Foundations of Tort Law

Foundations of
Tort Law

SAUL LEVMORE

New York   Oxford
Oxford University Press
1994

Oxford University Press

Oxford   New York   Toronto
Delhi   Bombay   Calcutta   Madras   Karachi
Kuala Lumpur   Singapore   Hong Kong   Tokyo
Nairobi   Dar es Salaam   Cape Town
Melbourne   Auckland   Madrid

and associated companies
in Berlin   Ibadan

Copyright © 1994 by Oxford University Press, Inc.

Published by Oxford University Press, Inc.
200 Madison Avenue, New York, New York 10016

Oxford is a registered trademark of Oxford University Press

Library of Congress Cataloging-in-Publication Data
Foundations of tort law / [edited by] Saul Levmore
p. cm.—(Interdisciplinary readers in law)
ISBN 0-19-508391-1 (cloth)
ISBN 0-19-508392-X (pbk.)
1. Torts—United States.   2. Torts—Economic aspects.
I. Levmore, Saul X.   II. Series.
KF1250.A2F68   1993
346.7303—dc20
[347.3063]   92-45580

9 8 7 6 5 4 3 2 1

Printed in the United States of America
on acid-free paper

Preface

This book is intended as a companion volume to the case materials used by law students taking a basic course in tort law. Although it makes no attempt to cover all the topics normally included in an introductory course, it does touch on a large number of issues and provides the reader with provocative, accessible, challenging, and diverse selections. None of the selections is beyond the reach of the novice. It can also be used as a sourcebook for students and lawyers interested in sampling the vast academic literature on torts. It might also be used in its entirety as a text in an advanced course or seminar. The introductions to each part and the notes and questions following each selection assume that the reader has some other exposure (or guide) to the law of torts. The issues raised in these introductory sections are meant to enhance the readings by posing a few provocative questions. For the most part, the readings speak for themselves. I have taken the liberty of including several of my own articles in order to obviate the need to infuse the notes and questions with personal viewpoints.

Two distinct aims are reflected in this volume. First, it is my hope to introduce the student to the value of academic writing. These selections may not help the lawyer organize files, structure depositions, or appeal to juries, but they will partially illuminate the black-letter law and perhaps encourage the reader to think of law not only as a means of earning a living but also as a subject of life-long fascination.

My second aim is to reflect the interdisciplinary possibilities of tort law. The study of law is not only an exploration of judges, legislators, and their pronouncements but also an inquiry into the behavioral effects as well as the evolution of legal rules. The "law and economics" literature has had an influence in torts that is far too great to be ignored, and this volume offers an opportunity to learn about and benefit from such a significant body of literature. Along with the economics materials, healthy doses of history and philosophy are also provided. Finally, there is some material on feminist theory, statistics, sociology, and political theory, for the aim is to show that the modern lawyer has many tools with which to understand the materials at hand.

The selections in this volume are of necessity extensively truncated versions of larger works. The reader is encouraged to refer to the original articles or essays for a more complete treatment of the arguments presented as well as for secondary references to other works and cases. I have indicated where such deletions have occurred and have eliminated virtually all citations and other footnote apparatus from the versions presented here.

The beginning law student who uses this book will want to know when to read various selections. The headings provide the best clues, but in order to simplify the process I have linked the names of some major cases with the parts or the selection titles in which they are explicitly—or implicitly—discussed.

I: All selections: *Stone v. Bolton, Vaughan v. Menlove.*

II: Posner: *Bird v. Holbrook, Katko v. Briney, Ploof v. Putnam;* Fletcher: *Vincent v. Lake Erie Transp. Co., Rylands v. Fletcher, Smith v. Lampe, Brown v. Kendall;* Levmore: *Vincent v. Lake Erie Transp. Co., Rylands v. Fletcher;* Schwartz: *Brown v. Collins, Rylands v. Fletcher, T. J. Hooper;* Calfee & Craswell: *City of Piqua v. Morris, U.S. v. Carroll Towing.*

III: Thomson: *Summers v. Tice, Vincent v. Lake Erie Transp. Co., Sindell v. Abbott Laboratories;* Levmore: *Sindell v. Abbott Laboratories, Haft v. Lone Palm Hotel;* Bishop: *Cattle v. Stockton Water Works, Union Oil v. Oppen;* Chamallas & Kerber: *Dillon v. Legg.*

IV: Grady: *U.S. v. Carroll Towing, Cooley v. Public Service Co., In re Polemis & Furness, Withy & Co., Berry v. Borough of Sugar Notch;* Sykes: *Ira S. Bushey & Sons v. U.S.*

VI: All selections: *Tarasoff v. Regents of University of California, Buch v. Amory Manufacturing Co.*

VII: Calabresi & Melamed: *Fountainbleau Hotel Corp. v. Forty-Five Twenty-Five, Inc., Rodgers v. Elliott, Ensign v. Wells, Boomer v. Atlantic Cement Co., Spur Industries v. Del E. Webb Dev. Co.;* Epstein: *Rylands v. Fletcher.*

VIII: Priest: *Greenman v. Yuba Power Products.*

Charlottesville, Va.   S.L.
January 1993

Contents

Torts and Bargains

So much of the study of tort law is concerned with the question of whether there should be liability for all injuries one causes (a strict liability rule) or only for those wrongfully caused (a negligence rule) that it is easy to pass over the question of whether liability is ever necessary. One of the great interdisciplinary starting points in torts is a famous article by Ronald Coase, winner of the Nobel Prize in economics. A lasting contribution of this article has been the suggestion that mishaps among strangers can be analyzed quite differently from transactions and disputes among parties that interact with one another. In particular, Coase taught law professors and their students to ask how parties might work out their own differences. And even when parties have litigated, the question remains as to what sort of agreement they might strike after the courts have spoken. Another way to think about Coase's contribution to tort law is to focus less on bargaining among parties and more on the reciprocal nature of harms. In many settings—and, some would say, in all circumstances—"it takes two to tort."

The Problem of Social Cost

RONALD H. COASE

The Problem to Be Examined

This selection is concerned with those actions of business firms which have harmful effects on others. The standard example is that of a factory the smoke from which has harmful effects on those occupying neighbouring properties. The economic analysis of such a situation has usually proceeded in terms of a divergence between the private and social product of the factory.

The conclusions to which this kind of analysis seems to have led most economists is that it would be desirable to make the owner of the factory liable for the damage caused to those injured by the smoke, or alternatively, to place a tax on the factory owner varying with the amount of smoke produced and equivalent in money terms to the damage it would cause, or finally, to exclude the factory from residential districts (and presumably from other areas in which the emission of smoke would have harmful effects on others). It is my contention that the suggested courses of action are inappropriate, in that they lead to results which are not necessarily, or even usually, desirable.

The Reciprocal Nature of the Problem

The traditional approach has tended to obscure the nature of the choice that has to be made. The question is commonly thought of as one in which A inflicts harm on B and what has to be decided is: how should we restrain A? But this is wrong. We are dealing with a problem of a reciprocal nature. To avoid the harm to B would inflict harm on A. The real question that has to be decided is: should A be allowed to harm B or should B be allowed to harm A? The problem is to avoid the more serious harm.

. . .

An example is afforded by the problem of straying cattle which destroy crops on the neighbouring land. If it is inevitable that some cattle will stray, an increase in the supply of meat can only be obtained at the expense of a decrease in the supply of crops. The nature of the choice is clear: meat or crops. What answer should be given is, of course, not clear unless we know the value of what is obtained as well as the value of what is sacrificed to obtain it.

. . .

Let us suppose that a farmer and a cattle-raiser are operating on neigh-bouring properties. Let us further suppose that, without any fencing between the properties, an increase in the size of the cattle-raiser's herd increases the total damage to the farmer's crops.

. . .

To simplify the argument, I propose to use an arithmetical example. I shall assume that the annual cost of fencing the farmer's property is $9 and that the price of the crop is $1 per ton. Also, I assume that the relation between the number of cattle in the herd and the annual crop loss is as follows:

Number in Herd (Steers)	Annual Crop Loss (Tons)	Crop Loss per Additional Steer (Tons)
1	1	1
2	3	2
3	6	3
4	10	4

Given that the cattle-raiser is liable for the damage caused, the additional annual cost imposed on the cattle-raiser if he increased his herd from, say, two to three steers is $3 and in deciding on the size of the herd, he will take this into account along with his other costs. That is, he will not increase the size of the herd unless the value of the additional meat produced (assuming that the cattle-raiser slaughters the cattle) is greater than the additional costs that this will entail, including the value of the additional crops destroyed. . . .

Given that the annual cost of fencing is $9, the cattle-raiser who wished to have a herd with four steers or more would pay for fencing to be erected and maintained. . . . When the fence is erected the marginal cost due to the liability for damage becomes zero. . . . But, of course, it may be cheaper for the cattle raiser not to fence and to pay for the damaged crops, as in my arithmetical example, with three or fewer steers.

It might be thought that the fact that the cattle-raiser would pay for all crops damaged would lead the farmer to increase his planting if a cattle-raiser came to occupy the neighbouring property. But this is not so. If the crop was previously sold in conditions of perfect competition, marginal cost was equal to price for the amount of planting undertaken and any expansion would have reduced the profits of the farmer. In the new situation, the existence of crop damage would mean that the farmer would sell less on the open market but his receipts for a given production would remain the same, since the cattle-raiser would pay the market price for any crop damaged.

. . .

I have said that the occupation of a neighbouring property by a cattle-raiser would not cause the amount of production, or perhaps more exactly the amount of planting, by the farmer to increase. In fact, if the cattle-raising has any effect, it will be to decrease the amount of planting. The reason for this is that, for any given tract of land, if the value of the crop damaged is so great that the receipts from the sales of the undamaged crop are less than the total costs of cultivating that tract of land, it will be profitable for the farmer and the cattle-raiser to make a bargain whereby that tract of land is left uncultivated. This can be made clear by means of an arithmetical example. Assume initially that the value of the crop obtained from cultivating a given tract of land is $12 and that the cost incurred in cultivating this tract of land is $10, the net gain from cultivating the land being $2. . . .

Now assume that the cattle-raiser starts operations on the neighbouring property and that the value of the crops damaged is $1. In this case $11 is obtained by the farmer from sale on the market and $1 is obtained from the cattle-raiser for damage suffered and the net gain remains $2. Now suppose that the cattle-raiser finds it profitable to increase the size of his herd, even though the amount of damage rises to $3; which means that the value of the additional meat production is greater than the additional costs, including the additional $2 payment for damage. But the total payment for damage is now $3. The net gain to the farmer from cultivating the land is still $2. The cattle-raiser would be better off if the farmer would agree not to cultivate his land for any payment less than $3. The farmer would be agreeable to not cultivating the land for any payment greater than $2. There is clearly room for a mutually satisfactory bargain which would lead to the abandonment of cultivation.

. . .

What needs to be emphasized is that the fall in the value of production . . . which would be taken into account in the costs of the cattle-raiser may well be less than the damage which the cattle would cause to the crops in the ordinary course of events. This is because it is possible, as a result of market transactions, to discontinue cultivation of the land. This is desirable in all cases in which the damage that the cattle would cause, and for which the cattle-raiser would be willing to pay, exceeds the amount which the farmer would pay for use of the land. In conditions of perfect competition, the amount which the farmer would pay for the use of the land is equal to the difference between the value of the total production when the factors are employed on this land and the value of the additional product yielded in their next best use (which would be what the farmer would have to pay for the factors). If damage exceeds the amount the farmer would pay for the use of the land, the value of the additional product of the factors employed elsewhere would exceed the value of the total product in this use after damage is taken into account. It follows that it would be desirable to abandon cultivation of the land and to release the factors employed for production elsewhere. A procedure which merely provided for payment for damage to the crop caused by the cattle but which did not allow for the possibility of cultivation being discontinued would result in too small an employment of factors of production in cattle-raising and too large an employment of factors in cultivation of the crop. But given the possibility of market transactions, a situation in which damage to crops exceeded the rent of the land would not endure. Whether the cattle-raiser pays the farmer to leave the land uncultivated or himself rents the land by paying the landowner an amount slightly greater than the farmer would pay (if the farmer was himself renting the land), the final result would be the same and would maximise the value of production. Even when the farmer is induced to plant crops which it would not be profitable to cultivate for sale on the market, this will be a purely short-term phenomenon and may be expected to lead to an agreement under which the planting will cease. The cattle-raiser will remain in that location and the marginal cost of meat production will be the same as before, thus having no long-run effect on the allocation of resources.

The Pricing System with No Liability for Damage

I now turn to the case in which . . . the damaging business is not liable for any of the damage which it causes. . . . I propose to show that the allocation of resources will be the same in this case as it was when the damaging business was liable for the damage caused. . . .

I return to the case of the farmer and the cattle-raiser. The farmer would suffer increased damage to his crop as the size of the herd increased. Suppose that the size of the cattle-raiser's herd is three steers (and that this is the size of the herd that would be maintained if crop damage was not taken into account). Then the farmer would be willing to pay up to $3 if the cattleraiser would reduce his herd to two steers, up to $5 if the herd were reduced to one steer, and would pay up to $6 if cattle-raising was abandoned. The cattle-raiser would therefore receive $3 from the farmer if he kept two steers instead of three. This $3 foregone is therefore part of the cost incurred in keeping the third steer. Whether the $3 is a payment which the cattle-raiser has to make if he adds the third steer to his herd (which it would be if the cattle-raiser was liable to the farmer for damage caused to the crop) or whether it is a sum of money which he would have received if he did not keep a third steer (which it would be if the cattle-raiser was not liable to the farmer for damage caused to the crop) does not affect the final result. In both cases $3 is part of the cost of adding a third steer, to be included along with the other costs. If the increase in the value of production in cattle-raising through increasing the size of the herd from two to three is greater than the additional costs that have to be incurred (including the $3 damage to crops), the size of the herd will be increased. Otherwise, it will not. The size of the herd will be the same whether the cattle-raiser is liable for damage caused to the crop or not.

It may be argued that the assumed starting point—a herd of three steers—was arbitrary. And this is true. But the farmer would not wish to pay to avoid crop damage which the cattle-raiser would not be able to cause. For example, the maximum annual payment which the farmer could be induced to pay could not exceed $9, the annual cost of fencing. And the farmer would only be willing to pay this sum if it did not reduce his earnings to a level that would cause him to abandon cultivation of this particular tract of land. Furthermore, the farmer would only be willing to pay this amount if he believed that, in the absence of any payment by him, the size of the herd maintained by the cattle-raiser would be four or more steers. Let us assume that this is the case. Then the farmer would be willing to pay up to $3 if the cattle-raiser would reduce his herd to three steers, up to $6 if the herd were reduced to two steers, up to $8 if one steer only were kept and up to $9 if cattle-raising were abandoned. It will be noticed that the change in the starting point has not altered the amount which would accrue to the cattle-raiser if he reduced the size of his herd by any given amount. It is still true that the cattle-raiser could receive an additional $3 from the farmer if he agreed to reduce his herd from three steers to two and that the $3 represents the value of the crop that would be destroyed by adding the third steer to the herd. Although a different belief on the part of the farmer (whether justified or not) about the size of the herd that the cattle-raiser would maintain in the absence of payments from him may affect the total payment he can be induced to pay, it is not true that this different belief would have any effect on the size of the herd that the cattle-raiser will actually keep. This will be the same as it would be if the cattle-raiser had to pay for damage caused by his cattle, since a receipt foregone of a given amount is the equivalent of a payment of the same amount.

It might be thought that it would pay the cattle-raiser to increase his herd above the size that he would wish to maintain once a bargain had been made, in order to induce the farmer to make a larger total payment. And this may be true. It is similar in nature to the action of the farmer (when the cattle-raiser was liable for damage) in cultivating land on which, as a result of an agreement with the cattle-raiser, planting would subsequently be abandoned (including land which would not be cultivated at all in the absence of cattle-raising). But such manoeuvres are preliminaries to an agreement and do not affect the long-run equilibrium position, which is the same whether or not the cattle-raiser is held responsible for the crop damage brought about by his cattle.

It is necessary to know whether the damaging business is liable or not for damage caused since without the establishment of this initial delimitation of rights there can be no market transactions to transfer and recombine them. But the ultimate result (which maximises the value of production) is independent of the legal position if the pricing system is assumed to work without cost.

The Problem Illustrated Anew

. . .

The problem of straying cattle and the damaging of crops which was the subject of detailed examination in the two preceding sections, although it may have appeared to be rather a special case, is in fact but one example of a problem which arises in many different guises. . . .

Let us first consider the case of *Sturges v. Bridgman* . . . In this case, a confectioner (in Wigmore Street) used two mortars and pestles in connection with his business (one had been in operation in the same position for more than 60 years and the other for more than 26 years). A doctor then came to occupy neighbouring premises (in Wimpole Street). The confectioner's machinery caused the doctor no harm until, eight years after he had first occupied the premises, he built a consulting room at the end of his garden right against the confectioner's kitchen. It was then found that the noise and vibration caused by the confectioner's machinery made it difficult for the doctor to use his new consulting room. . . . The doctor therefore brought a legal action to force the confectioner to stop using his machinery. The courts had little difficulty in granting the doctor the injunction he sought. . . .

The court's decision established that the doctor had the right to prevent the confectioner from using his machinery. But, of course, it would have been possible to modify the arrangements envisaged in the legal ruling by means of a bargain between the parties. The doctor would have been willing to waive his right and allow the machinery to continue in operation if the confectioner would have paid him a sum of money which was greater than the loss of income which he would suffer from having to move to a more costly or less convenient location or from having to curtail his activities at this location or, as was suggested as a possibility, from having to build a separate wall which would deaden the noise and vibration. The confectioner would have been willing to do this if the amount he would have to pay the doctor was less than the fall in income he would suffer if he had to change his mode of operation at this location, abandon his operation or move his confectionery business to some other location. The solution of the problem depends essentially on whether the continued use of the machinery adds more to the confectioner's income than it subtracts from the doctor's. But now consider the situation if the confectioner had won the case. The confectioner would than have had the right to continue operating his noise- and vibration-generating machinery without having to pay anything to the doctor. The boot would have been on the other foot: the doctor would have had to pay the confectioner to induce him to stop using the machinery. If the doctor's income would have fallen more through continuance of the use of this machinery than it added to the income of the confectioner, there would clearly be room for a bargain whereby the doctor paid the confectioner to stop using the machinery. That is to say, the circumstances in which it would not pay the confectioner to continue to use the machinery and to compensate the doctor for the losses that this would bring (if the doctor had the right to prevent the confectioner's using his machinery) would be those in which it would be in the interest of the doctor to make a payment to the confectioner which would induce him to discontinue the use of the machinery (if the confectioner had the right to operate the machinery). The basic conditions are exactly the same in this case as they were in the example of the cattle which destroyed crops. With costless market transactions, the decision of the courts concerning liability for damage would be without effect on the allocation of resources. It was of course the view of the judges that they were affecting the working of the economic system—and in a desirable direction. . . .

The judges' view that they were settling how the land was to be used would be true only in the case in which the costs of carrying out the necessary market transactions exceeded the gain which might be achieved by any rearrangement of rights. And it would be desirable to preserve [quiet on] Wimpole Street . . . (by giving nonindustrial users the right to stop the noise, vibration, smoke, etc., by injunction) only if the value [thereby] obtained was . . . greater than the value of cakes . . . lost. But of this the judges seem to have been unaware.

. . .

Bryant v. Lefever raised the problem of the smoke nuisance in a novel form. The plaintiff and the defendants were occupiers of adjoining houses, which were of about the same height. Before 1876 the plaintiff was able to light a fire in any room of his house without the chimneys smoking; the two houses had remained in the same condition some thirty or forty years. In 1876 the defendants took down their house, and began to rebuild it. They carried up a wall by the side of the plaintiff's chimneys much beyond its original height, and stacked timber on the roof of their house, and thereby caused the plaintiff's chimneys to smoke whenever he lighted fires. The reason, of course, why the chimneys smoked was that the erection of the wall and the stacking of the timber prevented the free circulation of air. In a trial before a jury, the plaintiff was awarded damages of £40. The case then went to the Court of Appeals where the judgment was reversed. Bramwell, . . . it is said, and the jury have found, that the defendants have done that which caused a nuisance to the plaintiff's house. We think there is no evidence of this. No doubt there is a nuisance, but it is not of the defendant's causing. They have done nothing in causing the nuisance. Their house and their timber are harmless enough. It is the plaintiff who causes the nuisance by lighting a coal fire in a place the chimney of which is placed so near the defendants' wall, that the smoke does not escape, but comes into the house.

. . .

What I shall discuss is the argument of the judges in the Court of Appeals that the smoke nuisance was not caused by the man who erected the wall but by the man who lit the fires. The novelty of the situation is that the smoke nuisance was suffered by the man who lit the fires and not by some third person. The question is not a trivial one since it lies at the heart of the problem under discussion. Who caused the smoke nuisance? The answer seems fairly clear. The smoke nuisance was caused both by the man who built the wall *and* by the man who lit the fires. Given the fires, there would have been no smoke nuisance without the wall; given the wall, there would have been no smoke nuisance without the fires. Eliminate the wall *or* the fires and the smoke nuisance would disappear. On the marginal principle it is clear that *both* were responsible and *both* should be forced to include the loss of amenity due to the smoke as a cost in deciding whether to continue the activity which gives rise to the smoke. And given the possibility of market transactions, this is what would in fact happen. Although the wall-builder was not liable legally for the nuisance, as the man with the smoking chimneys would presumably be willing to pay a sum equal to the monetary worth to him of eliminating the smoke, this sum would therefore become for the wall-builder a cost of continuing to have the high wall with the timber stacked on the roof.

The judges' contention that it was the man who lit the fires who alone caused the smoke nuisance is true only if we assume that the wall is the given factor. This is what the judges did by deciding that the man who erected the higher wall had a legal right to do so. . . .

Judges have to decide on legal liability, but this should not confuse economists about the nature of the economic problem involved. In the case of the cattle and the crops, it is true that there would be no crop damage without the cattle. It is equally true that there would be no crop damage without the crops. The doctor's work would not have been disturbed if the confectioner had not worked his machinery; but the machinery would have disturbed no one if the doctor had not set up his consulting room in that particular place. . . .

If we are to discuss the problem in terms of causation, both parties cause the damage. If we are to attain an optimum allocation of resources, it is therefore desirable that both parties should take the harmful effect (the nuisance) into account in deciding on their course of action. It is one of the beauties of a smoothly operating pricing system that, as has already been explained, the fall in the value of production due to the harmful effect would be a cost for both parties.

. . .

The economic problem in all cases of harmful effects is how to maximise the value of production. . . .

It has to be remembered that the immediate question faced by the courts is *not* what shall be done by whom *but* who has the legal right to do what. It is always possible to modify by transactions on the market the initial legal delimitation of rights. And, of course, if such market transactions are costless, such a rearrangement of rights will always take place if it would lead to an increase in the value of production.

The Cost of Market Transactions Taken into Account

The argument has proceeded up to this point on the assumption . . . that there were no costs involved in carrying out market transactions. This is, of course, a very unrealistic assumption. In order to carry out a market transaction it is necessary to discover who it is that one wishes to deal with, to inform people that one wishes to deal and on what terms, to conduct negotiations leading up to a bargain, to draw up the contract, to undertake the inspection needed to make sure that the terms of the contract are being observed, and so on. These operations are often extremely costly, sufficiently costly at any rate to prevent many transactions that would be carried out in a world in which the pricing system worked without cost.

In earlier sections, when dealing with the problem of the rearrangement of legal rights through the market, it was argued that such a rearrangement would be made through the market whenever this would lead to an increase in the value of production. But this assumed costless market transactions. Once the costs of carrying out market transactions are taken into account it is clear that such a rearrangement of rights will only be undertaken when the increase in the value of production consequent upon the rearrangement is greater than the costs which would be involved in bringing it about. When it is less, the granting of an injunction (or the knowledge that it would be granted) or the liability to pay damages may result in an activity being discontinued (or may prevent its being started) which would be undertaken if market transactions were costless. In these conditions the initial delimitation of legal rights does have an effect on the efficiency with which the economic system operates. One arrangement of rights may bring about a greater value of production than any other. But unless this is the arrangement of rights established by the legal system, the costs of reaching the same result by altering and combining rights through the market may be so great that this optimal arrangement of rights, and the greater value of production which it would bring, may never be achieved. The part played by economic considerations in the process of delimiting legal rights will be discussed in the next section. In this section, I will take the initial delimitation of rights and the costs of carrying out market transactions as given.

It is clear that an alternative form of economic organisation which could achieve the same result at less cost than would be incurred by using the market would enable the value of production to be raised. . . . The firm represents such an alternative to organising production through market transactions. Within the firm individual bargains between the various cooperating factors of production are eliminated and for a market transaction is substituted an administrative decision. The rearrangement of production then takes place without the need for bargains between the owners of the factors of production. A landowner who has control of a large tract of land may devote his land to various uses taking into account the effect that the interrelations of the various activities will have on the net return of the land, thus rendering unnecessary bargains between those undertaking the various activities. Owners of a large building or of several adjoining properties in a given area may act in much the same way. In effect, using our earlier terminology, the firm would acquire the legal rights of all the parties and the rearrangement of activities would not follow on a rearrangement of rights by contract, but as a result of an administrative decision as to how the rights should be used.

· · ·

The firm is not the only possible answer to this problem. The administrative costs of organising transactions within the firm may also be high, and particularly so when many diverse activities are brought within the control of a single organisation. In the standard case of a smoke nuisance, which may affect a vast number of people engaged in a wide variety of activities, the administrative costs might well be so high as to make any attempt to deal with the problem within the confines of a single firm impossible. An alternative solution is direct government regulation. Instead of instituting a legal system of rights which can be modified by transactions on the market, the government may impose regulations which state what people must or must not do and which have to be obeyed. Thus, the government (by statute or perhaps more likely through an administrative agency) may, to deal with the problem of smoke nuisance, decree that certain methods of production should or should not be used (e.g., that smoke-preventing devices should be installed or that coal or oil should not be burned) or may confine certain types of business to certain districts (zoning regulations).

The government is, in a sense, a super-firm (but of a very special kind) since it is able to influence the use of factors of production by administrative decision. But the ordinary firm is subject to checks in its operations because of the competition of other firms, which might administer the same activities at lower cost and also because there is always the alternative of market transactions as against organisation within the firm if the administrative costs become too great. The government is able, if it wishes, to avoid the market altogether, which a firm can never do. The firm has to make market agreements with the owners of the factors of production that it uses. Just as the government can conscript or seize property, so it can decree that factors of production should only be used in such and such a way. Such authoritarian methods save a lot of trouble (for those doing the organising). Furthermore, the government has at its disposal the police and the other law enforcement agencies to make sure that its regulations are carried out.

It is clear that the government has powers which might enable it to get some things done at a lower cost than could a private organisation (or at any rate one without special governmental powers). But the governmental administrative machine is not itself costless. It can, in fact, on occasion be extremely costly. Furthermore, there is no reason to suppose that the restrictive and zoning regulations, made by a fallible administration subject to political pressures and operating without any competitive check, will necessarily always be those which increase the efficiency with which the economic system operates. Furthermore, such general regulations which must apply to a wide variety of cases will be enforced in some cases in which they are clearly inappropriate. From these considerations it follows that direct governmental regulation will not necessarily give better results than leaving the problem to be solved by the market or the firm. But equally there is no reason why, on occasion, such governmental administrative regulation should not lead to an improvement in economic efficiency. This would seem particularly likely when, as is normally the case with the smoke nuisance, a large number of people are involved and in which therefore the costs of handling the problem through the market or the firm may be high.

There is, of course, a further alternative, which is to do nothing about the problem at all. And given that the costs involved in solving the problem by regulations issued by the governmental administrative machine will often be heavy (particularly if the costs are interpreted to include all the consequences which follow from the government engaging in this kind of activity), it will no doubt be commonly the case that the gain which would come from regulating the actions which give rise to the harmful effects will be less than the costs involved in government regulation.

The discussion of the problem of harmful effects in this section (when the costs of market transactions are taken into account) is extremely inadequate. But at least it has made clear that the problem is one of the choosing the appropriate social arrangement for dealing with the harmful effects. All solutions have costs and there is no reason to suppose that government regulation is called for simply because the problem is not well handled by the market or the firm.

Notes and Questions

1. Imagine a small amusement park with a driving range for golfers located adjacent to a home-improvement center that sells windows. Occasionally the golfers' drives fly far out of bounds into the stacked windows, prompting the question: What can be done about these losses? Assume that the annual loss as a result of golf balls smashing windows adds up to $400, that the windows could be moved to another part of the property at a projected annual cost to the store owner of $1,200, and that the store owner could erect a fence to protect the windows at an annual cost of $300. In turn, the owner of the amusement park could either pay for the occasional broken window (a cost we already know to be $400), build a fence for $300, or modify her use of the property by constructing a miniature golf course. This last option is expected to reduce annual profits by $1,500 (including the cost of construction). If a court requires the amusement park owner to pay for the broken windows, what arrangement might eventually transpire between the parties? If, instead, the amusement park owner is not held liable for these "accidental" losses, what result would you expect?

2. Continuing with the previous example, what if the fence costs not $300 to build and maintain but rather $2,000? What will happen if the amusement park owner is held liable? What will happen if she is held not liable? Do you see why the Coase Theorem is sometimes sketched with the statement that if parties can communicate easily with one another, then the legal rule really does not matter?

3. If the home-improvement center is an enormous enterprise, while the amusement park is a small, struggling one, do you expect the bargaining to be different than if the enterprises were of similar size and profitability? What if the fence is the cheapest option and the costs of moving the windows and switching to miniature golf are not $1,200 and $1,500 but rather $1,200 (for the store) and $10,000 (for the amusement park)? In laboratory experiments where players were given unequal amounts of money and offered an opportunity to bargain for "contracts" that would increase their combined rewards (at the expense of the experimenter), most players were inclined to limit their bargaining to agreements calling for an equal split of the proceeds. It is thus possible that many people stick to preconceived norms about fair bargaining or outcomes. (See Elizabeth Hoffman and Matthew L. Spitzer, The Coase Theorem: Some Experimental Tests, 25 *Journal of Law and Economics* 73 (1982).) How might this contrived experiment be analogous to real conflicts? If there is some norm of equal sharing, would you expect the amusement park owner and the owner of the home-improvement center to be more likely to split the cost of the fence, the profits from their two stores, or the gains from the fence? Is there some other norm they might converge on in order to avoid haggling?

4. If the amusement park owner loses in court or, in Coase's examples, cattleraisers are made to pay farmers, what would you expect to happen over time to the

profitability and the number of driving-range (and cattleraising) businesses? This question will be considered in the next selection.

5. Coase's insight appears to be based on the notion that valuations are, in an important sense, reciprocal, so that the price one would pay in order to gain a right not presently enjoyed (such as raising cattle or operating a driving range without liability) is the same as the price one would need to be paid in order to give up a preexisting right to engage in these activities. Can you think of circumstances in which this symmetry is unlikely to hold true? This question is considered in the last selection in this part.

6. Consider the position of a judge who, having read, comprehended, and agreed with Coase, must now decide the case arising out of the conflict between the home-improvement store and the driving range. What should this judge decide? What if the judge feels that although the parties are neighbors, they are unlikely to communicate easily because of a language barrier or perhaps because of a long history of unpleasant disputes?

When Does the Rule of Liability Matter?

HAROLD DEMSETZ

. . .

The questions with which we shall be concerned are whether and under what conditions a legal decision about liability affects the uses to which resources will be put and the distribution of wealth between owners of resources. If ranchers are held liable for the damage done by their cattle to corn fields, how will the outputs of meat and corn be affected? If drivers or pedestrians, alternatively, are held liable for automobile-pedestrian accidents, how will the accident rate be affected? What implications for extortion (an extreme form of wealth redistribution) are found in the decision about who is liable for damages?

Recent developments in this area began with an article by Professor R. H. Coase. Coase's work presented a penetrating criticism of the conventional treatment by economists of divergences between private and social cost. The social cost of furthering an economic activity is the resulting reduction in the value of production that is obtainable from other activities. Such reductions occur because the resources required to further an activity are scarce and must be diverted from other possible uses. According to the view that Coase challenged, social cost, being the sum total of the costs incurred to carry on any activity, might very well differ from private cost. For example, the social cost of running steam locomotives properly includes the fire damage done to surrounding farm crops by sparks from the locomotive. If the railroad is not

Abridged and reprinted without footnotes by permission from 1 *Journal of Legal Studies* 13 (1972). Copyright © 1972 by The University of Chicago Press.

required to pay for these damages, perhaps through a tax per train, then, according to the conventional analysis, the railroad would not take account of crop damage costs in deciding how many trains to run. In the absence of a specific public policy to intervene, the rate at which an activity is carried forth, which is determined solely by private cost, would diverge from the optimum rate, which is determined by social cost. In the present example, the private cost of running additional trains, being less than the social cost because crop damage is not taken into account, would encourage the railroad to run too many trains per day. The conventional economic analysis called for the levy of a tax per train, equivalent to the damages, in order to bring social cost and private cost into equality.

Coase demonstrated that the imposition of such a tax could, in some circumstances, aggravate the difficulty; but two other aspects of his work are of more concern to us. Coase (1) showed that powerful market forces exist that tend to bring private and social cost into equality without the use of a tax, and (2) discussed the conditions under which the legal position toward liability for damages would and would not alter the allocation of resources. Coase discusses an interaction between two productive activities, ranching and farming, in the context of a competitive regime in which the cost of transacting (or negotiating) is assumed to be zero. His analysis concludes that social cost and private cost will be brought into equality through market negotiations—and this regardless of which party is assigned the responsibility for bearing the cost that results from the proximity of ranching and farming.

The law, reasoning that crops *stand in the way* of a neighbor's cattle, can leave the farmer to bear the cost of crop damage; alternatively, reasoning that cattle *stray errantly* across farm fields, the law can assign liability for crop damages to ranchers. Coase's work demonstrates that either legal position will result in the same resource allocation—i.e., in the same quantities of corn and meat—and, also, that negotiations between the parties to the damage will, with either legal position, eliminate any divergence between private and social cost.

. . .

The significance of Coase's work quickly led to . . . criticism centered around two allegations—that the Coase theorem neglects long-run considerations that negate it and that the spirit of the work endorses the use of resources for the undesirable purpose of "extortion."

. . .

The question of long-run considerations has been raised because it would seem that different liability rules would alter the profitability of remaining inside or outside each industry. It is alleged that if farmers are left to bear the cost that arises from proximity to cattle, the rate of return to farming will fall and resources will therefore leave the farming industry. Alternatively, if ranchers are left to bear the cost, the resulting reduction in rate of return to ranching will lead to the exit of resources from that industry. Hence, even if transaction cost is zero, the market will allocate resources differently in the long run depending upon which rule of liability is chosen.

But short-run versus long-run considerations should have no bearing on the Coase theorem, which is based on the proposition that an implicit cost (the forgone payment from the farmer) is just as much a cost as is an explicit cost (the liability damage), and this proposition surely must hold in the long run as well as in the short run. One way of demonstrating this is by allowing the two activities to be merged under a common owner. . . .

If there is no special cost to operating a multiproduct firm, the costly interaction between farming and ranching will be fully brought to the owner's attention in his operation of a farming-ranching enterprise. The mix of output that he produces will be that which maximizes his earnings. The rule of liability that is chosen can have no effect on his decisions because the owner of such a firm must bear the interaction cost whichever legal rule is adopted. The cost interdependence is a technical-economic interdependence, not a legal one. Since such merged operations are possible, the rule of liability is rendered irrelevant to the choice of output mix.

. . .

If owners of farmland bear the cost of crop damage, what must be the cost conditions that are associated with an equilibrium allocation of land to farming, ranching, and other uses? For such damage to arise there must be a sufficient scarcity of land to force farms and ranches into proximity. Marginal farm acreage (acreage that just "breaks even") must earn revenue sufficient to cover all cost, including the cost of crop damage done by straying cattle. Suppose the proximity of ranching and farming reduces the net return to the owner of farmland by $100 as compared with what could be earned were there no neighboring ranch. If ranching continues on the neighboring acres, it must be true that the net return to the owner of the ranchland exceeds $100, for otherwise the farmer would have been able to purchase the removal of cattle from neighboring land by offering $100 to the neighboring ranchers. The reduction in net return to farming brought about because of crop damage is thus implicitly taken into account by the owner of the ranchland when he refuses the offer of the owner of the neighboring farmland. The money offered by the farmer is refused by the rancher precisely because the continued use of the land for grazing brings in additional net revenue in excess of $100.

Land that is submarginal as farm or ranchland (land that cannot be profitably farmed or ranched) is unable to earn revenue sufficient to cover the explicit $100 damage cost if it is put to the plough or the $100 implicit cost if it is employed in ranching. Were this land to be employed in farming, its owner would suffer losses attributable in part to the damage done to his crops by straying cattle, whereas if it were employed in ranching its owner would suffer the implicit loss of forgoing a $100 payment from the neighboring farmer (the owners of farmland bearing the cost of crop damage). Submarginal land by definition can be neither farmed nor ranched profitably.

Now let the rule of liability be changed so that ranchers become liable for crop damage. If there is to be a long-run effect, it must be true that the cost interrelationships change in a manner that causes either the conversion of ranchland to farmland or submarginal land to farmland. But neither conversion can be made profitable by the change in liability.

Acreage that was marginally profitable in ranching must remain ranchland because it had been earning net revenues in excess of the $100 damages done by cattle to surrounding farmland. If a producing ranch were to be switched from ranching to farming to avoid the new liability its owner would forgo revenues (in excess of $100) that exceeded the resulting reduction in liability ($100). The owner of what was marginal ranchland, therefore, will continue to employ his land in ranching under the new liability rule.

The land that previously was submarginal must remain submarginal. The changed liability rule will not attract this land into farming. Submarginal lands under the original rule of liability earned insufficient revenues in farming to cover the $100 cost of crop damage. Under the new rule of liability, neighboring ranchers will succeed in negotiating with the owners of this land to keep it out of farming. Operating ranches, under the old rule of liability, had been yielding net revenues in excess of $100; therefore it will be possible and profitable for ranchers, in order to avoid the $100 crop damage that otherwise would result, to offer an amount to the owners of submarginal land that is sufficient to keep the land out of farming.

There is a temptation at this point in the argument to believe that an error has been made. Suppose that the farmer suffers damages equal to $100 and the rancher enjoys a net return equal to $110. If the rancher is not liable he will choose to continue ranching and to refuse a payment from a neighboring farmer of $100 to stop ranching. But if the rule of liability is reversed, he will continue ranching only if his $110 net return is sufficient to cover the cost imposed on the farmer ($100) *plus* the payment required to keep submarginal land (which can be assumed to border on another boundary of the ranch) out of farming. There has been no error in the argument, but there is an error in introducing the second neighbor halfway through the analysis. With the rancher not liable, he would have elected to remain in ranching only if the net return to ranching exceeded the payments to leave ranching offered to him by *both* his neighbors. If the rancher finds it remunerative to remain in ranching in the face of both these offers he must earn a sufficient net return from ranching, after the rule of liability is changed and he is held liable, to be able to pay damages to his neighboring farmer *and* to pay the owner of neighboring submarginal land to keep that land out of production.

The change in the rule of liability does not lead to a conversion of ranchland or submarginal land to farming. The use of land that maximized returns before the change in liability rule continues to maximize returns after the new rule is adopted, and the mix of output is unaffected by the choice of liability rule even when long-run considerations are analyzed. To understand the effect of altering the rule of liability it is important to recognize that the owner of a resource who finds it in his interest to employ that resource in a particular way when he bears the cost of an interaction will be paid to employ that resource *in the same way* when the rule of liability is reversed. What can happen, and in this case does happen, when the rule of liability is changed is that present owners of land having a comparative advantage in ranching suffer a *windfall* loss in the value of their land while owners of farmland enjoy a *windfall* gain. But this redistribution of wealth cannot alter the uses of these lands.

. . .

The problem of "extortion" arises when a change in liability gives rise to a redistribution in wealth. In the farmer-rancher case, the relative values of nearby farm and ranchlands will be changed when the rule of liability is altered. Under one rule of liability, with farmers required to bear the cost of crop damage, farmers will need to pay ranchers to reduce herd size; under the other rule ranchers will have to pay farmers for damages or for any alteration in the quantity of corn grown nearby. The change in the direction of payments must affect the rents that can be collected by owners of these lands and thus the market values of these lands.

In these cases the owner of the specialized resource, ranchland or farmland, that is not required to bear the cost of the interaction may threaten to increase the intensity of the interaction in an attempt to get his neighbor to pay him a larger sum than would ordinarily be required to obtain his cooperation in adjusting the intensity of the interaction downward. The owner of ranchland, if he is not liable for crop damage done by straying cattle, might, in the absence of a neighboring farmer, raise only 1,000 head of cattle. With proximity between farming and ranching, a neighboring owner of farmland might be willing to pay the rancher the sum required to finance a 200-head reduction in herd size. However, if the owner of ranchland *threatens* to raise 1,500 head, he may be able to secure more than this sum from the farmer because of the additional crop damage that would be caused by the larger herd size. With or without this "extortion" threat, the size of the herd will be reduced to 800 because that is the size, by assumption, that maximizes the total value of both activities. Given the interrelationship between the two activities, that is the herd size that will maximize the return to the farmer and, indirectly, the sum available for possible transfer to the rancher. What is at issue is the sharing of this maximum return.

To the extent that there exist alternative farm sites, the ability of the owner of ranchland to make such a threat credible is compromised. Competition among such owners will reduce the payment that farmers make to ranchers to that sum which is just sufficient to offset the revenue forgone by ranchers when herd size is reduced. No rancher could succeed in a threat to increase herd size above normal numbers because other ranchers would be willing to compete to zero the price that farmers are asked to pay to avoid abnormally large herd sizes. Abnormally large herd size, in itself, will generate losses to owners of ranchland and, for this reason, competition among such owners will reduce the price that owners of farmland must pay to avoid such excessive herd sizes to zero.

But if a ranchland owner has a locational monopoly, in the sense that there are no alternative sites available to farmers, then the rancher may succeed in acquiring a larger sum from his neighboring farmer in order to avoid abnormally large herd sizes. The acquisition of a larger sum by the owner of ranchland generally will require him to incur some cost to make his threat credible, perhaps by actually beginning to increase herd size beyond normal levels. If the cost of making this threat credible is low relative to the sum available for transfer from the owner of farmland, the rancher will be in a good position to accomplish the transfer. The sum available for transfer will be the amount by which the value of the neighboring land when used as farmland exceeds its value in the next best use. If the rancher were to demand a larger payment from his neighbor, the neighboring land would be switched to some other use.

The temptation to label such threats extortion or blackmail must be resisted by economists for these are legal and not economic distinctions. The rancher merely attempts to maximize profits. If his agreements with neighboring farmers are marketed in competition with other ranchers, profit maximization constrained by competition implies that an agreement to reduce herd size can be purchased for a smaller payment than if effective competition in such agreements is absent. The appropriate economic label for this problem is nothing more nor less than monopoly. It takes on the cast of such legal classifications as extortion only because the context seems to be one where the monopoly return is received by threatening to produce something that is not wanted—excessively large herds. The conventional monopoly problem involves a reduction or a threat to reduce the output of a desired good. In the unconventional monopoly problem presented here, there is a threat to increase herd size beyond desirable levels. But this difference is superficial. The conventional monopoly problem can be viewed as one in which the monopolist produces more scarcity than is desired, and the unconventional monopoly problem discussed here can be considered one in which the monopolist threatens to produce too small a reduction in crop damage. Any additional sum that the rancher succeeds in transferring to himself from the farmer is correctly identified as a monopoly return.

The temptation to resolve this monopoly problem merely by reversing the rule of liability must be resisted. Should the liability rule be reversed and the owner of ranchland now be held liable for damage done by his cattle to surrounding crops, the specific monopoly problem that we have been discussing would be resolved. But if the farmer enjoys a locational monopoly such that the rancher has nowhere else to locate, the shoe will now be on the other foot. The farmer can threaten to increase the number of bushels of corn planted, and hence the damage for which the rancher will be liable, unless the rancher pays the farmer a sum greater than would be required under competitive conditions. The potential for monopoly and the wealth redistribution implied by monopoly is present in principle whether or not the owner of ranchland is held liable for damages. Both the symmetry of the problem and its disappearance under competitve conditions refute the allegation that Coase's analysis implicitly endorses the use of resources in undesirable activities. . . .

The costly interaction between farming and ranching is not properly attributed to the actions of either party individually, being "caused," instead, by resource scarcity, the scarcity of land and fencing materials. If transaction cost is negligible, it would seem that the choice of liability rule cannot depend on who "causes" the damage since both jointly do, or on how resource allocation will be altered, since no such alteration will take place, but largely on judicial or legislative preferences with regard to wealth distribution.

Once significant transacting or negotiating cost is admitted into the analysis, the choice of liability rule will have effects on resource allocation, and it no longer follows that wealth distribution is the main or even an important consideration in choosing the liability rule. The assumption of negligible transacting cost can be only a beginning to understand the economic consequences of the legal arrangements that underlie the operations of the economy, but little more can be done here than to illustrate the nature of the considerations.

The most obvious effect of introducing significant transacting cost is that negotiations will not be consummated in those situations where the expected benefits from exchange are less than the expected cost of exchanging. . . .

Notes and Questions

1. When *does* the rule of liability matter? What does the author mean by "matter," and in what other ways might the rule of liability make a difference?

2. Can you draw the line between extortion and bargaining? One of the mysteries of criminal law is to understand why some forms of extractions are criminal. Does this reading provide an answer to that question?

3. Can we always intuit the author's (sometimes difficult) results by asking what a single owner would do with both properties and businesses under his or her control?

4. A possible criticism of the Coase Theorem (a term often applied to the middle of the first reading) is that it ignores parties' incentives to bargain strategically. Thus it may sometimes be profitable to refuse to reach a deal, or both parties may bargain so fiercely that no deal is reached. In these situations, does the rule of liability matter?

Consumption Theory, Production Theory, and Ideology in the Coase Theorem

MARK KELMAN

The Coase Theorem is the most significant legal-economic proposition to gain currency since the early utilitarians identified the maximization of individual satisfaction with consumer freedom from conscious state regulation. The Theorem is simple in structure and can be easily understood by working through a straightforward example.

Abridged from Mark Kelman, Consumption Theory, Production Theory, and Ideology in the Coase Theorem 52 *S. Cal. L. Rev.* 669 (1979). Reprinted without footnotes with permission of the *Southern California Law Review*.

Problems of entitlement arise whenever two parties interact, e.g., a manufacturer (P) who pollutes a stream, and water users (U) downstream from the manufacturer. In all such interactions, each party is the "but for" cause of harm to the other party. Although the water users could have clean water but for the polluter, so could the polluter spew forth pollutants without costly (and therefore harmful) abatement devices but for the desire of the downstream actors to use clean water. Legal systems establish entitlements to determine which party will be deemed to be the legally responsible cause of harm. In economic terms, these rules force one producer to internalize the external costs he imposes on others: either the legally "real" cost of production of the manufacturer's products includes the cost to the users of having to live with or clean up the dirtier water, or the legally real cost of clean water for the users includes the cost of making manufacture more difficult. If, for instance, U is entitled to a clear stream, P is liable in tort for polluting the stream. Thus, P will bear the burden of "bribing" U to permit him to pollute. If, on the other hand, P is entitled to use water in any manner he chooses, including as a disposal site, U would have to bribe him to abate the pollution.

. . .

Coase attempts to demonstrate, through simple marginal utility analysis detailing the process through which the parties will trade to equilibrium, that the same substantive result will be reached, regardless of who is liable to whom, as long as there are no transaction (negotiation) costs. Assume there are two feasible states that the parties would be entitled to: U could be entitled to pure water (state A) or P could be entitled to render the water as filthy as he wants (state C). There are also intermediate states, e.g., B, where some pollution abatement is used, but not enough to make the water pure. Coase postulates that there are fixed subjective values that the user places on moving between the states, e.g., if B is worth nothing, A is worth $100. The cost of abatement to the polluter is $200 to get from B to A.

Given this assumption, Coase correctly argues that state A will never result, regardless of the liability rule. Even if the users are entitled to pure water, the polluter will still pay more than the $100 pure water is worth to the users to avoid the $200 outlay necessary to create it. Both sides will be better off if the users receive any amount between $100 and $200. If the polluters are entitled to use the water for dumping, U, willing to spend only the $100 that state A is worth to him, will not offer enough to induce a change that costs P $200.

Likewise, state C may not be stable even if the polluter is "entitled" to dump at will. If moving from C to intermediate state B is worth $50 to the user, and it costs the polluter only $10 to make the change, C is obviously unstable: the users will bribe the polluter to abate the pollution by paying the $10 abatement costs and throwing in some of the difference between $10 and $50 to make the polluter want to do it. If the polluter is legally obligated to pay the user for causing the water to be less than pure, he would be unwilling to pay the bribe. P would save only $10 in production costs, while spending $50 for a bribe. Only in a state where marginal cost and marginal benefit are equal, i.e., where it is not worth more to one party to change the state than it costs the other, will there be an equilibrium, and that substantive equilibrium point, where a fixed amount of pollution is generated, will result regardless of whether users are entitled to pure water or polluters are entitled to use water as they will. The only difference will be distributive: who will bribe and who will receive the bribe.

. . .

An Empirical Criticism of Coase's Hypothesis

This section . . . demonstrates that viewing the Coase Theorem as a positive, i.e., empirical and hypothetically falsifiable, hypothesis is wrong. Consumers do not behave in a way such that the theorem holds true. A digression at the end of the section contains speculation on the reasons behind the consumer behavior discussed, although these reasons are not important in falsifying the Coase hypothesis.

Testing the Coase Theorem

Obviously, there are not perfect tests for the Coase Theorem because there are no real world situations in which there are absolutely no transaction costs. Further, there have been very few shifts in entitlements, historically, such that the "legal" cause of harm in a mutual interaction situation suddenly changes, making it possible to see if distribution alone, not substance, is affected by a shift.

However, the behavior of individuals in situations where the Coase problem is implicitly posed can be observed, and it can be seen that the individuals do not behave as they would were his hypothesis to prove true. First, casual empirical examples, which the reader can certainly verify as within his own experience, are examined, before turning to survey evidence. Finally, one of the few real world, macroeconomic cases in which the hypothesis can actually be tested is discussed.

Casual Empirical Evidence.   a. A fully rational individual, a professor at a business school, buys a bottle of imported wine for $5. After its value increases, a wine dealer with whom he regularly deals offers him $100 for the bottle of wine. Although he has never purchased a bottle of wine for $100, in fact, he has never paid more than $35 for one and would not do so now, the professor drinks the wine rather than sell it. What is observed here is a dramatic divergence between the treatment of opportunity cost income and realized income. *If* Coase's hypothesis were true, the individual must treat the $95 in opportunity cost income, i.e., the amount he would realize *were he to sell* the wine, just like the received income that he actually possesses. However, the individual does not do this. Although he would not spend $95 of realized income on wine, he "spends" $95 of opportunity income on it.

The divergence can be seen equally well if the case is viewed as one of "valuing" moves between two states—life without the imported wine (state A) and life with the wine (state B). Although he would not pay $95 to move from A to B (it is not worth $95 to him to *attain* state B), he will not accept a $95 bribe—the selling price of the wine on the market—to bring him back from B to A (it seems to be worth $95 to him to *remain* at B).

The case is surely not one of simple transaction costs. Obviously, one would expect the wine-holder to demand more to part with the wine then he would pay to buy a similar bottle because there is some cost to selling it. The *net receipts* from sale are lower than the *price* at sale. Of course, there are some costs to purchase as well, so that *net price* exceeds nominal purchase price, but the divergence between the price the consumer would pay to purchase and that he would demand to give up the wine should not be any greater than the amount the actual cost of selling (locating a seller, delivering the wine) exceeds the cost of buying. In the instant case, the wine dealer offered $100 for the bottle; because the greatest transaction cost involves locating a purchaser, the $65 divergence between the maximum received income offer for the wine ($100) and the price of a similar bottle of wine ($35) is clearly greater than the transaction cost.

The relevance of this to our paradigm case should be obvious. At a given point in time, the polluter must pay the user for encroachments on perfectly pure water (rule A). Some equilibrium is established through bargaining. Subsequently, the liability rules change so that the user must pay the polluter for encroachments on the polluter's right to dispose of waste in whatever manner he wishes (rule B). Coase hypothesizes that the amount of pollution will remain the same whether rule A or rule B is used: the user will now pay out the same amount for the departure from a no-abatement state to the old equilibrium state as he previously demanded to remain at the old equilibrium state. This is doubtless untrue; the opportunity cost income that the user had enjoyed in the rule A setting, where he could sell his right to clean water for cash, is spent more readily than the received income that he must actually spend to attain that state. Income that the person could realize is worth less than the income that the person already has realized. Viewed in the other manner by which the Coase Theorem is being examined, the value to the user of moving from the initial equilibrium state attained under rule A to a situation of no-abatement (represented by the possibility of payment of money from P to U) is perceived to be worth less to U than the value of moving from a no-abatement state to the old equilibrium situation (this move being represented by the payment of money from U to P).

Alternatively, assume that pure water is "first in time" and that polluters then make their appearance on the social scene, so that a liability rule must be picked to govern the user-polluter interaction. If rule B were chosen, U would have to spend received income to keep pure water. He would be more reluctant to do this than to forego the opportunity income to which he would be entitled if rule A were chosen.

b. A consumer buys a new color television and decides to keep his old black-and-white set for which he could *realize* $50. He makes the decision after accounting for all transaction costs—an offer for $50 has been received, and he will incur no additional costs in disposing of the television. If that second television were destroyed, he would not pay $50 for a second television; neither would he pay out $50 to ship the first set to a new home. Again, the $50 of opportunity income is spent on preserving the status quo, keeping the television, although $50 of received income would not be spent to get to the same substantive two-television state. In other words, the consumer does not feel it is worth *as much as* $50 to go from state *A* (one television) to state *B* (two televisions), but once at state *B,* he feels that it is worth *more* than $50 not to move to state *A,* and he will not accept a $50 bribe to make the move.

. . .

Explaining Behavior of Economic Actors

For the purposes of this selection it does not matter whether the behavior manifested by the economic actors that have been discussed is rational or irrational, explicable or inexplicable. Insofar as the Coase Theorem is designed to undercut the notion that the liability rules that determine who is legally accountable for harm in cases of mutual interaction will affect substantive results—therefore to deny the inevitablity of political decisions on the contours of the policy—it simply fails; liability rules will affect substance.

One could reasonably argue that the Coase result will still obtain so long as both the interactors are "producers," rather than "consumers," who are personally indifferent to the end-state except as a means for attaining the lowest cost of production. Thus, in the farmer-railroad interaction,* where the farmer values crops *only* as a marketable commodity, and their market value is unaffected by *where* on the field they are grown, the Coase Theorem will still hold. In fact, though, this is not a significant case. In the classic nuisance situations, where advocates of one liability rule or the other claim to have a *political* stake in the choice of rules, there will inevitably be a "consumer" on one side. Insofar as farmers are indifferent as producers to where the crops go, there will be political indifference to the liability rule. *If,* however, the farmer has some aesthetic (pastoralist) attachment to his fields (and this attachment is mirrored politically by pastoral advocates), he *becomes* a consumer in the interaction. His willingness to buy pastoralism will depend on whether he must pay out for it or simply resist bribes.

Likewise, many "producers" sell products when the "consumer" issues that have been discussed are clearly relevant. For example, assume that our downstream landowner is a landlord, not an occupant. He will generally treat opportunity and realized income identically; having no attachment (and perhaps no knowledge) of the current level of water quality, there is no reason for him to differentiate between two kinds of income-enhancing offers: bribes from the polluter to dirty the water and higher rent offers from tenants to live

*[The reference is to the discussion in the preceding selection—Ed.]

on cleaner water. Still, however, the liability rule will matter. Suppose the polluter moves to town while renters are paying $200 for an apartment on the clean water. If the liability rule disfavors the polluter, he may offer $50 per unit to dirty the water. The landlord will refuse this offer unless tenants are willing to pay $150 for an apartment on dirty water. The tenants, however, may well be unwilling to live next to dirty water for as little as $50 in *saved* rent. But if the landlord is liable to the polluters, he will be unable to buy clean water unless the tenants will pay $250 per month to attain that state, and they may be unwilling to spend $50 per month to get the clean water.

The Coase Theorem already contains a deliberately unrealistic assumption—that there are no transaction costs. Although the asymmetry of treatment of opportunity and realized income has not been explored, or criticized, by the Chicago school proponents, it is conceivable that they would attempt to deal with the difficulties presented by the asymmetry of treatment by adding a new assumption to the Coase Theorem—that people treat opportunity and realized income in the same way. If this were done, the tautological boundaries of the Coase Theorem would become apparent. The theorem would then become the exciting and daring proposition that people will always bargain to the same substantive position regardless of liability rules so long as they are people who bargain to the same position regardless of liability rules. The assumption of no "transaction costs," although unrealistic, at least does not reduce the theorem to a definition.

The asymmetry of treatment of opportunity and realized income that has been observed cannot be explained in terms of transaction costs. Calabresi defined the Coase Assumption as follows: " '[N]o transaction costs' must be understood extremely broadly as involving both perfect knowledge and the absence of any impediments or costs of negotiating." Even this "extremely broad" reading of "transaction costs" does not include the purely *internal* aversion to sale after negotiation is complete. Transaction costs, unlike this aversion, can be reduced through simple *technical* progress. Markets may be organized and information spread more cheaply when scientific techniques advance (e.g., computer terminals in the home listing "rights for sale"), but if people remain privately and psychically oblivious to *marketing*, the technical advances will not matter.

It might be argued, however, that the general presence of transaction costs prevents people from considering sale, even in particular cases where those transaction costs have been eliminated. For example, the *ordinary* absence of a costless resale market for wine or television sets may contribute to the fact that people avoid resale, even when resale would be costless. Thus, presumptive (though on particular occasions nonexistent) transaction costs may be *one* of the mechanisms that create asymmetries in treatment of "opportunity" and "realized" income, and thereby facilitate the positive effect of liability rules on behavior.

Consider the growth of garage sales. These sales suggest three factors that may increase the likelihood that our hypothetical consumer will sell his extra television: First, in the manner already recognized by Coase advocates, the general existence of these sales will typically lower transaction costs, enabling the seller to locate a broader base of consumers at less cost. Second, in line with the hypothesis just discussed, sales may become more probable if the prevalence of garage sales causes a shift in the *presumption* against resale, which tends to make people overestimate the costs of resale even when those costs are not present in particular cases. Third, and perhaps most significant, the social practice of garage sales may simply serve to redefine "static states" from which departures are unwelcome. Once the seller becomes aware that he is foregoing "ordinary" income receipts, he will be more likely to wish to resell.

The second and third explanations of the consumer's behavior are distinct. In the second situation, the presumption against resale has been shifted because consumers have more information and are less likely to overestimate the costs of resale. The consumer who chooses not to resell is simply misusing information that is accessible to him; he fails to perceive a "costless" transaction and continues to apply the presumption that large transaction costs are present. In the third situation, the consumer who decides not to sell is simply indifferent to transactions. He is aware that he is foregoing "ordinary" income receipt but, nonetheless, he does not wish to resell.

Consider a parallel phenomenon: the growth of a black market in babies. As the black market develops, three distinct things happen: (1) the net amount of money *realized* by sale will, on the average, increase as each seller locates buyers at lower cost and transactions become routinized; (2) sellers may take more seriously particular offers that are equivalent to those they received before the market developed, having believed somehow that the earlier above-average offers did not represent the amount they could realize on a particular sale; and (3) babies will start to be viewed as a commodity so that the failure to sell them will seem like an economic decision. The parent who would not have sold a child for $10,000 (and perhaps not even for *any* price) before the market developed may now do so, not so much because, in general, more people can realize $10,000 on sale or because he or she perceives that $10,000 can really be made on this sale, but because babies are now perceived as being part of the world of marginalism and calculation: Further facilitating market transactions by legalizing the sale of children would not merely increase the money realized by sellers (affecting the *quantity supplied*) but would alter sellers' willingness to sell at any particular price (affecting *supply functions*).

It is beyond the scope of this selection to explain in great detail economic behavior inconsistent with the false assumptions of the Coase Theorem, and, as noted, refutation of the theorem does not depend on whether the behavior is explicable. A few further comments, however, in explanation of this behavior may be fruitful.

The Perception That Treating Received Income Unlike Lost Opportunity Income Is Rational: "Could Be" v. "Is".   There is probably no appealing definition of rationality that requires a person to treat unrealized, opportunity income the same as realized income. In terms of ordinary fact-value distinctions, the asymmetry of treatment of the two types of income seems like pure "taste," unrelated to instrumental, fact-grounded rationality. Certainly, people often care less about things that could be than about things that are immediate and tangible. For instance, there seems to be less concern, even in the Catholic Church, with protecting "potential life" (by banning contraception) than with protecting realized life (by protecting the fetus or the newborn child). Consider also, that a person will probably feel worse when he loses a $50 bill than when he fails to walk from one store to the next to save $50 in the purchase of a washing machine. Such behavior, however, hardly qualifies a person for the loony bin; spending one's life fixated on "what would have been" if one had followed through all opportunities seems a far surer path to the psychiatrist's couch. Neoclassical economists traditionally satisfy their urge to label all behavior as rational by recourse to tautology: the "psychic cost" of treating opportunity income as received income is positive. The [preceding] explanation of the divergent treatment of lost opportunities and ordinary income reflects a rational accounting for this "psychic cost."

Growth of Income.   Implicit in the [previous] examples is the proposition that a rise in income matters less to people than a fall in income. Although this is certainly true, it is not the only claim. Although the two-television-set-owner's behavior might be partially explained as reflecting relative indifference to a rise in income, it would seem plausible that if he got a $50 raise the same day he bought the new television, and the second set were later destroyed, he would not buy a second television with the "new income."

There appears to be a nexus between the income-bearing entitlement (e.g., possession of the wine, possession of a right to clean water) and the decision on disposition of the good associated with the entitlement (e.g., to drink the wine or not, to "spend" the "opportunity income" from the entitlement on clean water or not). This nexus transcends the question of a shift in income position. In Kahneman and Tversky's work on gambling preferences, must subjects perferred a certain $500 gain to a fifty-fifty chance of winning $1,000, while most of the subjects preferred to take a fifty-fifty chance of losing $1,000 rather than to give up $500 certainly. Even if a subject had just received a separate $500 windfall gain, he might have made the same choice. The choice would seemingly depend on whether the windfall were one that he emotionally connected with the decision to gamble. In Las Vegas, for example, people do treat first-day gambling winnings as money that can be freely gambled away during the remainder of a stay; that increment to income affects their view of whether subsequent losses are deemed "losses" from some psychological base position. Whether they would so treat a raise that they learned about while on the Las Vegas vacation seems very doubtful. An inheritance that came through on the trip would probably be a closer question, and the treatment of it might depend in part on whether it was long-expected (and therefore already incorporated into the fixed asset conception from which downward departures were painful) or not.

Nonetheless, consumers probably do treat rises in total income differently than they do the remainder of their income. For example, the most acceptable theory of savings behavior indicates that most savings occur in disequilibrium periods when income is going up and people have not yet become accustomed to spending all that they have. It has also been observed that people tend to defer or avoid having children until the expenditures that the children will entail can come out of income that they are not yet used to spending. This dogged attachment to protecting current levels of income, coupled with a relative indifference to growth of income, may reflect the peculiar nature of consumption in a class society where exogenous needs, which are presumably either met or unmet at fixed levels of *absolute* income regardless of whether that level represents a rise or fall in income, are subordinated to ever-expandable addictive needs to maintain a current standard of living. The attachment may also reflect the more universal, less socially contingent, fact that people enjoy most what they have learned to enjoy. It is a difficult question whether people's attachment to their current economic state is a reflection of acquiring new "consumption skills," or instead is the result of an unwanted, regrettable *anxiety* about material deprivations in a society in which self-image and possessions are intertwined.

"Closing Transactions."   The behavior that is being described is conceptually related to another form of behavior that traditional economists, like Coase, find puzzling: the refusal of consumers to ignore sunk costs. For example, in deciding to play tennis at the club every week despite a bad elbow, very few people would consider it irrelevant that they had paid $100 to join the club even though the $100 was spent in the past and was therefore unrecoverable. Economists generally find this hard to explain. If the consumer would not play if the tennis were free, it is hard to understand why playing tennis is any cheaper (more desirable) when one has already paid out money. The consumer, however, feels he would be wasting the $100 if he does not play; he ignores the fact that no funds will be demanded in the future and asks whether, in the past, he has received value for the $100 he has spent. Consumers try to "close" transactions: $100 was spent on tennis, and the consumer wants $100 of tennis value.

The hypothetical cases that [have been] discussed involve an event that is psychologically related to closing transactions. Just as some transactions are *not* closed because value has not yet been obtained (and sunk costs are, therefore, accounted for), other transactions *are* closed once value is obtained (and future opportunity income from the right obtained in the initial transaction is, therefore, ignored).

Consider the case where the family found itself with two televisions. It is necessary to explain why the family treats retention of the television it has long owned as free, rather than costing the money that would be realized by selling the set. In part, this feeling reflects the general gap between realized and unrealized events and in part the sense that extra income is worth less to individuals than lost income. In terms of the sunk-cost argument, however, if the consumer feels that he has already received adequate value from the original purchase, then the transaction vis-a-vis that old television can be psychically closed. The television's future-oriented, economic impact, as a possible source of income, is not important; there is no need to open up the "case of the television" again. If the set were destroyed, however, and a decision to buy a new one had to be made, the case would be inexorably opened and a future-oriented decision would be unavoidable.

In summary, there is a desire to withdraw spheres of activity from the realm of marginalism and calculation: We may wish to extend the non-commodity relationship, which most people still have with babies, to many traditional material concerns (like holding on to possessions that are formally alienable) because the commodity relationship itself is undesirable.

Although worrying about whether everything one owns is now worth its resale value may be perceived as obsessive, a presumptive waste of time, and an unnecessary capitulation to unwanted "trader's" status, the assumption that a person ought to act as if he is throwing out money when sunk costs are ignored may be quite functional. Cases that at first appear to be sunk-costs cases may not be. For example, the decision not to drop out of college during a semester because one has already spent $2,500 on tuition may seem entirely past-oriented at first, but if one will eventually want to re-do the semester, and thus spend an additional $2,500, then the decision proves not to be entirely past-oriented. Even in the tennis club case, one may argue that some-day a person is going to play a fixed, finite amount of tennis with tennis elbow. If one will later pay an additional $100 to do so, you might as well play with tennis elbow now. Thus, the presumption that one should ignore his initial perception of costs as sunk, and instead attempt to complete transactions, may, therefore, be a good rule of thumb.

On a deeper level, the desire to complete transactions, to refuse to treat costs as sunk, facilitates planning and the integration of the contingent self into a whole, willed personality. If a person goes to the theatre when he has bought tickets, in part because he does not like to see money "thrown out," that part of the person, perhaps with a longer, wiser perspective, that wants to go to the theatre, even on a rainy night, can win out more readily over the shorter-term personality who finds going to the theatre more inconvenient than not.

Conservatism.   One argument that should be rejected as an explanation of the behavior that has been observed is a "risk aversion" or "conservatism" approach. The argument would basically be as follows: Once the wine owner or television owner possesses the commodity, he becomes fixed on it. The obverse side of the coin is that buying a new bottle of wine or a new television is always risky, and the value of the purchase is inevitably discounted by the possibility of "failure" or disappointment. This account of nonsymmetrical buying and selling behavior seems to depend on the "protected interest" (e.g., clean water) being more familiar than alternative states. Although the user will be *most* familiar with the particularities of his present circumstances (i.e., *this* living situation with *this* body of water), he may well be relatively familiar with a vast range of water-quality states, having seen pictures or having lived with different water qualities. His refusal to sell the clean water, unless paid more than he would pay to attain it were the water dirty, would not, however, rest on the difficulty of getting him to surrender the familiar because both clean and dirty water may be, broadly speaking, "familiar" to him. Rather, the refusal would rest on the cost of getting people to marginalize, calculate, and put everything in their life up for sale. Our two-television man refuses to sell although that refusal involves living in an *unfamiliar* state; he has always been a one-television man in the past.

The User-Polluter Paradigm.   The reasons that liability rules will affect substance in light of the above terms may be explained as follows: Assume, once more, that in a world where pure water came first, a liability rule must be selected once the polluter comes into existence. Suppose the cost to P of maintaining the water in its pure state is \$100; if P has to pay for the right to dirty the water (he will offer up to \$100), the water may stay clean (U will refuse this offer) whereas if U has to pay for cleanliness, he will not pay out \$100. Perhaps, as in the "could be" versus "is" argument, U simply ignores money that he "could" make; the thought of selling his new-found right is no more on his mind, even when raised by P, than wondering whether he ought to sell Avon products on his lunch hour. Perhaps, as in the growth of income argument, the positive increment of \$100 is less pleasurable to him than the loss of \$100 is painful; having income grow is not as urgent as avoiding income losses.

Finally, at the intersection of the above arguments is the argument on "closing transactions." U may consider his life in a general equilibrium state, where all goods (including clean water) are purchased as he wishes before the liability rule. It is relatively difficult and expensive to disturb this equilibrium. If the polluter is disfavored by the liability rule, U can easily remain in that equilibrium simply by turning down offers. If the user is disfavored, the equilibrium state is inexorably shattered: If the user pays out the money, he must give up other goods that have been part of his equilibrium state; if he does not, he will lose the clean water. Once this equilibrium is shattered, the terms on which the user will trade off clean water with other goods will be different from the state where he would be trading clean water for goods not part of his equilibrium packet. Because all goods at issue for U are "familiar" or "on the market" when he is obliged to pay for clean water (or, alternatively, because the water transaction cannot be considered psychically closed), pure water is lower in value than when it is traded off against unfamiliar or nonmarketed goods (or, alternatively and preferably, when the transaction involving it must be psychically opened by the necessity of paying out money and foregoing other purchases).

Substantive Outcomes May Be Altered by the Sociolegal Impact of Liability Rules

For the Coase Theorem to be true, the parties' tastes must be completely exogenous, i.e., they must not be affected by the legal rules. There are two ways in which this assumption is relevant.

First, if the consumer in our paradigm case values pure water more highly *because* he is entitled to it, because it is a legally protected interest, then the substantive outcome will obviously be affected by the liability rule. He will not give up his *right* to pure water for the $100 cost of abatement because he values the right too highly. But pure water, a good like any other on the market that he can buy or not, may not be worth $100 to him. Perhaps society learns what to value in part through the legal system's descriptions of our protected spheres. Second, if the polluter considers tortfeasance morally difficult, i.e., if he does not view overstepping *U*'s rights as something that he is entitled to do as long as he pays damages, he will likely spew forth more pollution when he is not committing a tort then when he is.

Consider the case of rape. In the Coase Theorem world, the amount of rape is unaffected by whether the victim must bribe the attacker to refrain from raping her or whether the attacker must compensate the victim for the rape. For this proposition to be true, one would have to assume that the rule *against* rape had *no* effect on either the attacker's taste for forced intercourse or on the woman's view of the outrageousness of forced sexual contact.

Naturally, it is difficult to determine in such cases what the independent effect of legal, rather than moral, precepts really is and whether the moral rules better retain their force when legal rules reinforce them. In the absence of empirical proof, however, the Coase presumption—that tastes are independent of sociological stamps of approval and disapproval—hardly seems convincing.

Notes and Questions

1. The differential in valuation and behavior emphasized in the beginning of this reading is sometimes referred to as the "offer-asking" difference, or problem; people will sometimes ask (sell) at a different price than they would offer (buy). When would you expect this differential to be most significant?

2. The examples offered by Coase and Demsetz involve profit-making businesses, whereas Professor Kelman focuses on valuations made by individuals. Is there an offer-asking differential for a business in valuing an opportunity or asset it has or wishes to acquire? Put differently, how does a business decide how much it will pay for an asset, including a legal right? Does this calculation change if the business is large or if it has been profitable in the past?

Consider again the example offered in the Notes and Questions following the first reading in this part, in which an amusement park owner and a home-improvement store owner disputed liability for the damage caused by errant golf balls. How does the

Kelman reading affect your view of this dispute? Is it likely that Coase and Kelman would dispute the analysis or ultimate judicial decision in this case?

3. One of the examples in this selection concerned the question of selling a used television set. Why might someone retain a used television set when just a day earlier that person would not have purchased one? Why is it difficult for a consumer to realize much money when selling used goods? In turn, might it make sense for consumers to get in the habit of retaining rather than selling old appliances?

4. A fairly advanced topic in tort law concerns the treatment of parties who have voluntarily "come to a nuisance," or problem. It is sometimes said that first in time is first in right. Thus courts might be disinclined to hold a polluter liable for damages experienced by a plaintiff who moved next door to the polluting business, built a house, and then sued to shut down the polluter's business or to collect for diminution in the (potential) value of the house. Imagine that in such a lawsuit the late-coming property owner loses on all claims and then sells the house to another private party who is fully aware of the history of the property and the legal dispute. If this second owner sues the polluter despite the first owner's loss in court, do you see any offer-asking problem, or might we say that because the plaintiff bought the property at a price that reflected the prevailing legal rule, such a problem does not exist? When might we expect (or wish) courts to be sympathetic to a latecomer who sues?

The Role of Fault

II

In much of tort law a precondition to liability is some kind of fault, wrongdoing, negligence, or defective manufacture or design. There remain, however, subsets of tort law (and many other areas of law) in which fault plays no role in the assessment of liability. A recurring topic in this part is understanding the division of the terrain between fault and stricter liability.

One way to think about the choice of a standard for liability is to adopt a functional approach. For example, a negligence rule may be less (or more) expensive to administer than any reasonable alternative. Strict liability may be thought to provide parties with the appropriate incentives to consider the true costs imposed by their actions. Such functional considerations are discussed in the first, fourth, and sixth selections.

A very different approach asks when it is appropriate, as a moral or philosophical matter, to have one party compensate another. Rather than viewing tort liability through an economic or utilitarian lens, this approach is based on corrective justice or on the notion that it is simply right for everyone to pay for what he or she "causes." Causation, a vexing problem, especially in light of Coase's insight (discussed in part I), is explored in part III. Two of the readings in part II, by Professor Fletcher and Judge Posner, take up the corrective justice approach and compare it with the functional, or economic, method.

A third way of thinking about the choice between negligence and strict liability (or even no liability) involves the possibility that historically judges were part of a movement to encourage certain activities or to disfavor subsets of citizens or enterprises. The penultimate reading in this part discusses and takes issue with the best known of these suggestions—that tort law served as an accessory to the industrial revolution by declining to make emerging businesses pay the full costs they imposed.

Killing or Wounding to Protect a Property Interest

RICHARD A. POSNER

A farmer in Iowa named Briney had a farmhouse in which he stored old furniture and odds and ends, including some antiques of undisclosed value. After several thefts, Briney rigged a spring gun* in the farmhouse. At his wife's suggestion, he pointed the gun so that it would hit an intruder in the legs—not, as Briney had initially planned, in the stomach. A man named Katko, who had previously stolen goods from the place, broke in, triggered the spring gun, and was badly wounded in the leg. He was initially charged with burglary but later permitted to plead guilty to petty larceny, fined $50, and given a suspended jail sentence. He brought a civil suit against the Brineys, charging that his wounding was a battery, and won a jury award of $20,000 in compensatory and $10,000 in punitive damages. The verdict occasioned a public outcry, and proposed legislation (modeled on a recent Nebraska law) that would explicitly have authorized the use of deadly force in defense of property was narrowly defeated in the state legislature. At this writing the case is awaiting decision in the Supreme Court of Iowa on defendants' appeal.

Spring guns were something of a cause célèbre in early nineteenth-century England, but since that time the reported cases have been few. Cases involving the use of deadly force to defend property in person have been few too, in part because self-defense is so often an issue when the defendant is present. Perhaps the Iowa case, when viewed against a background of mounting public concern over crime, signifies a resurgence of the problem. What makes the deadly-force issue worth discussing, however, is not its topicality but its theoretical interest, which I believe to be considerable. Involving as it does a conflict between the preservation of life and the protection of property interests, the privilege (if there is a privilege) to use deadly force to protect property cannot fail to raise fundamental issues of legal policy. It also presents interesting questions concerning the allocation of law enforcement authority between the public and private sectors. The approach of conventional legal scholarship has been unsatisfactory, and here is another source of interest, since this failure illuminates characteristic deficiencies of such scholarship, especially as it is embodied in the American Law Institute's *Restatements*. I am led to explore an alternative approach, an economic approach, whose

---

*A gun (usually a shotgun) rigged to fire when a string or other triggering device is tripped by an intruder.

Abridged and reprinted without original footnotes by permission from 14 *Journal of Law and Economics* 201 (1971). Copyright © 1971 by The University of Chicago Press.

utility in helping to answer questions of legal policy and to interpret opaque
and apparently conflicting judicial decisions is a major theme of this paper. As
we shall see, an economic approach is useful even though it is not usually
thought of as an especially apt tool for resolving such basic value questions as
life versus property. Finally, although our privilege may seem far from the
mainstream of contemporary concern with tort law, we shall see that it is in
fact paradigmatic of a wide spectrum of tort questions and illuminates the
central policies of that law.

                                              . . .

The failure of generations of distinguished scholars to give an adequate
account of the law on what is after all both an old and a narrow question
reflects, I believe, limitations inherent in a certain type of legal scholarship
and in the attempt to restate the common law in code form. Limitations of the
first kind include a propensity to compartmentalize questions and then consider each compartment in isolation from the others; a tendency to dissolve
hard questions in rhetoric (for example, about the transcendent value of
human life); and, related to the last, a reluctance to look closely at the
practical objects that a body of law is intended to achieve. Codification, as in
the *Restatements,* would hardly counteract these tendencies. Indeed, the pre-
occupation with completeness, conciseness, and exact verbal expression natural to a codifier would inevitably displace consideration of fundamental issues
and obscure the flexibility and practicality that characterize the common law
method.

Perhaps these failures of scholarship stem ultimately from a tendency to
confuse what should be distinct levels of discourse. I expect that most judges,
before deciding a case, conceive it in highly practical terms. I do not mean by
this that they consider which party's plight is more desperate, which more
engages their sympathies. I mean that they consider the probable impact of
alternative rulings on the practical concerns underlying the applicable legal
principles. . . .

Why judges, having made a practical decision, so often embody it in the
pompous, stilted, conclusionary prose that the layman derides as legalistic is
something of a mystery. But what seems clear is that the task of legal scholarship is to get behind the prose and back to the practical considerations that
motivated the decision. Yet scholars often seem mesmerized by the style,
terminology, and concepts of the judicial opinion; they confuse their function
with the judicial.

A possible way of avoiding this danger is to take an economic approach to
questions of legal interpretation. One who tries to explain cases in economic
terms may expose himself to many pitfalls, but they will not include the pitfall
of attempting to analyze cases in the conceptual modes employed in the
opinions themselves. An economic approach is especially plausible with regard to tort law, since the subject of economics is how society meets the
conflicting wants of its members and tort cases, as we shall see, are plainly
concerned with arbitrating such conflicting wants.

The nature of an economic approach to our problem can be illustrated by reference to an old English case, *Bird v. Holbrook*. The defendant owned a valuable tulip garden located about a mile from his house. It was surrounded by a wall 7–8 feet high on one side and somewhat lower (how much lower is not indicated) on the other sides. After some of his tulips were stolen, the defendant rigged a spring gun. One day a neighbor's peahen escaped and strayed into the garden. A young man (the plaintiff in the case) tried to retrieve the bird for its owner, tripped the spring gun, and was badly injured. The incident occurred during the daytime, and there was no sign warning that a spring gun had been set.

The case involved two legitimate activities, raising tulips and keeping peahens, that happened to conflict. Different rules of liability would affect differently the amount of each activity carried on. A rule that the spring-gun owner was not liable for the injuries inflicted on the plaintiff would promote tulip raising but impose costs on (and thereby tend to contract) peahen keeping, for knowing that efforts to retrieve straying fowl from neighbors' yards might invite serious (and uncompensable) injuries, keepers of peahens would keep fewer fowl, or invest in additional measures to keep the birds from straying, or do both. The opposite rule, one that recognized no privilege ever to use spring guns in defense of property, would benefit peahen keeping but burden tulip growing. The wall surrounding the garden had not been effective in preventing theft. The garden was too far from the defendant's home for him to watch over it himself. Raising the wall or hiring a watchman may have been prohibitively costly.

One would have to know a good deal about tulip growing and peahen keeping, and about the likelihood and character of other trespasses to the garden, in order to design a rule of liability that maximized the (joint) value of both activities, net of any protective or other costs (including personal injuries). And if one wanted a rule that applied to still other crops and straying creatures one would have to know a lot more. But what seems reasonably clear without extended inquiry is that the economically sound rule will be found somewhere in between the extreme possibilities of making the spring-gun owner never liable or always liable. At the minimum, someone in the defendant's position should be required to post notices that anyone entering his garden might be shot: the cost of doing so would be less than the cost (in medical expenses, loss of earnings, and suffering) likely to be incurred by someone who strayed into the garden on an innocent mission. (Of course, a daring and ingenious thief, alerted by the notices, might be able to avoid or disarm the spring gun.) It is possible to go further and suggest a plausible rule that avoids the extremes of blanket prohibition and blanket permission. Given that the expenses of protecting the defendant's valuable tulips other than by a spring gun would probably have been high, that a theft was most likely to be attempted at night, that domestic animals are usually confined then, and that people (other than burglars) do not customarily climb walls at night, the defendant should have been permitted to set a spring gun only at night and after posting appropriate notification. The actual decision in the case is consistent with such a rule. The Court of Common Pleas held for the plaintiff, stressing the absence of notices and the fact that the incident occurred in the daytime.

The reader may object that an analysis which focuses exclusively on the value of the interfering activities is too narrow and in one respect he will be clearly right: it improperly ignores the costs of administering different rules of law. A complex rule, one carefully tailored to relevant differences among the situations to which it might be applied, may do better in terms of maximizing the joint value of the interfering activities than a simple and crude rule yet be inferior because the additional costs of administering the complex rule exceed the additional value of the activities. The complex rule may require lengthier (and hence more costly) litigation or settlement negotiations; or it may be more uncertain and the uncertainty may have a dampening effect on productive activity. Because the costs of different types of legal rule have never (to my knowledge) been seriously studied, it is very difficult to introduce the element of administrative expense into the economic calculus but I assume that judges attempt to do so in a rough way. Our law is replete with instances where judges explicitly rejected a more complex in favor of a simpler rule because the costs of administering the former were thought to outweigh its benefits. That is what the debate over per se rules in antitrust law—to take one of many examples—is all about.

. . .

We need to consider a rule of liability that will have a more general application than the rule suggested for *Bird v. Holbrook* and will be more firmly grounded in a discussion of the relevant considerations—including the value of a human life.

Some people express shock at the idea of weighing personal injury and death in the same balance with purely economic costs and benefits, but it is done all the time. Individuals who work at hazardous jobs for premium pay are exchanging safety for other economic goods. And where life is taken or injury inflicted in an involuntary transaction, such as an automobile accident, society often attempts to approximate the loss in monetary terms. It goes without saying that the task of approximation is an extremely difficult one. Some dimensions of the loss—such as the anguish to family and friends— cannot even be approximated by the methods available to the courts and are therefore usually ignored. But it is out of the question to ban all hazardous activities on these grounds.

A difficult problem of analysis is created where, as will often be the case when deadly force is used to defend property, the person killed or injured is a criminal. One could argue that burglary and other thefts involving trespass to land are risky activities and that someone who engages in them is no different from a man who agrees to drive a dynamite truck for extra pay: he assumes the risk of being killed. Or one could argue that society should place only a small value on the lives of people who engage in antisocial conduct. These arguments could be debated endlessly; it is sufficient to note that they ignore important practical considerations. If a burglar is injured, his injuries will be tended, if need be at the expense of the state; and if he is disabled, he will not be left to starve. The costs of treating and maintaining him are no less real costs to society than the costs of treating and maintaining the innocently injured and disabled. The interest in minimizing such costs cannot be ignored in the design of a proper rule of liability. Furthermore, a rule that greatly increased the hazards of certain property crimes might disrupt a more or less carefully calibrated scheme of criminal penalties. One reason for not punishing all crimes with equal severity is to preserve an incentive for criminals to commit less serious in preference to more serious crimes. If robbery were punished as severely as murder, there would be fewer robberies but more occasions on which the robber killed everyone who might be a witness. If the burglar of an unoccupied building ran the same risk of being killed or maimed as a burglar of an occupied dwelling, there might be more burglaries of occupied dwellings and hence a greater danger to personal safety than under legal arrangements that made burglaries of unoccupied buildings safer for burglars.

It does not follow that an appropriate rule of liability would be one under which a burglar injured by a spring gun set in an unoccupied building could always recover damages. It is one thing to attempt to graduate punishment in accordance with the gravity of different crimes and another to adopt policies that make the punishment, when discounted by the probability of escaping apprehension, a negligible deterrent. One can imagine situations, for example the storage of valuable property in a remote location, where the likelihood of preventing theft or apprehending the thief afterward without using armed watchmen or spring guns would be so small that even nominally quite severe criminal penalties would not deter. In cases such as these deadly force may be an appropriate, because it is the only practical, deterrent.

These observations reinforce the point made earlier in connection with *Bird v. Holbrook* that neither blanket permission nor blanket prohibition of spring guns and other methods of using deadly force to protect property interests is likely to be the rule of liability that minimizes the relevant costs. What is needed is a standard of reasonableness that permits the courts to weigh such considerations as the value of the property at stake, its location (which bears not only on the difficulty of protecting it by other means but also on the likelihood of innocent trespass), what kind of warning was given, the deadliness of the device (there is no reason to recognize a privilege to kill when adequate protection can be assured by a device that only wounds), the character of the conflicting activities, the trespasser's care or negligence, and the cost of avoiding interference by other means (including storing the property elsewhere). The enumeration of the relevant criteria is simple enough. The real challenge is to fashion them, on the basis of scanty information, into a rule of liability that will maximize the value of the affected activities, subject to the constraint that any rule chosen be simple enough to be understood by those subject to the rules and to be applied by courts (our administrative-cost point). I offer the following as a plausible such rule:

1. Deadly force should not be privileged in situations where the owner of property has an adequate legal remedy (as in the typical boundary dispute), or

where the threatened property loss is small. In these cases the costs of protection in human life or limb exceed the value being protected. However, in the computation of value, the economic status of the owner should be considered, since property that would be of no moment to a person of average means might be extremely valuable to a poor person.

2. There should be no privilege to set deadly contrivances such as spring guns in heavily built-up residential and business areas. The protection of property by means of alarms, watchmen, or the police should normally be feasible in such areas and is much to be preferred in view of the undiscriminating character of the mechanical devices. To be sure, one can imagine cases where a spring gun might seem an appropriate measure in such areas: an old lady living alone in a high-crime-rate area; a house full of priceless paintings. But even in such cases (and note the self-defense element in the first) the dangers inherent in the use of the device seem inordinate. The old lady might die in her sleep; her house would be a death trap. A fire might break out in the house containing the paintings; the firemen would trigger the spring gun. The likelihood of beneficent, or at least innocent, intrusions—by public officers, concerned neighbors, mischievous boys, meter readers, and the like— seems greatest in a built-up area, the very situation where alternative protective measures are most likely to be relatively effective at reasonable cost. In contrast, in remote areas the alternative protective measures are less feasible and at the same time noncriminal intrusions are less frequent. (To be sure, in a remote area the victim is also less likely to receive prompt aid.)

Because the stationing of armed watchmen involves fewer dangers, it need not be confined to remote locations. The watchman can discriminate between the harmful and harmless intruder and can usually prevent a theft or apprehend the thief without actually harming him.

Although the undiscriminating character of the spring gun, as I have indicated, is a matter of legitimate concern, it has at least one redeeming grace. One danger of recognizing any privilege to kill is that it may be used as a shield for unjustified killing. *A* hates *B,* shoots him, and then claims it was self-defense. *B* cannot dispute the point because he is dead. Spring guns are at least devoid of any personal animus—though so are most watchmen. The privilege to kill in self-defense is more prone to abuse than a properly limited privilege to kill in defense of property. The latter privilege is ordinarily asserted against strangers; it is harder to use against a personal enemy. (Periods or places where racial or other tensions create a danger that armed watchmen will kill total strangers merely because they belong to a disliked group require special rules.)

3. The privilege to use deadly force in defense of property should be forfeited if the user fails to take reasonable precautions to minimize the danger of accidental injury both to innocent and to criminal intruders. If theft is likely only at night the spring gun should not be set during the day. Signs should be posted with explicit and credible warnings. The defendant should be liable if he left his door open or his land unfenced—thereby virtually inviting intrusion—or if he declared that he had not set a spring gun when he

had. Lethal calibers, or, in the case of a shotgun, lethal shot, should be avoided in spring guns or other devices since ordinarily the wounding of an intruder is adequate to prevent intrusion. A watchman should not be subjected to this requirement, because his personal safety might be endangered. But he should be required to warn a thief before shooting at him, at least where the thief is clearly not armed; and, consideration of his own safety permitting, he should be required to shoot to wound rather than to kill.

4. In property not sufficiently enclosed to keep out straying animals, children, and youths, the privilege to set spring guns should be limited to the nighttime.

5. Where the use of deadly force is permissible under the foregoing precepts:

a. An adult intruder killed or injured in an attempt to steal or destroy property should not be permitted to recover damages. This result is appropriate in order to prevent serious property losses due to theft in circumstances where, as discussed earlier, other means of deterrence may be impracticable.

b. An innocent intruder should be denied recovery if carelessness on his part contributed materially to the accident. This part of the rule is designed to minimize the joint cost (which is another way of saying, maximize the value) of legitimate but interfering activities by placing responsibility on the participant who could have avoided the interference at least cost. Suppose that a watchman has been stationed, or a spring gun set, and all reasonable precautions observed. A birdwatcher comes along, climbs a high fence—ignoring a clear warning notice in plain view—and triggers the spring gun against which the notice warned; or he ignores the repeated warnings of an armed watchman who reasonably believes that the theft or destruction of valuable property is being attempted. One could prevent the accident by forbidding spring guns and armed watchmen in all circumstances, but this may be a very costly means of prevention. The method of averting accidents likely to minimize the relevant costs is one that encourages the intruder to take a few precautions himself by barring recovery of damages otherwise.

The reader may question whether it is realistic to suppose that the denial of damages will deter an intruder not already deterred by fear of being killed or maimed. Perhaps people do not, in general, take greater precautions (other things being equal) against those hazards that are not compensable, such as being struck by lightning, then against those that are. It is hard to believe they do not. When a person takes out accident insurance, he is reducing the likelihood not of an accident but only of an uncompensated accident; that there is a market for such insurance indicates that people are influenced by considerations of compensability as well as by fear of injury itself. Rules of liability could also influence conduct more subtly. A rule that owners of property were strictly liable for any injuries accruing to intruders might be taken by the latter to imply that they need not be careful. They might assume that the completeness of the landowner's liability would impel him to eliminate any hazard. Or they might think the rule was based on a finding that all accidents were caused by the carelessness of landowners rather than of intruders.

6. An accident may occur even though neither the landowner nor the intruder was demonstrably careless. The warning sign may have been sturdily fixed to the fence but then stolen before the innocent intruder chanced on the scene. In such a case there is no clear basis on economic grounds for preferring one rule of liability to another. But I incline to making the landowner liable (though, for reasons explained earlier, only to the innocent trespasser). He is in control of the premises and so in a better position, in the usual case, to anticipate and avoid contingencies that increase the hazards created by the employment of deadly force. Stated otherwise, there may be reason to suspect that in most cases where an accident occurs and the intruder was not careless, the landowner was—though we cannot prove it.

. . .

The decisions in which courts have been asked to recognize a privilege to kill or wound to protect property compose a pattern that seems broadly consistent with an economic approach, and with the specific rule of liability that I have suggested. Thus, the courts have refused to sanction the use of deadly force to repel merely technical trespasses that cause no loss or damage, as when a property owner shoots a hole in a boat that has strayed into the owner's part of the lake; and they have rejected any privilege to use deadly force in support of a legal claim asserted in a boundary or other property dispute. A dispute differs from theft or vandalism in that there are well developed judicial remedies—temporary restraining orders, preliminary injunctions, bonds, and the like—by which a person can avert loss or destruction of substantial property values without having to resort to force. This, incidentally, would seem to be the explanation of why the courts have held the poisoning of trespassing animals to be wrongful even when the owner of the animals was forewarned. The victim of the trespass has adequate remedies (including the right to impound the animals) that do not entail the destruction of valuable property. If he had time to warn the animals' owner he also had time to obtain temporary injunctive relief and the cost of so proceeding would in the usual case be smaller than the value of the animals killed.

The courts have likewise refused to recognize a privilege to use deadly force to avert the loss of property having little value, as by killing a thief with a spring gun in order to protect goods worth no more than $6 or shooting a drunken man because he refused to return a bottle of whiskey belonging to his assailant. At the same time, it has been implied that the amount of force permissible is, up to a point, proportional to the value of the property at stake.

In a number of cases where a claim of privilege has been rejected, the defendant exhibited carelessness in his use of deadly force, as by failing to give adequate warning or by using an excessively lethal weapon. In one case the defendant saw a 15-year-old boy stealing watermelons from his watermelon patch, shot to frighten him—and hit him. One defendant set a spring gun and neglected to notify an employee, who was killed by it. In one case the court pointed out that the defendant, before setting a spring gun, should have erected a higher fence around his property to prevent cattle (and their keep-ers) from straying on to the property. Another defendant who had set a spring gun placed a vague warning on two pieces of paper—"Dangerous, don't go in this patch. Go back out"—and the plaintiff, a 14-year-old boy who testified that he had not seen the notices and thought his family owned the watermelon patch, was seriously wounded when he triggered the gun. In another case a policeman was killed when he tried the door of the defendant's store on his nightly rounds to see whether it was locked and the door swung open, trigger-ing a spring gun. The defendant, who knew that the police tried the door on their rounds, had not told them about the spring gun and had neglected to fasten the door securely.

The language of the opinions is not always consistent with an economically sensible rule of liability. In particular, the courts are prone to say that the infliction of injury by a spring gun is privileged when the defendant would have been privileged to inflict the same injury in person, and not otherwise. . . .

This is how the privilege to use deadly mechanical devices was stated in section 85 of the *Restatement of Torts*. Such a formulation is inconsistent with our rule of liability in two respects. In part 2 of our rule, we explain why the privilege to use deadly force by means of an armed watchman should in some circumstances be greater than the privilege to kill or wound using a spring gun. And it was implicit in part 5b that it should be narrower in other circum-stances. We said that a careless victim should be denied recovery if the use of a spring gun was otherwise privileged, but of course an armed watchman would not be privileged to shoot an intruder merely because the intruder was care-less; the watchman must actually and reasonably believe that the intruder was an adult about to steal or destroy valuable property and that he could not be stopped in any gentler way.

To repeat an earlier point, the task of legal scholarship is to get behind the prose of the opinions and when we do this we find much less support for the "indirectly" principle than the *Restatement* would lead us to believe exists. The courts have appeared to recognize a broader privilege for the armed watchman in circumstances where our rule of liability would dictate a broader privilege. The relevance of the victim's conduct is less clear in the cases. An early case barred recovery on the ground that, although the defendant was negligent in having set the spring gun, the victim was careless too. In two more recent cases the victim's apparent carelessness was not given any weight. In one the plaintiff climbed a fence and broke two locks to get into the house where the spring gun was set (his motives in doing so were found to have been innocent); in the other the plaintiff (again with innocent motives) climbed over a fence at night into a watermelon patch where the defendant had set a spring gun. But perhaps both cases should be explained as resting on the absence of any notice that a spring gun had been set.

A fairly recent case appears to illustrate part 6 of our rule (although again the absence of notice may have been a factor in the court's decision). The defendant fastened two locks on an unoccupied building in which he had set a spring gun. Someone broke the locks and when the plaintiff, whose motives were completely innocent, came along the door was unlocked. The court held the defendant liable.

. . .

The reader will not have failed to notice that my approach and particular conclusions assume that the dominant purpose of rules of liability is to channel people's conduct, and in such a way that the value of interfering activities is maximized. Even those who find the analysis plausible may wish to quarrel with the premise. They may argue that judges do not think in economic terms. No doubt very few judges would articulate their grounds of decision in the precise terms used in this paper. But they could easily hit on the approach intuitively. The adversary process forces them to consider the impact of a ruling on both parties, and therefore on both interfering activities. The fact that the incomes of parties and other such factors bearing on their relative deservedness are excluded from the consideration of judge and jury (except as bearing on a claim for punitive damages) also helps to keep the focus on the *activities* affected by the rule of liability. Such factors are not excluded from legislative judgments, which is one reason for expecting the legislative product to be different. What is truly unlikely is that the process of judicial reasoning is exhausted in the conceptual categories exhibited in judicial opinions.

One might also question the assumption that the rules of liability prescribed by the tort law actually affect conduct. As Professor Coase has shown, where transactions between interfering parties can be effected without cost, the market will bring about an optimum adjustment between the interfering activities regardless of the rule of liability initially prescribed by the law. Although the cost of transacting is never zero, Coase's point has force whenever it is low. But in the cases that we have been considering the cost of transacting is normally prohibitive. It is not feasible for the landowner to contract with the potential trespasser or the potential trespasser with the landowner. There are exceptions—the reader will recall the case where an employee was killed by his employer's spring gun—but they seem rare. As I have mentioned elsewhere, Coase's insight, were it taken seriously, might lead to a redefinition of the boundaries of tort law that excluded all sorts of accidents and injuries incidental to a contractual relationship. Most of the deadly-force cases, however, would remain inside the tort boundary.

Our basic premise will also be challenged by anyone who believes that the dominant purpose of the law of torts is to compensate people for wrongs suffered rather than to shape people's conduct; the latter is the proper sphere, it is sometimes argued, for criminal and other regulatory laws. In support of this argument one might cite the criminal penalties for excessive use of deadly force. Many of the cases discussed in the preceding part were in fact criminal cases. (Their inclusion in a discussion of tort law is justified by the fact that the criminal and tort standards governing the propriety of using deadly force to protect property are basically identical.)

Although the issue is too large for adequate discussion here, I will venture the suggestion that a compensation theory of the law of torts has little content.

The need for compensation is typically independent of the nature of the accident giving rise to the need. A man killed by lightning suffers the same loss as if he had been killed by a careless driver. The law of torts decrees compensation only where there is "wrongful" conduct, and the criteria of wrongfulness are not self-evident. One plausible meaning that can be assigned the term is conduct that society wishes to deter in order to increase the joint value of two (or more) interfering activities, consistently with protecting certain sunk costs in order to encourage adequate investment (our "fairness" point).

Because tort law is concerned, as I would argue, with shaping conduct, it does not follow that we need no other machinery of deterrence. There are good reasons for supplementing tort with criminal sanctions in certain areas. Tort law will not deter a judgment-proof individual while the threat of imprisonment may, and it is not a fully effective deterrent where, for one reason or another, many victims will not sue at all (burglars and other thieves may be reluctant to institute tort suits) and others, who do sue, may be barred from recovering damages by their own carelessness. Furthermore, where tortious conduct involves killing or maiming, a tort judgment, for reasons touched on earlier, may well undervalue the true social cost of the conduct and hence fail to deter it sufficiently for the future, in which case an additional, penal sanction may be appropriate. It does not follow that we should place exclusive reliance on criminal sanctions. Considering how overburdened the institutions of criminal law enforcement seem at present, we should be seeking ways of increasing rather than of diminishing the scope and effectiveness of tort law in deterring socially harmful behavior.

All things considered, the approach to tort questions sketched here seems decidedly superior to the "method of maxims"—the pseudo-logical deduction of rules from essentially empty formulas such as "no man should be permitted to do indirectly what he would be forbidden to do directly" or "the interest in property can never outweigh the value of a human life"—that plays so large a role in certain kinds of legal scholarship. And the present study provides a good point of departure for investigation of other areas of tort law. Distant as our subject may seem from the dominant concerns of modern tort law, on closer examination it is seen to be curiously central. We mentioned in passing one group of cases where the reasoning underlying the privilege to defend property by deadly force was invoked, and properly so, to solve a different problem: the destruction of trespassing domestic animals. Another doctrine with a strong affinity to the deadly-force cases is that of "private necessity." In *Ploof v. Putnam,* the plaintiff and his family were sailing their boat on a lake when a storm came up. They moored at a dock owned by the defendant. The defendant's employee unmoored their boat, it ran aground, and several of the occupants were injured—a sequence the employee should have anticipated. The court held that the plaintiff's trespass had been justified by necessity and that the defendant's employee had acted wrongfully in casting him off. It could as well have viewed the case as one where deadly force—which is what the employee used, in effect, in repelling the trespass—was manifestly unjustified in defense of a property right. Had the plaintiff's act in mooring his boat to the dock damaged the dock, the defendant could have obtained damages from the plaintiff. The plaintiff was not a criminal against whom legal remedies would probably have been unavailing, so there was no occasion to endanger human safety.

Just as the *Ploof* case might have been decided by reference to the limitations on the privilege to use deadly force in defense of property, so *Holbrook* might conceivably have been decided under the doctrine of necessity. A valuable fowl had strayed into the defendant's garden and the defendant had no privilege (in the circumstances) to use deadly force to prevent the plaintiff from recovering it. In *Ploof,* to be sure, the trespass was necessary to avert danger to human safety, but the doctrine of necessity has also been invoked to excuse trespasses committed solely in order to avert property losses, often in the context of deviations from highways—where the doctrine (naturally) is called by a different name.

As noted, the privilege to commit a trespass to avert serious injury to life or property does not relieve the trespasser from the obligation to pay for the harm that his trespass inflicts. This principle comports with the economic objectives that I have argued best explain the course of decisions in these areas. It not only protects sunk costs but forces the individual contemplating a trespass to weigh the injury he will cause by committing the trespass against the injury that would result from refraining and to choose the course that maximizes the joint value of the interfering activities; we do not want people trampling on tulips to save peahens if the damage to the tulips would exceed the value of the peahen. In addition, as Clarence Morris has suggested, the right to recover damages may incline the landowner to cooperate with the trespasser in situations where cooperation is likely to minimize the social costs of the intrusion. When a boat unexpectedly moors at a stranger's dock in a storm, the owner of the dock will have a greater incentive to assist with fresh rope if he knows that any injury to the dock is fully compensable.

An old case where compensation was not allowed, *Mouse's Case,* is the exception that proves the rule. The parties were passengers on a ferry that began to sink in a storm. The defendant cast a valuable chest belonging to the plaintiff overboard in order to lighten the craft. The plaintiff sued the defendant for the value of the chest and lost. The court found that, but for the defendant's action, the boat would have sunk. Therefore the defendant wasn't really responsible for the loss of the chest—it would have been lost anyway. Moreover, the defendant should be entitled to offset the value of the plaintiff's life, which his action was instrumental in saving, against the value of the plaintiff's goods. These are good grounds but the ground I would stress is that the denial of compensation served the same purpose as the grant of compensation does in the usual necessity case: to encourage the value-maximizing course of conduct. We do not want each of the passengers of a sinking ship to hesitate in casting off excess baggage in the hope that another one will act first and save him from tort liability.

As these examples and I hope the whole paper suggest, there are far more conceptual pigeonholes in the law of torts—the privilege to use deadly force to protect property, the privilege to use such force to prevent certain crimes, the privilege in cases of arrest, rules about animals, the doctrine of necessity, rules governing deviations from highways onto private land, liability for engaging in ultrahazardous activities—than there are useful distinctions. By utilizing the approach to tort questions sketched here, legal scholarship has an opportunity to effect a drastic and necessary simplification of doctrine and to place the analysis of tort law on a more functional basis.

Notes and Questions

1. Consider the well-known Learned Hand negligence formula, as stated in *United States v. Carroll Towing,* 159 F.2d 169, 173 (2d Cir. 1947): one who causes injury is negligent only if the cost of an untaken precaution is less than the probability of the expected loss occurring in the absence of such a precaution times the magnitude of such loss. Is the author simply suggesting that the rules regarding violence in defense of property comport with this formulaic recitation of the law of negligence? If so, it would appear that the common law, at least in this area, reached an efficient result. What might explain this (possible) efficiency of the common law? If judges do not think in these terms, what could possibly cause legal doctrines to evolve in a way that satisfies the efficiency concerns of modern economists? See Richard A. Posner, *Economic Analysis of Law* 229–33, 505–16 (3d ed. 1986).

2. What does Posner suggest ought to be the rule if a watchman shoots a criminal intruder, even though there is every reason to believe that the watchman's own life is not in danger (as when the watchman is protected by a barrier or the intruder is plainly unarmed and on the verge of running away with stolen goods)? Do you suppose that this is the legal rule?

3. What does the reading suggest about the familiar aphorism, "a man's home is his castle?" What if a homeowner observes a criminal intrusion and faces the choice of (1) standing ground and shooting or (2) running out the rear door in order to escape any possible personal harm? What might explain the law's sympathy for a homeowner who stays and wounds an intruder?

4. Consider the discussion of *Mouse's Case* at the very end of the selection. Was the defendant negligent in throwing over the plaintiff's valuable chest? If the defendant were held (strictly) liable, do you really think that the defendant would hesitate in the future to jettison heavy objects? What if the defendant tossed the plaintiff overboard as a means of saving other passengers?

Fairness and Utility in Tort Theory

GEORGE P. FLETCHER

. . .

Tort theorists tend to regard the existing doctrinal framework of fault and strict liability as sufficiently rich to express competing views about fairly shifting losses. This conceptual framework accounts for a number of traditional beliefs about tort law history. One of these beliefs is that the ascendancy of fault in the late nineteenth century reflected the infusion of moral sensibility into the law of torts. That new moral sensibility is expressed sometimes as the principle that wrongdoers ought to pay for their wrongs. . . . The underlying assumption . . . is that negligence and strict liability are antithetical rationales of liability. This assumed antithesis is readily invoked to explain the ebbs and flows of tort liability. Strict liability is said to have prevailed in early tort history, fault supposedly held sway in the late nineteenth century, with strict liability now gaining ground.

These beliefs about tort history are ubiquitously held, but to varying degrees they are all false or at best superficial. There has no doubt been a deep ideological struggle in the tort law of the last century and a half. But, as I shall argue, it is not the struggle between negligence and fault on the one hand, and strict liability on the other. Rather, the confrontation is between two radically different paradigms for analyzing tort liability. . . .

Of the two paradigms, I shall call the first the paradigm of reciprocity. According to this view, the two central issues of tort law—whether the victim is entitled to recover and whether the defendant ought to pay—are distinct issues, each resolvable without looking beyond the case at hand. Whether the victim is so entitled depends exclusively on the nature of the victim's activity when he was injured and on the risk created by the defendant. . . . Further, according to this paradigm, if the victim is entitled to recover by virtue of the risk to which he was exposed, there is an additional question of fairness in holding the risk-creator liable for the loss. This distinct issue of fairness is expressed by asking whether the defendant's creating the relevant risk was excused on the ground, say, that the defendant could not have known of the risk latent in his conduct. To find that an act is excused is in effect to say that there is no rational, fair basis for distinguishing between the party causing harm and other people. Whether we can rationally single out the defendant as the loss-bearer depends on our expectations of when people ought to be able to avoid risks. . . .

Abridged and reprinted without footnotes by permission from 85 *Harvard Law Review* 537 (1972).

As part of the explication of the first paradigm of liability, I shall propose a specific standard of risk that makes sense of the *Restatement*'s emphasis on uncommon, extra-hazardous risks, but which shows that the *Restatement*'s theory is part of a larger rationale of liability. . . . The general principle expressed in all of these situations governed by diverse doctrinal standards is that a victim has a right to recover for injuries caused by a risk greater in degree and different in order from those created by the victim and imposed on the defendant—in short, for injuries resulting from nonreciprocal risks. Cases of liability are those in which the defendant generates a disproportionate, excessive risk of harm, relative to the victim's risk-creating activity. . . . Conversely, cases of nonliability are those of reciprocal risks, namely those in which the victim and the defendant subject each other to roughly the same degree of risk. . . .

The conflicting paradigm of liability—which I shall call the paradigm of reasonableness—represents . . . a commitment to the community's welfare as the criterion for determining both who is entitled to receive and who ought to pay compensation. Questions that are distinct under the paradigm of reciprocity—namely, is the risk nonreciprocal and was it unexcused—are collapsed in this paradigm into a single test: was the risk unreasonable? The reasonableness of the risk thus determines both whether the victim is entitled to compensation and whether the defendant ought to be held liable. Reasonableness is determined by a straightforward balancing of costs and benefits. If the risk yields a net social utility (benefit), the victim is not entitled to recover from the risk-creator; if the risk yields a net social disutility (cost), the victim is entitled to recover. The premises of this paradigm are that reasonableness provides a test of activities that ought to be encouraged and that tort judgments are an appropriate medium for encouraging them.

The function of both of these paradigms is to distinguish between those risks that represent a violation of individual interests and those that are the background risks that must be borne as part of group living. The difference between the two paradigms is captured by the test provided by each for filtering out background risks. The paradigm of reciprocity holds that we may be expected to bear, without indemnification, those risks we all impose reciprocally on each other. If we all drive, we must suffer the costs of ordinary driving. The paradigm of reasonableness, on the other hand, holds that victims must absorb the costs of reasonable risks, for these risks maximize the composite utility of the group, even though they may not be mutually created background risks.

The paradigm of reasonableness bears some resemblance to present-day negligence, but it would be a mistake to associate the two paradigms, respectively, with strict liability and negligence. As I shall argue, the paradigm of reciprocity cuts across strict liability, negligence and intentional torts, and the paradigm of reasonableness accounts for only a subset of negligence cases. . . .

The Paradigm of Reciprocity

The Victim's Right to Recover

Our first task is to demonstrate the pervasive reliance of the common law on the paradigm of reciprocity. The area that most consistently reveals this paradigm is the one that now most lacks doctrinal unity—namely, the disparate pockets of strict liability. We speak of strict liability or "liability without fault" in cases ranging from crashing airplanes to suffering cattle to graze on another's land. Yet the law of torts has never recognized a general principle underlying these atomistic pockets of liability. The *Restatement's* standard of ultra-hazardous activity speaks only to a subclass of cases. In general, the diverse pockets of strict liability represent cases in which the risk is reasonable and legally immune to injunction. They are therefore all cases of liability without fault in the limited sense in which fault means taking an unreasonable risk. Beyond these characteristics distinguishing strict liability from negligence, there is no consensus of criteria for attaching strict liability to some risks and not to others.

I shall attempt to show that the paradigm of reciprocity accounts for the typical cases of strict liability—crashing airplanes, damage done by wild animals, and the more common cases of blasting, fumigating and crop dusting. To do this, I shall consider in detail two leading, but seemingly diverse instances of liability for reasonable risk-taking—*Rylands v. Fletcher* and *Vincent v. Lake Erie Transportation Co.* . . .

In *Rylands v. Fletcher* the plaintiff, a coal-mine operator, had suffered the flooding of his mine by water that the defendant had pumped into a newly erected reservoir on his own land. The water broke through to an abandoned mine shaft under the defendant's land and thus found its way to the plaintiff's adjoining mine. . . . Though the defendant's erecting and maintaining the reservoir was legally permissible, the Exchequer Chamber found for the plaintiff, and the House of Lords affirmed. Blackburn's opinion in the Exchequer Chamber focused on the defendant's bringing on to his land, for his own purposes, "something which, though harmless whilst it remain there, will naturally do mischief if it escape." Lord Cairns, writing in the House of Lords, reasoned that the defendant's activity rendered his use of the land "non-natural"; accordingly, "that which the Defendants were doing they were doing at their own peril." . . . The fact was that the defendant sought to use his land for a purpose at odds with the use of land then prevailing in the community. He thereby subjected the neighboring miners to a risk to which they were not accustomed and which they would not regard as a tolerable risk entailed by their way of life. Creating a risk different from the prevailing risks in the community might be what Lord Cairns had in mind in speaking of a non-natural use of the land. . . .

A seemingly unrelated example of the same case law tradition is *Vincent v. Lake Erie Transportation Co.*, a 1910 decision of the Minnesota Supreme Court. The dispute arose from a ship captain's keeping his vessel lashed to the plaintiff's dock during a two-day storm when it would have been unreasonable, indeed foolhardy, for him to set out to sea. The storm battered the ship against the dock, causing damages assessed at $500. The court affirmed a judgment for the plaintiff even though a prior case had recognized a ship captain's right to take shelter from a storm by mooring his vessel to another's dock, even without consent. The court's opinion conceded that keeping the ship at dockside was justified and reasonable, yet it characterized the defendant's damaging the dock as "prudently and advisedly [availing]" himself of the plaintiff's property. Because the incident impressed the court as an implicit transfer of wealth, the defendant was bound to rectify the transfer by compensating the dock owner for his loss. . . . The critical feature of both cases is that the defendant created a risk of harm to the plaintiff that was of an order different from the risks that the plaintiff imposed on the defendant. . . .

Without the factor of nonreciprocal risk-creation, both cases would have been decided differently. Suppose that Rylands had built his reservoir in textile country, where there were numerous mills, dams, and reservoirs, or suppose that two sailors secured their ships in rough weather to a single buoy. In these situations each party would subject the other to a risk, respectively, of mundation and abrasion. Where the risks are reciprocal among the relevant parties, as they would be in these variations of *Rylands* and *Vincent,* a rule of strict liability does no more than substitute one form of risk for another—the risk of liability for the risk of personal loss. Accordingly, it would make little sense to extend strict liability to cases of reciprocal risk-taking, unless one reasoned that in the short run some individuals might suffer more than others and that these losses should be shifted to other members of the community.

Expressing the standard of strict liability as unexcused, non-reciprocal risk-taking provides an account not only of the *Rylands* and *Vincent* decisions, but of strict liability in general. It is apparent, for example, that the uncommon, ultra-hazardous activities pinpointed by the *Restatement* are readily subsumed under the rationale of nonreciprocal risk-taking. If uncommon activities are those with few participants, they are likely to be activities generating nonreciprocal risks. Similarly, dangerous activities like blasting, fumigating, and crop dusting stand out as distinct, nonreciprocal risks in the community. They represent threats of harm that exceed the level of risk to which all members of the community contribute in roughly equal shares.

The rationale of nonreciprocal risk-taking accounts as well for pockets of strict liability outside the coverage of the *Restatement*'s sections on extra-hazardous activities. For example, an individual is strictly liable for damage done by a wild animal in his charge, but not for damage committed by his domesticated pet. Most people have pets, children, or friends whose presence creates some risk to neighbors and their property. These are risks that offset each other; they are, as a class, reciprocal risks. Yet bringing an unruly horse into the city goes beyond the accepted and shared level of risks in having pets, children, and friends in one's household. If the defendant creates a risk that exceeds those to which he is reciprocally subject, it seems fair to hold him liable for the results of his aberrant indulgence. Similarly, according to the latest version of the *Restatement,* airplane owners and pilots are strictly liable for ground damage, but not for mid-air collisions. Risk of ground damage is nonreciprocal; homeowners do not create risks to airplanes flying overhead. The risks of mid-air collisions, on the other hand, are generated reciprocally by all those who fly the air lanes. Accordingly, the threshold of liability for damage resulting from mid-air collisions is higher than mere involvement in the activity of flying. To be liable for collision damage to another flyer, the pilot must fly negligently or the owner must maintain the plane negligently; they must generate abnormal risks of collision to the other planes aflight.

Negligently and intentionally caused harm also lend themselves to analysis as nonreciprocal risks. As a general matter, principles of negligence liability apply in the context of activities, like motoring and sporting ventures, in which the participants all normally create and expose themselves to the same order of risk. These are all pockets of reciprocal risk-taking. Sometimes the risks are grave, as among motorists; sometimes they are minimal, as among ballplayers. Whatever the magnitude of risk, each participant contributes as much to the community of risk as he suffers from exposure to other participants. To establish liability for harm resulting from these activities, one must show that the harm derives from a specific risk negligently engendered in the course of the activity. Yet a negligent risk, an "unreasonable" risk, is but one that unduly exceeds the bounds of reciprocity. Thus, negligently created risks are nonreciprocal relative to the risks generated by the drivers and ballplayers who engage in the same activity in the customary way.

If a victim also creates a risk that unduly exceeds the reciprocal norm, we say that he is contributorily negligent and deny recovery. The paradigm of reciprocity accounts for the denial of recovery when the victim imposes excessive risks on the defendant, for the effect of contributory negligence is to render the risks again reciprocal, and the defendant's risk-taking does not subject the victim to a relative deprivation of security.

Thus, both strict liability and negligence express the rationale of liability for unexcused, nonreciprocal risk-taking. The only difference is that reciprocity in strict liability cases is analyzed relative to the background of innocuous risks in the community, while reciprocity in the types of negligence cases discussed above is measured against the background of risk generated in specific activities like motoring and skiing. To clarify the kinship of negligence to strict liability, one should distinguish between two different levels of risk-creation, each level associated with a defined community of risks. Keeping domestic pets is a reciprocal risk relative to the community as a whole; driving is a reciprocal risk relative to the community of those driving normally; and driving negligently might be reciprocal relative to the even narrower community of those driving negligently. The paradigm of reciprocity holds that in all communities of reciprocal risks, those who cause damage ought not to be held liable.

. . .

All of these manifestations of the paradigm of reciprocity . . . express the same principle of fairness: all individuals in society have the right to roughly the same degree of security from risk. By analogy to John Rawls' first principle of justice, the principle might read: we all have the right to the maximum amount of security compatible with a like security for everyone else. This means that we are subject to harm, without compensation, from background risks, but that no one may suffer harm from additional risks without recourse for damages against the risk-creator. Compensation is a surrogate for the individual's right to the same security as enjoyed by others. But the violation of the right to equal security does not mean that one should be able to enjoin the risk-creating activity or impose criminal penalties against the risk-creator. The interests of society may often require a disproportionate distribution of risk. Yet, according to the paradigm of reciprocity, the interests of the individual require us to grant compensation whenever this disproportionate distribution of risk injures someone subject to more than his fair share of risk.

Excusing Nonreciprocal Risks

If the victim's injury results from a nonreciprocal risk of harm, the paradigm of reciprocity tells us that the victim is entitled to compensation. Should not the defendant then be under a duty to pay? Not always. For the paradigm also holds that nonreciprocal risk-creation may sometimes be excused, and we must inquire further, into the fairness of requiring the defendant to render compensation. We must determine whether there may be factors in a particular situation which would excuse this defendant from paying compensation.

Though the King's Bench favored liability in its 1616 decision of *Weaver v. Ward,* it digressed to list some hypothetical examples where directly causing harm would be excused and therefore exempt from liability. One kind of excuse would be the defendant being physically compelled to act, as if someone took his hand and struck a third person. Another kind would be the defendant's accidentally causing harm, as when the plaintiff suddenly appeared in the path of his musket fire. The rationale for denying liability in these cases, as the court put it, is that the defendant acted "utterly without . . . fault."

If a man trespasses against another, why should it matter whether he acts with "fault" or not? What the King's Bench must have been saying is that if a man injures another without fault on his part, there is no rational and fair basis for charging the costs of the accident to him rather than to an arbitrary third person. The inquiry about fault and excusability is an inquiry about rationally singling out the party immediately causing harm as the bearer of liability. Absent an excuse, the trespassory, risk-creating act provides a sufficient basis for imputing liability. Finding that the act is excused, however, is tantamount to perceiving that the act is not a factor fairly distinguishing the trespassing party from all other possible candidates for liability.

. . .

The hypotheticals of *Weaver v. Ward* correspond to the Aristotelian excusing categories of compulsion and unavoidable ignorance. Each of these has spawned a line of cases denying liability in cases of inordinate risk-creation.

The excuse of compulsion has found expression in the emergency doctrine, which excuses excessive risks created in cases in which the defendant is caught in an unexpected, personally dangerous situation. In *Cordas v. Peerless Transportation Co.,* for example, it was thought excusable for a cab driver to jump from his moving cab in order to escape from a threatening gunman on the running board. In view of the crowd of pedestrians nearby, the driver clearly took a risk that generated a net danger to human life. It was thus an unreasonable, excessive, and unjustified risk. Yet the overwhelmingly coercive circumstances meant that he, personally, was excused from fleeing the moving cab. An example of unavoidable ignorance excusing risk-creation is *Smith v. Lampe,* in which the defendant honked his horn in an effort to warn a tug that seemed to be heading toward shore in a dense fog. As it happened, the honking coincided with a signal that the tug captain expected would assist him in making port. Accordingly the captain steered his tug toward the honking rather than away from it. That the defendant did not know of the prearranged signal excused his contributing to the tug's going aground. Under the facts of the case, the honking surely created an unreasonable risk of harm. If instantaneous injunctions were possible, one would no doubt wish to enjoin the honking as an excessive, illegal risk. Yet the defendant's ignorance of that risk was also excusable. Under the circumstances he could not fairly have been expected to inform himself of all possible interpretations of honking in a dense fog.

As expanded in these cases, the excuses of compulsion and unavoidable ignorance added dimension to the hypotheticals put in *Weaver v. Ward.* In *Cordas* and *Smith* we have to ask: What can we fairly expect of the defendant under the circumstances? Can we ask of a man that he remain in a car with a gun pointed at him? Can we require that a man inform himself of all local customs before honking his horn? Thus the question of rationally singling out a party to bear liability becomes a question of what we can fairly demand of an individual under unusual circumstances. Assessing the excusability of ignorance or of yielding to compulsion can be an instrumentalist inquiry. As we increase or decrease our demands, we accordingly stimulate future behavior. Thus, setting the level of excusability could function as a level of social control. Yet one can also think of excuses as expressions of compassion for human failings in times of stress—expressions that are thought proper regardless of the impact on other potential risk-creators.

Despite this tension between thinking of excusing conditions in an instrumentalist or noninstrumentalist way, we can formulate two significant claims about the role of excuses in cases decided under the paradigm of reciprocity. First, excusing the risk-creator does not, in principle, undercut the victim's right to recover. In most cases, it is operationally irrelevant to posit a right to recovery when the victim cannot in fact recover from the excused risk-creator. Yet it may be important to distinguish between victims of reciprocal, background risks and victims of nonreciprocal risks. The latter class of victims— those who have been deprived of their equal share of security from risk— might have a claim of priority in a social insurance scheme. Further, for a variety of reasons, one might wish in certain classes of cases to deny the availability of particular excuses, such as insanity in general or immaturity for teenage drivers. Insanity has always been a disfavored excuse; even the King's Bench in *Weaver v. Ward* rejected lunacy as a defense. However, it is important to perceive that to reject the excuse is not to provide a rationale for recovery. It is not being injured by an insane man that grounds a right to recovery, but being injured by a nonreciprocal risk—as in every other case applying the paradigm of reciprocity. Rejecting the excuse merely permits the independently established, but previously unenforceable, right to prevail.

Secondly, an even more significant claim is that these excuses—compulsion and unavoidable ignorance—are available in all cases in which the right to recovery springs from being subjected to a nonreciprocal risk of harm. We have already pointed out the applicability of these excuses in negligence cases like *Cordas* and *Smith v. Lampe*. What is surprising is to find them applicable in cases of strict liability as well; strict liability is usually thought of as an area where courts are insensitive to questions of fairness to defendants. Admittedly, the excuses of compulsion and unavoidable ignorance do not often arise in strict liability cases, for men who engage in activities like blasting, fumigating, and crop dusting typically do so voluntarily and with knowledge of the risks characteristic of the activity. Yet there have been cases in which strict liability for keeping a vicious dog was denied on the ground that the defendant did not know, and had no reason to know, that his pet was dangerous. And doctrines of proximate cause provide a rubric for considering the excuse of unavoidable ignorance under another name. In *Madsen v. East Jordan Irrigation Co.,* for example, the defendant's blasting operations frightened the mother mink on the plaintiff's farm, causing them to kill 230 of their offspring. The Utah Supreme Court affirmed a demurrer to the complaint. In the court's judgment, the reaction of the mother mink "was not within the realm of matters to be anticipated." This is precisely the factual judgment that would warrant saying that the company's ignorance of this possible result was excused, yet the rubric of proximate cause provided a doctrinally acceptable heading for dismissing the complaint. . . .

Recognizing the pervasiveness of nonreciprocity as a standard of liability, as limited by the availability of excuses, should provide a new perspective on tort doctrine and demonstrate that strict liability and negligence as applied in the cases discussed above are not contrary theories of liability. Rather, strict liability and negligence appear to be complementary expressions of the same paradigm of liability.

The Paradigm of Reasonableness

Until the mid-nineteenth century, the paradigm of reciprocity dominated the law of personal injury. . . . In the course of the nineteenth century, however, the concepts underlying the paradigm of reciprocity gradually assumed new contours. . . . The new paradigm challenged the asssumption that the issue of liability could be decided on grounds of fairness to both victim and defendant without considering the impact of the decisions on the society at large. It further challenged the assumption that the victim's right to recovery was distinguishable from the defendant's duty to pay. In short, the new paradigm of reasonableness represented a new style of thinking about tort disputes.

The core of this revolutionary change was a shift in the meaning of the word "fault." At its origins in the common law of torts, the concept of fault served to unify the medley of excuses available to defendants who would otherwise be liable in trespass for directly causing harm. As the new paradigm emerged, fault came to be an inquiry about the context and the reasonableness of the defendant's risk-creating conduct. Recasting fault from an inquiry about excuses into an inquiry about the reasonableness of risk-taking laid the foundation for the new paradigm of liability. It provided the medium for tying the determination of liability to maximization of social utility, and led to the conceptual connection between the issue of fault and the victim's right to recover. . . .

The reasonable man became a central, almost indispensable, figure in the paradigm of reasonableness. . . . Reasonable men, presumably, seek to maximize utility; therefore, to ask what a reasonable man would do is to inquire into the justifiability of the risk. If the risk-running might be excused, say by reason of the emergency doctrine or a particular defect like blindness or immaturity, the jury instruction might specify the excusing condition as one of the "circumstances" under which the conduct of the reasonable man is to be assessed. If the court wished to include or exclude a teenage driver's immaturity as a possible excusing condition, it could define the relevant "circumstances" accordingly. . . .

No single appellate decision ushered in the paradigm of reasonableness. It derived from a variety of sources. If there was a pivotal case, however, it was *Brown v. Kendall,* decided by the Massachusetts Supreme Judicial Court in 1850. . . . In an effort to separate two fighting dogs, Kendall began beating them with a stick. Brown was standing nearby, which Kendall presumably knew; and both he and Brown moved about with the fighting dogs. At one point, when he had just backed up to a position in front of Brown, Kendall raised his stick, hitting Brown in the eye and causing serious injury. Brown sought to recover on the writ of trespass, whereby traditionally a plaintiff could establish a prima facie case simply by proving that his injuries were the direct result of the defendant's act—a relationship which clearly existed in the case. . . . Chief Justice Shaw's opinion created possibilities for an entirely new and powerful use of the fault standard, and the judges and writers of the late nineteenth and early twentieth centuries responded sympathetically.

Shaw's revision of tort doctrine made its impact in cases in which the issue was not one of excusing inadvertent risk-creation, but one of justifying risks of harm that were voluntarily and knowingly generated. Consider the following cases of risk-creation: (1) the defendant operates a streetcar, knowing that the trains occasionally jump the tracks; (2) the defendant police officer shoots at a fleeing felon, knowing that he thereby risks hitting a bystander; (3) the defendant undertakes to float logs downriver to a mill, knowing that flooding might occur which could injure crops downstream. All of these victims could receive compensation for their injuries under the paradigm of reciprocity, as incorporated in the doctrine of trespassory liability; the defendant or his employees directly and without excuse caused the harm in each case. Yet as *Brown v. Kendall* was received into the tort law, the threshold of liability became whether, under all the circumstances, the defendant acted with ordinary, prudent care. . . . The test for justifying risks became a straightforward utilitarian comparison of the benefits and costs of the defendant's risk-creating activity. The assumption emerged that reasonable men do what is justified by a utilitarian calculus, that justified activity is lawful, and that lawful activities should be exempt from tort liability. . . .

The Interplay of Substance and Style

The conflict between the paradigm of reasonableness and the paradigm of reciprocity is, in the end, a struggle between two strategies for justifying the distribution of burdens in a legal system. The strategy of utility proceeds on the assumption that burdens are fairly imposed if the distribution optimizes the interests of the community as a whole. The paradigm of reciprocity . . . takes as its starting point the personal rights of individuals in society to enjoy roughly the same degree of security, and appeals to the conduct of the victims themselves to determine the scope of the right to equal security. By interpreting the risk-creating activities of the defendant and of the victim as reciprocal and thus offsetting, courts may tie the denial of liability to the victim to his own waiver of a degree of security in favor of the pursuit of an activity of higher risk.

. . .

The major divergence is the set of cases in which a socially useful activity imposes nonreciprocal risks on those around it. These are the cases of motoring, airplane overflights, air pollution, oil spillage, sonic booms—in short, the recurrent threats of modern life. In resolving conflict between those who benefit from these activities and thos who suffer from them, the courts must decide how much weight to give to the net social value of the activity. . . .

On the whole, however, the paradigm of reasonableness still holds sway over the thinking of American courts. The reasonable man is too popular a figure to be abandoned. The use of litigation to pursue social goals is well entrenched. Yet the appeal to the paradigm might well be more one of style than of substance.

In assessing the reasonableness of risks, lawyers ask many seemingly precise questions: What are the consequences of the risk, its social costs and social benefits? What specific risks are included in the "ambit of the risk"? One can speak of formulae, like the Learned Hand formula, and argue in detail about questions of costs, benefits and trade-offs. This style of thinking is attractive to the legal mind. . . .

The paradigm of reciprocity, on the other hand, for all its substantive and moral appeal, puts questions that are hardly likely to engage the contemporary legal mind: When is a risk so excessive that it counts as a nonreciprocal risk? When are two risks of the same category and thus reciprocally offsetting? It is easy to assert that risks of owning a dog offset those of barbecuing in one's backyard, but what if the matter should be disputed? There are at least two kinds of difficulties that arise in assessing the relationship among risks. The first is that of protecting minorities. Does everyone have to engage in crop dusting for the risk to be reciprocal, or just half the community? A tempting solution to the problem is to say that as to someone not engaged in the activity, the risks are per se nonreciprocal. But the gains of this simplifying stroke are undercut by the assumption necessarily implicit in the concept of reciprocity that risks are fungible with others of the same "kind." Yet how does one determine when risks are counterpoised as species of the same genus? If one man owns a dog, and his neighbor a cat, the risks presumably offset each other. But if one man drives a car, and the other rides a bicycle? Or if one plays baseball in the street and the other hunts quail in the woods behind his house? No two people do exactly the same things. To classify risks as reciprocal risks, one must perceive their unifying features. . . . Determining the appropriate level of abstraction is patently a matter of judgment.

Notes and Questions

1. Professor Fletcher claims that without the factor of nonreciprocal risk, the *Vincent* case would have been decided differently. What do you suppose would happen over time to the number of docks on the lake if the case had gone against the plaintiff dock owner? Do you think that most ship captains and owners were rooting for the plaintiff or the defendant in *Vincent?* If they were pulling for the plaintiff, is this an independent explanation for the result in the case or a restatement of Fletcher's point about strict liability for damage caused by nonreciprocal risks?

2. Are you confident that you can identify reciprocal risks? If you think that most risks (such as automobile driving) are reciprocally imposed, would you favor a requirement that insurance companies charge equal premiums to all insureds? How might Fletcher restate his argument in a context with unequal and therefore only partially reciprocal risks?

3. Consider the claim that "the paradigm of reciprocity holds that in all communities of reciprocal task, those who cause damage ought not to be held liable." Thus, keeping domestic pets is a reciprocal task relative to the community as a whole, while driving negligently is said to be a reciprocal risk relative to the narrow community of those driving negligently. Does this mean that a negligent driver whose victim happens to be a negligent driver need not pay? Do you suppose more people own pets than occasionally drive too fast?

The Concept of Corrective Justice in Recent Theories of Tort Law

RICHARD A. POSNER

For the last 100 years, which is to say since the publication of Holmes's *The Common Law,* most tort scholars have thought that tort doctrines were, and should be, based on utilitarian (or, more recently, economic) concepts. This was the view of Holmes, of Ames, and of Terry; of the draftsmen of the first and second *Restatement of Torts;* and of the legal realists who thought the focus of tort law should be on loss spreading rather than on assessment of fault. It is also the view of economic analysts of tort law such as Guido Calabresi and myself. Writing in 1972 about tort scholarship, George Fletcher declared that "the fashionable questions of the time are instrumentalist: What social value does the rule of liability further in this case? Does it advance a desirable goal, such as compensation, deterrence, risk-distribution, or minimization of accident costs?" It would be easy to show that the goals which he listed were regarded by their advocates as utilitarian, or sometimes economic, goals. Since he wrote, the economic approach to tort law has developed apace; at the same time, several scholars have joined Fletcher both in questioning the proposition that tort doctrines are or should be based on utilitarian or economic ideas, and in arguing that the tort law should be, and perhaps already is, based on the idea of corrective justice.

. . .

The advocates of the corrective justice approach contend variously that corrective justice is and that it should be the basis of tort law. While ordinarily the difference between positive and normative analysis is of great importance in discussions of law, it is in this instance irrelevant. My argument is not that the theory of corrective justice provides either an inaccurate description of or an unsound guide to principles of tort liability, but that those who believe it is necessarily a rival to the economic approach are mistaken; the Aristotelian concept, at least, is not. . . .

Aristotle's Concept of Corrective Justice

In Book V, Chapter 4, of the *Nicomachean Ethics,* Aristotle develops the concept of corrective justice. He had discussed, in Chapter 3, distributive justice—that is, justice in the distribution by the state of money, honors, and other things of value—saying that such awards should be made according to

Abridged and reprinted without footnotes by permission from 10 *Journal of Legal Studies* 187 (1981). Copyright © 1981 by The University of Chicago Press.

merit (*kat' axian*). Chapter 4 discusses a contrasting concept of justice, the rectificatory or corrective (*diorthōtikos*—literally "making straight"), which he says applies to transactions (*sunallagmata*), both voluntary (*hekosia*) and involuntary (*akosia*); the distinction is roughly that between contracts and torts. The crucial passage in Chapter 4 is the following:

> [I]t makes no difference [from a corrective justice standpoint] whether a good man has defrauded a bad man or a bad man a good one, nor whether it is a good or a bad man that has committed adultery; the law looks only to the distinctive character of the injury, and treats the parties as equal, if one is in the wrong and the other is being wronged, and if one inflicted injury and the other has received it.

As far as remedy is concerned, Aristotle says that

> the judge tries to equalize things by means of the penalty, taking away from the gain of the assailant. For the term 'gain' [*kerdos*] is applied generally to such cases—even if it be not a term appropriate to certain cases, e.g., to the person who inflicts a wound—and 'loss' [*zēmia*] to the sufferer; at all events, when the suffering has been estimated, the one is called loss and the other gain. . . . Therefore the just . . . consists in having an equal amount before and after the transaction.

There is more, but the chapter is short, and the part I have summarized and especially the passages I have quoted contain the gist of Aristotle's concept.

The *Nicomachean Ethics,* consisting as they do of classroom notes of Aristotle's lectures, are notoriously obscure and there is a fair amount of exegetic literature on Book V, Chapter 4. But there is little controversy with regard to the basic features of his concept of corrective justice, summarized above. As paraphrased by Joachim,

> If, for example, the thief was a gentleman and the injured party a beggar—a member of an inferior class in the State—this difference of rank is nothing to the law. . . . All that the law is concerned with is that, of two parties before it, one has got an unfair advantage and the other has suffered an unfair disadvantage. There is, therefore, a wrong which needs redress—an inequality which needs to be equalized.

Three points should be noted about Aristotle's concept of corrective justice:

1. The duty to rectify is based not on the fact of inquiry but on the conjunction of injury and wrongdoing. The injurer must do wrong (*adikei*) as well as do harm (*eblapsen*), and the victim must be wronged (*adiketei*) as well as harmed (*beblaptai*). Not all departures from distributive justice call for correction. Someone who voluntarily makes a bad bargain may end up worse off than the principles of distributive justice would, but for the bad bargain, dictate. But he has not been wronged, and he is not entitled to rectification. Moreover, what is wrongful or unjust—*adikos*—is not defined in Chapter 4; it

is assumed. In Chapter 8 of Book V we learn that "Whether an act is or is not one of injustice (or of justice) is determined by its voluntariness or involuntariness." But even within the class of voluntary acts, only those that are deliberate can be acts of injustice. Those done by misadventure (where "the injury takes place contrary to reasonable expectation") or by mistake (where, for example, "he threw not with intent to wound but only to prick") are not.

2. The idea that distributive considerations do not count in a setting of corrective justice ("it makes no difference whether a good man has defrauded a bad man or a bad man a good one . . .") is a procedural principle. It is not equivalent to saying that distributive notions should not affect the definition of rights or even that they should not enter into the determination of what sorts of acts are unjust or wrongful. The point, rather, is that the judge is interested only in the character—whether it is wrongful—of the injury, rather than in the character of the parties apart from that of the injury: "the moral worth of persons . . . is ignored."

3. Aristotle was writing against the background of the Athenian legal system of his day, where even suits to redress crimes were (with rare exceptions) instituted and prosecuted by private individuals, the victim or a member of his family, rather than by the state. So he naturally assumed that redress for wrongful injuries was by means of private actions. But there is no indication in Chapter 4 that he thought there could be only one mode of rectification consistent with the concept of corrective justice—namely, a tort action, in which the judge orders the wrongdoer to pay a damages judgment to the victim. In fact . . . it is not even certain that Aristotle required that rectification involve full compensation of the victim.

To summarize, the main point in Chapter 4 is that if someone injures another wrongfully, he has behaved unjustly irrespective of his merit, relative to the victim's, evaluated apart from the wrongful injury itself. Chapter 4 is thus a corollary to Chapter 3, which discusses distributive justice. Chapter 4 makes clear that the rights of the superior individual do not include the right to injure an inferior person through wrongful conduct. This idea of "impartial legal correction" is important, but it is more limited than the corrective justice concepts of recent tort scholars, to which I turn next. It is limited because it is part of what Aristotle called "particular" justice (in Chapter 2 of Book V), in contrast to universal justice which he equates to virtue in general.

Modern Tort Scholars on Corrective Justice

Professor Fletcher analyzes tort law under two competing "paradigms"—the "paradigm of reciprocity" and the "paradigm of reasonableness." The former is derived from notions of corrective justice that Fletcher locates in Book V, Chapter 4 of the *Nicomachean Ethics,* and the latter from utilitarian ideas. The paradigm of reasonableness corresponds in a rough way to the negligence standard, with its implicit (sometimes explicit) balancing of the costs and benefits of risky activity. Under the paradigm of reciprocity, in contrast, "a victim has a right to recover for injuries caused by a risk greater in degree and different in order from those created by the victim and imposed on the defendant," irrespective of the social value of the defendant's or the plaintiff's activity giving rise to the injury. The choice between the two paradigms depends on "whether judges should look solely at the claims and interests of the parties before the court . . . without looking beyond the case at hand" (as corrective justice requires, according to Fletcher)—in which event they should choose the paradigm of reciprocity—or whether judges should "resolve seemingly private disputes in a way that serves the interests of the community as a whole," in which event they should choose the paradigm of reasonableness.

Fletcher's suggested rule of reciprocity has no basis in the concept of corrective justice expounded by Aristotle, the only authority on corrective justice to whom Fletcher refers. Nowhere does Aristotle suggest that the concept of wrongful or unjust conduct excludes consideration of the social value of conduct. To be sure, the Aristotelian judge is not to look "beyond the case at hand," but only in the sense that he is not to consider whether the defendant is a better man than the plaintiff, evaluated apart from the character of the injury; it does not follow that the social utility of the defendant's conduct that gave rise to the injury is irrelevant to whether the injury was wrongful. There is no basis in the Aristotelian concept of corrective justice for Fletcher's conclusion that negligence is an inappropriate standard when the victim's conduct is less dangerous to the injurer than the injurer's is to the victim.

. . .

Professor Jules Coleman has written a series of articles on tort law emphasizing what he calls compensatory or sometimes rectificatory justice, a term equivalent to Aristotle's corrective justice. Coleman states that "compensatory justice is concerned with eliminating undeserved or otherwise unjustifiable gains and losses. Compensation is therefore a matter of justice because it protects a distribution of wealth—resources or entitlements to them—from distortion through unwarranted gains and losses. It does so by requiring annulment of both." Coleman recognizes that a duty of corrective justice is compatible with a substantive concept of unjust conduct based on economics or utilitarianism. The "distortion" of which he speaks comes about because the injurer has violated a standard of conduct and the standard could be "one of maximizing social utility." This is an important point that is easily missed. If one equates retributive justice to punishment based on a nonutilitarian theory of desert, in the manner of Kant, and then equates retributive to corrective justice, then corrective justice will indeed seem necessarily inconsistent with utilitarianism. But even if the Kantian concept of retributive justice is accepted, the further step of equating retributive and corrective justice is unwarranted; Aristotle himself rejects retribution as a basis for punishment in Chapter 5 of Book V.

The twist that Coleman gives the concept of corrective justice is to emphasize the victim's deserts more than the injurer's guilt. If an injury is wrongful, the victim is entitled to be compensated, but not necessarily by the injurer; if the injurer did not gain from his wrongful act, corrective justice does not require that he be the source of the victim's compensation. But there is a problem: if the injurer is not the source of the compensation, then someone else, who is innocent, must be, and why is not that innocent party a victim of the wrongdoer's injurious conduct?

Defending no-fault automobile accident compensation plans against arguments based on corrective justice notions, Coleman argues that the victim of an accident in which the injurer was at fault is entitled to compensation and receives it under a no-fault system, but the injurer is not required as a matter of justice to be the source of the compensation because he does not gain by his wrongful act, as he would if we were speaking of a theft rather than an accident. Both propositions—that the victim is compensated, and that the wrongdoer does not gain—can be questioned. Take the second first. The injurer avoids the costs of taking care. This cost saving is a gain to him; if his conduct (driving too fast, for example) is wrongful, it is a wrongful gain. Negligence under the Hand formula is a failure to take cost-justified precautions, and this failure involves a cost savings to the injurer which is a wrongful gain to him. Coleman has made not only a mistake in economics but a mistake about Aristotle, who used "gain" and "loss" to describe the relation between injurer and victim even when the term "gain" was (he thought) not quite appropriate, as in the case of a wounding.

And is the victim really compensated? He receives the insurance proceeds, and let us assume they are sufficient to make him whole: but he paid for the insurance, so he just receives what is his. Potential victims as a class are clearly harmed by people who cause accidents; accident insurance premiums will be higher the higher the accident rate, and there will be no compensation by the wrongdoers for these higher premiums. Therefore, no-fault automobile accident compensation plans, which amount to eliminating liability and compelling potential victims to insure (at their own cost) against being hurt in automobile accidents, would appear to violate corrective justice because they do not redress injuries caused by wrongdoing.

It does not follow that allowing people to buy *liability* insurance is inconsistent with the Aristotelian concept of corrective justice. It might appear that the effect of liability insurance is to shift the victim's costs resulting from the wrongdoer's action to the other members of the wrongdoer's risk pool, who become in effect uncompensated victims of his action. But they are compensated, albeit ex ante, by the opportunity which insurance affords them to shift some of their accident costs to other members of the risk pool. There is nothing in Aristotle to preclude this mode of compensation, indirect as it may seem; it is the principle of rectification, rather than the form it takes, that Aristotle insists on. . . .

The Economic Basis of Corrective Justice

Once the concept of corrective justice is given its correct Aristotelian meaning, it becomes possible to show that it is not only compatible with, but required by, the economic theory of law. In that theory, law is a means of bringing about an efficient (in the sense of wealth-maximizing) allocation of resources by correcting externalities and other distortions in the market's allocation of resources. The idea of rectification in the Aristotelian sense is implicit in this theory. If A fails to take precautions that would cost less than their expected benefits in accident avoidance, thus causing an accident in which B is injured, and nothing is done to rectify this wrong, the concept of justice as efficiency will be violated. The reason is discovered by considering the consequences of doing nothing. Since A does not bear the cost (or the full cost) of his careless behavior, he will have no incentive to take precautions in the future, and there will be more accidents than is optimal. Since B receives no compensation for his injury, he may be induced to adopt in the future precautions which by hypothesis (the hypothesis that the accident was caused by A's wrongful conduct, in an economic sense of "wrongful") are more costly than the precaution that A failed to take. B's precautions will reduce the number of accidents, thus partially offsetting the adverse consequences of A's continuing failure to take the precaution, but aggregate social welfare will be diminished by this allocation of care between the parties.

The substantive concept of "wrongful" conduct in this example is of course different from Aristotle's substantive concept of wrongful conduct as set forth in Chapter 8 of Book V. He did not consider negligence the kind of wrongful conduct that triggers a duty of rectification, because negligence is not a deliberate wrong: the negligent injurer does not desire to cause an injury. But the idea of corrective justice as redress for wrongful injury (Chapter 4) is logically separable from the idea of wrongful injury as *deliberately* wrongful (Chapter 8). By the same token, the act of injustice that triggers the duty of corrective justice could be defined more broadly than Aristotle, or a normative economist, would define it.

Although the economic theory of justice requires rectification in the above example and thus implies the Aristotelian concept of corrective justice, the precise mode of rectification remains, for the economist as for Aristotle, a secondary question having to do with the practical advantages and disadvantages of alternative modes. Aristotle assumed that the method of rectification would involve private actions, mainly for damages, because that was how things were done in his day. The situation today remains much the same and economists have presented arguments why the private damage action continues to be the cornerstone of the system of redress in most tort (and contract) settings. But where private tort remedies are infeasible, as where injurers have no assets to levy on—not even what they wrongfully took from the victim—there is nothing in Aristotle to imply that an alternative mode of rectification, such as criminal punishment, would be unjust.

A more difficult case is where tort remedies, while feasible, are more costly than the alternatives. Suppose the advocates of no-fault automobile accident compensation plans are correct that a combination of criminal penalties for dangerous driving and compulsory accident insurance for potential victims would be a more efficient method of accident control, considering all relevant social costs—the costs of accidents, the costs of accident avoidance, and the costs of administering the accident-control system itself—than the present tort system. If the criminal penalties deterred all negligent driving, there would be no victims of wrongful conduct and so no problem with the abolition of liability. But not all negligent injuries would be deterred, so some victims of wrongful injury would go uncompensated. Would the no-fault system therefore violate corrective justice? Not necessarily. The concept of ex ante compensation, introduced earlier, is again relevant. If the no-fault system is really cheaper, potential victims (who are also drivers) may prefer to buy accident insurance and forgo their tort rights in exchange for not having to buy liability insurance.

But there is a simpler route to the conclusion that a no-fault plan would not necessarily violate the concept of corrective justice. If there are good reasons, grounded in considerations of social utility, for abolishing the wrong of negligently injuring another, then the failure to compensate for such an injury is not a failure to compensate for *wrongful* injury. To repeat an earlier point, corrective justice is a procedural principle; the meaning of wrongful conduct must be sought elsewhere.

Let us consider another example of arguable conflict between corrective justice and economics. Suppose a favorite idea of economists was adopted, and a very high fine was set for some offense coupled with a very low probability of apprehension and conviction. Say the fine was $100—although the social cost of the offense was only $1—and the probability of apprehending and convicting an offender was set at 1 percent. Then in 99 out of 100 cases the offender would go scot-free. Would such a penalty scheme, though economically optimal, violate the principles of corrective justice? I think not. The expected cost of the offense is $1, which we said was its social cost. The offender has paid for the offense—in advance. To be sure, ex post there will be unequal treatment of offenders; ex post, some really will get off scot-free; but unless the ex ante perspective is inconsistent with the Aristotelian idea of corrective justice (and why should it be?), a failure of redress ex post is not necessarily a failure to do corrective justice.

Book V, Chapter 4 of the *Nicomachean Ethics* makes the point not only that a wrongful and injurious act requires rectification in some unspecified form—a point perilously close to being a tautology—but also, and more interestingly, that the duty of rectification is unaffected by the relative merit of injurer and victim considered apart from the injury; unaffected, that is, by distributive considerations. Distributive neutrality is also required by the economic analysis of law. Consider two otherwise identical accident cases, but in one the injurer and the victim have incomes of 100 (in present-value terms) and in the other the victim's income is only 60. The accident is the result of a wrong (in the economic sense of a failure to take a cost-justified precaution) by the injurer, and the victim is totally disabled from gainful work by the accident but not otherwise injured. Under the economic approach as under the Aristotelian, and assuming rectification takes the form of private damages actions, the first victim would be entitled to damages of 100 and the second to

60. If the second victim's damages were reduced by a further 60 percent—say on the ground that he is only 60 percent as meritorious as the injurer—there would be underdeterrence of accidents from an economic standpoint, because the injurer would not bear the full social costs of his accidents. Similarly, it would be wrong as a matter both of economics and corrective justice to award the same damages to both victims—notwithstanding the difference in their incomes—on the ground, for example, that they were in some sense equally good people. To adjust the compensation according to the relative merit of the injurer and the victim as persons would be contrary to Aristotle's concept of corrective justice, and it would also be inefficient because it would induce an inefficient level of precautions by one of the victims, or by both, or even by the injurer. For example, if victim A receives only 60 in damages, and victim B also 60, A will be undercompensated and will be led to take excessive precautions. If A and B each receive 100, there will be overdeterrence of injurers; in addition, B will have an incentive to act carelessly since he profits from being disabled. If each receives 80 (one-half their combined loss), then A will be undercompensated and B overcompensated, with inefficient results as just described. Thus, the distributive neutrality of the economics analysis of torts is not a merely adventitious characteristic of that analysis. Neutrality is required as a matter of justice, where justice is defined in terms of economic efficiency.

I am not arguing that Aristotle anticipated the economic analysis of law. He did not. Not only was his substantive concept of wrongful conduct too narrow from an economic standpoint, because limited to deliberate wrongs, but it is not clear whether his idea of corrective justice required that the victim of wrongful conduct be correctly (from an economic standpoint) compensated. The problem is most acute in the case where the wrongdoer's gain is less than the victim's loss. One commentator has suggested that if the wrongdoer gained three and the victim lost seven, Aristotle would have wanted the judge to award damages of five. Another commentator thinks Aristotle would have required full compensation. The first suggestion would involve giving the victim an incentive to overinvest in safety.

· · ·

To all that I have said in this section two possible responses remain to be considered. The first is that I have limited discussion to Aristotle's concept of corrective justice and other concepts might lead to other results, perhaps inconsistent with the economic approach. This is of course possible, but while Aristotle's concept is not always followed, . . . no alternative concept has, to my knowledge at least, been elaborated. Second, it may be argued that while both the Aristotelian concept and economic analysis result in the same or at least similar systems of redress, they do so for different reasons: the Aristotelian to carry out some ideal of justice, and the economic to maximize the wealth of society. But Aristotle did not explain *why* he thought there was a duty of corrective justice; he merely explained what that duty was. Economic analysis supplies a reason why the duty to rectify wrongs, and the corollary principle of distributive neutrality in rectification, is (depending on the cost of rectification) a part of the concept of justice. Corrective justice is an instrument for maximizing wealth, and in the normative economic theory of the state—or at least in that version of the theory that I espouse—wealth maximization is the ultimate objective of the just state.

To summarize, my argument is not that Aristotle advocated an economic approach to law; it is that the concept of corrective justice in Book V, Chapter 4 of the *Nicomachean Ethics* is, and must be, a component of the economic theory of law.

Notes and Questions

1. When a wrongdoer gains three and the victim loses seven as a result, what *should* the wrongdoer pay the victim in your view? What theory of corrective justice do you think is intuitively correct to most observers?

2. When a wrongdoer gains three and the victim loses seven as a result, how much should the wrongdoer be required to pay in order to deter future wrongdoing? Should the focus be on the wrongdoer's gain or the victim's loss? Do you see why the former, or restitutionary approach, might prove difficult to calibrate in many cases? What if the wrongdoer always gains three from a certain kind of negligence, and there is a 50 percent chance that this behavior causes no damage and a 50 percent chance that it causes a victim fourteen in damage? Can we still deter such negligence with a corrective justice rule aimed at the wrongdoer's unjust deserts?

Rethinking Comparative Law: Variety and Uniformity in Ancient and Modern Tort Law

SAUL LEVMORE

Ancient codes and customary laws have a great deal of subject matter in common with one another and with contemporary legal systems. It could hardly be otherwise. Any system of rules that seeks to regulate behavior or resolve disputes in a community that is too large to rely solely on informal sanctions, but that is large enough for its rules to have been recorded in a way that allowed survival to the present day, must deal with murder, theft, accident prevention, insolvent debtors, and other circumstances in which, for various reasons, purely private agreements and enforcement mechanisms are likely to be inferior to more formal rules and arrangements. Indeed, nearly every legal system of which we are aware, from ancient Babylon to our own, not only contains

Abridged and reprinted without footnotes by permission from 61 *Tulane Law Review* 235 (1986).

rules about murder and theft but also deals with inheritance, marital obligations, and those mundane matters that are the stuff of private law.

Traditional comparativists, be they lawyers or philologists, explain uniformity among legal systems as a product of direct borrowing, imposition, or
common inheritance. Most anthropologists seem more flexible and sophisticated in their willingness to explore parallel evolution, comparable cultural
adaptation, and even similarities in social problems and institutional functions
across different societies. . . .

In this essay, I introduce an alternative or supplementary explanation of
variety and uniformity in legal systems. I begin with a belief or conjecture that
many legal rules serve to channel behavior and I argue that we should find
more uniformity across legal systems when theory tells us that a rule matters.
For example, since it is easy to predict the deterioration of the social and
economic fabric of any society if there are no deterrents to theft, we should
expect to find thieves liable at least for what they have taken, and probably
more. For many of the same reasons we should expect negligent behavior to
be discouraged as well. . . . In fact, although it is easy to quibble over the
definition of uniformity among penalties, I know of not a single legal system
that fails to discourage theft and the negligent infliction of harm.

On the other hand, there are other legal rules that are not compelled by
behavioral effects and realities. Many procedural rules fall into this category;
it is unlikely, for example, that a society would undergo great change simply
by altering its rules governing the admissibility of certain kinds of evidence or
the number of persons contained on its juries. It is therefore easy to predict
that such rules will not be uniform in different societies because such uniformity would be the product of happenstance (or detailed imitation). More
interestingly, some fundamental substantive rules fall into the category of
uncompelled rules. In this Article, I discuss one of the most familiar of these
rules, the choice between strict liability and negligence in tort law.

Substantial literature exists on the difference, if any, between liability
standards. For example, if T saves his $500 boat from certain destruction by
taking a 50 percent chance with V's $400 dock—and actually destroys the
dock—then, under a negligence rule, T will presumably not be liable. If the
situation should repeat itself, T will have no reason to do anything differently;
he will again attach his boat to V's fragile dock. And even if a rule of strict
liability prevails and T must pay V $400 because he has "caused" V damage, T
will still dock because an expected loss of $200 (50 percent of $400) is more
attractive than the certain loss of a $500 boat. The difference between the two
rules is one of relative wealth between T and V. . . . [But] as far as immediate
behavioral effects are concerned, both negligence and strict liability rules
discourage the sacrifice of, say, a $3,000 dock to save a $500 boat, but do not
discourage the sacrifice of a $400 dock for this same purpose.

In studying different legal systems, we might therefore expect to see the
adoption of a negligence rule in some systems, the use of a strict liability rule
in others, and even the adoption of something in between in some systems, for

a nonnegligent T would still continue to dock in an efficient way even if he had to pay some portion of V's damage. In short, we can look for uniformity when the rules matter in a direct behavioral way, and variety among rules when they do not matter. . . .

In order to demonstrate the remarkable degree of variety in the tort area and the uniform deterrence of negligent acts, I have focused on material that has survived from ancient legal systems. Although ancient legal systems obviously could have influenced one another, it is more useful to compare old codes far enough apart to suggest independent origins than to compare modern legal systems that almost necessarily share influences. . . .

The literature on this subject is so important and extensive that little further rehearsal is necessary. If A negligently, but not maliciously, injures B, then A must be made to pay B's damages if the community or legal system wishes to discourage A from such activity in the future. If A's boat is worth $199 and the expected damage to B's dock is $200, A must be made to pay at least $200 or he will dock his boat. A rule that requires A to pay B more than B's damages raises familiar problems of moral hazard, overdeterrence, and magnification of fact-finding errors. On the other hand, if A destroys B's property in a nonnegligent fashion, as when the boat he docks is worth $201, then whether we make A pay nothing (a negligence rule) or all of B's damage (a strict liability rule), A may just as well continue his activity. The rule thus does not matter in an immediate behavioral sense, even if A and B are unable to bargain, for the $201 boat, but not the $199 boat, will continue to be docked regardless of the rule.

. . .

The implications of these theoretical sketches of basic tort law rules for understanding variety and uniformity are straightforward. Just as it is almost unimaginable that a society could be successfully organized without some sanction to discourage theft, so too it is unlikely that it could prosper without discouraging negligent behavior. Whether it does this with a strict liability rule or a negligence rule, it almost surely requires some rule to prevent the inefficient exploitation of neighbors' resources. In fact, I have been unable to find even a single example of a system of laws in which negligent actors regularly avoid liability. Oaths, warnings, curses, and other penalties may be assigned to actors who are nonnegligent or only possibly negligent, but clearly negligent tortfeasors are either singled out for liability or held liable along with other causal agents. Either way, there is uniform imposition of liability for (and hence discouragement of) negligent acts.

It is more difficult to theorize about uniformity regarding the magnitude of this liability for negligently imposed injury. Considerations of moral hazard*

*["Moral hazard" refers to the possibility that some provision—such as a rule of law or a contractual promise or the issuance of insurance—will actually exacerbate an underlying problem. For example, multiple damages may decrease the incentive for potential victims to be careful, so that more accidents may result when potential victims know they will collect multiple damages from tortfeasors than when they expect single damages.—Ed.]

and factfinding error often argue against multiple damages. Since these considerations are unlikely to disappear in other societies, we might expect a uniform rule of single damages for negligent tortfeasors. On the other hand, we might expect damages to be multiplied when there is little fear of factfinding error and there is reason to think that many negligent injuries are difficult to trace or otherwise part of an underenforcement problem. As far as deterrence is concerned, the prospect of paying treble damages one-third of the time is obviously equal to paying full, single damages each time. Such underenforcement might be the case for different wrongs in different societies or might simply be a general problem to a different degree in different societies. It is, in short, easy to predict that negligent tortfeasors will be liable for damages in all societies but less easy to predict that this uniformity will extend to the remedy of single damages.

· · ·

The variety of rules governing nonnegligent injurers is so striking in primitive legal systems that only a few words of introduction are necessary. In our own common-law system, we find, as might be expected, oscillation between a negligence and strict liability rule. Although some observers favor a nearly universal strict liability rule, it is fair to describe current tort law as built on a negligence foundation with exceptional cells of strict liability. Generally speaking, these cells are easily explained as solutions to specific problems in the application of a negligence rule. It is, for example, not at all surprising that in our own system the doctrine of res ipsa loquitur and other presumptions practically turn negligence into strict liability for certain encounters, such as those between passers-by and falling objects. If, under a negligence rule, plaintiff-pedestrians would win 99 percent of all litigated cases, then society is probably better off letting such plaintiffs win automatically, avoiding the burden of the negligence suits. The cost of getting 1 percent of all such cases wrong, including losses from the underdeterrence of the few wild pedestrians who actually cause accidents, is very likely far outweighed by the savings in litigation costs. This rule can be refined, as it is in our own system, to allow a defendant to show that his case is abnormal (that is, that he was not negligent).

Somewhat similarly, blasters are said to be governed by a rule of strict liability. This cell of strict liability is quite explicable. For a negligence rule to succeed (in deterring antisocial activity), the potential negligent actor must expect to be detected and forced to pay at least those damages his actions generate. But an explosion will often destroy all the evidence that a victim could use to prove negligence. This must have been especially true before the development of modern arson squads and forensic techniques, when the blasting rule was developed. If plaintiffs are often unable to prove negligence when their damages were in fact negligently caused, then a negligence rule will not succeed in deterring antisocial behavior (unless damages are multiplied). Fortunately, since strict liability and negligence are almost equivalent in their effect upon behavior, the law is able to employ the strict liability rule where the negligence rule would fail. That the early common law also imposed strict liability for a spreading fire, which also destroys the evidence of its origin, reinforces this explanation. The blasting rule (and other cells of strict liability) is further explained by noting that alternative blasting techniques, safety steps, and the like are sufficiently technical, time-specific, and obscure that potential victims are much less able to bargain for alternatives with the blaster than is the blaster able to initiate bargaining with them to protect valuable objects or take other precautionary steps. Again, the rule is quite sensible in allowing the blaster to convince a perceptive court that plaintiff might better have avoided (or bargained to avoid) the injury in question. Thus, in one splendid case, a blaster was not held liable for losses suffered by a mink rancher whose mink killed their offspring when frightened by defendant's blasting. The decision speaks in terms of intervening causes, but it is fair to argue that most blasters are unaware of the unusual proclivities of mink and that the best route to the socially desirable result is for mink ranchers who are aware of nearby blasting to separate mother mink from their kittens.

There are, of course, other such cells of (and exceptional subcells to) strict liability, but the argument can proceed without a comprehensive survey. The point is simply that just as in our own system we find some oscillation between strict liability and negligence, so too we ought to expect such variety in other systems. We ought to expect uniformity with respect to things like theft and negligent blasting or firesetting. We might even expect uniformity regarding nonnegligent blasting or firesetting if many societies face the problem of evidence destruction. In contrast, there is no social need for, and therefore no reason to expect, uniform liability or freedom from liability for most non-negligent behavior; as illustrated earlier, such behavior will normally continue whether or not some liability is imposed. The rule does not matter (or does not matter much) so that whether it is the product of assimilation, randomness, cultural influences, or conquest, it will not lead to unsettling behavioral consequences. Somewhat similarly, since it is not clear which rule best encourages socially desirable behavior when numerous misbehaving parties are concerned, there is no reason to expect uniformity among legal systems in this context. Just as the variety of rules across jurisdictions in our own legal system can be said to reflect the fact that we are unsure whether contributory negligence, comparative negligence, or any single modification of these rules dominates or is dominated by the other possible rules, so too the lack of uniformity that we will see presently among primitive legal systems dealing with multiple tortfeasors or contributorily negligent victims may reflect the fact that no rule emerged as clearly inferior or superior.

It is possible, of course, that although the choice between no liability (a negligence rule) and liability (a strict liability rule) for nonnegligent acts is not affected by behavioral implications, at least in the short run, all legal systems would choose the same rule. Thus, as a theoretical matter, uniformity does not reflect utilitarian necessity nearly as much as variety reflects the lack of such necessity or the presence of conflicting goals. . . .

The Code of Hammurabi

The tort rules found in the Code of Hammurabi are most easily described by reference to our own. Although the common-law tort system that we know (modified by occasional legislative decisions) is negligence-based, there are areas that are carved out for the application of a strict liability rule. As noted earlier, strict liability is often employed when negligent behavior is either hard to discern and prove, or so likely, that the litigation costs associated with a negligence rule are best avoided.

The Code of Hammurabi describes a variety of situations in which an intentional wrongdoer, such as one who strikes another in a dispute, must pay a penalty. Wrongful or negligent behavior consistently calls for single damages or another penalty suitable in the circumstances. Thus, a careless boatman must replace the boat and cargo that he loses. A boat builder must repair a boat that becomes disabled within one year of his building it. Presumably, boats required substantial maintenance so that the warranty period lasted but one year. By way of contrast, a house builder's liability toward injured persons and property (and the reconstruction of the house itself) if his building collapses is not limited in time. It becomes clear from the constant emphasis on the poor work done by the builder that the code assumes he has been negligent. The owner of a regoring ox who has not blunted its horns or otherwise taken precautions must also pay when damage is done. . . . This code requires no payment for an unpredictable goring. The wetnurse who suckles a second child without informing its parents that a prior child died in her service is severely punished. Finally, the landowner who has been too lazy to strengthen the bank of an irrigation canal must pay for the flood damage that occurs.

There are, however, a few cells of strict liability in the Code of Hammurabi. A shepherd is strictly liable for cattle that become diseased while in his care. This treatment is probably akin to res ipsa loquitur, the lawmaker assuming that disease only strikes a herd in the charge of a negligent herdsman. A second rule that has at least a flavor of strict liability concerns the bursting of irrigation canals. In Babylon, irrigation would have been necessary during the summer, and there is substantial evidence that in Hammurabi's time there were extensive canal systems and land reclamation projects. Indeed, the potential gains from such public works may have been the force behind Hammurabi's success in uniting the region. The code is especially attuned to the problems of flooding on land controlled by tenants.

> 53. If a man has been too lazy to strengthen his dyke, and has not strengthened the dyke, and a breach has opened in the dyke, and the ground has been flooded with water; the man in whose dyke the breach has opened shall reimburse the corn he has destroyed.
>
> 54. If he has not corn to reimburse, he and his goods shall be sold for silver, and it shall be divided among those whose corn has been destroyed.

> 55. If a man has opened his irrigation ditch, and, through negligence, his neighbour's field is flooded with water, he shall measure back corn according to the yield of the district.
>
> 56. If a man has opened the waters, and the plants of the field of his neighbor the waters have carried away, he shall pay ten *gur* of corn per *gan*.

Section 56 presents an interpretative, or hermeneutical problem. It is quite clear that sections 53 through 55—while establishing in passing a pro-rata distribution rule in bankruptcy—codify the rule that a negligent party must pay (single) full damages. It is difficult to know whether there is an underenforcement problem in this setting. It is possible that neighbors always knew the state of one another's canal banks so that when flooding occured evidence quickly developed regarding negligent behavior. Alternatively, it may be a res ipsa situation; breaches of the sort mentioned in section 53 may rarely occur unless there has been negligent behavior. In section 55, the bank has not crumbled but the tenant seems to have mismanaged the floodgate. The section specifies damages either because the victim's land has so eroded that planting will be impossible (although one would think that liability should then be for lost net profit and not for gross yield) or because the flooding has destroyed evidence of what was lost and corroborating evidence is difficult to obtain. The question, then, is whether section 56 provides a negligence rule, a strict liability rule, or a splitting rule. The compensation it calls for is equal to one year's rent, or one-third of the typical crop yield. It is thus possible that section 56 is different from section 55 in that section 55 specifies negligent behavior, leaving us to conclude that section 56 provides for one-third damages when flooding is caused by nonnegligent behavior. This would be a clear splitting rule, although not down the middle.

Alternatively, it is possible that the precondition of negligence found in section 55 is meant to carry over to section 56, and that section 56 envisages the possibility that the injured party will be able to replant his field or use the payment to rent another field. It is also possible that section 55 refers to an intentional act, but inasmuch as the damages are light and, as we have seen, the code elsewhere extracts more serious penalties from intentional wrongdoers, this interpretation seems improbable. Finally, it is possible that both interpretations are to be combined: section 56 is concerned with nonnegligent behavior and with a field that can be replanted. Under this last interpretation, section 56 announces a strict liability rule. This interpretation is not unreasonable in light of the explicit reference to negligence in section 55. Such a reference implies that flooding can occur both through the negligent and the nonnegligent opening of a floodgate. Section 55 provides the result for the first case and section 56 for the second one.

It is easy to see why strict liability might be chosen for the management of floodgates. When there is negligence, there will sometimes be witnesses and sometimes confessions, but often it will be hard for victims to prove that a floodgate was mismanaged. Farming in that part of the world is to this day a lonely business, and it is easy to imagine that the owner of a flooded field would not know whether to blame his neighbor or simply to bemoan his bad luck. Just as the common law generally holds a blaster strictly liable—perhaps because it is difficult for victims to develop proof of wrongdoing on his part—so, too, the Code of Hammurabi may have chosen in section 56 to legislate some disincentive for those who, but for a strict liability rule, might open floodgates knowing that a negligence-based rule may result in no liability because their action would be difficult for others to discern and prove.

Finally, the code imposes penalties, or liability, on unsuccessful surgeons.

> 218. If a doctor has treated a man with a metal knife for a severe wound, and has caused the man to die, or has opened a man's tumour with a metal knife, and destroyed the man's eye; his hands shall be cut off.
>
> 219. If a doctor has treated the slave of a plebeian with a metal knife for a severe wound, and caused him to die; he shall render slave for slave.

Although I do think that the code creates a cell of strict liability in the operating room, two contextual comments need to be made. First, there is reason to accept the view of some commentators that in this and in other ancient societies talionic or retaliative penalties, such as that calling for amputation of the surgeon's hands, were generally commutable at the parties' option into monetary compensation. This is especially easy to accept in the standard tort (as opposed to the more public criminal) case, for if A strikes out B's eye, B will generally prefer compensation to A's joining him in misery. That we have so many records of the period, and especially of Hammurabi's Babylon, and yet no evidence that citizens walked around without hands and eyes, reinforces this view of talionic or in-kind justice as a metaphor or, at most, as a powerful spur to negotiations.

Moreover, given the septic conditions of surgical practices before the modern era, amputation would have often meant death, and there is little reason to suppose that ancient laws were intended to produce the unnecessary deaths of tortfeasors and careless contractors. Indeed, there is every reason to think that these ancient lawmakers sought to decrease the amount of violent retribution in their societies. Traditions of blood revenge and of long-standing and spiraling vengeance among clans were almost surely a major concern of the talionic decrees in the Code of Hammurabi and other ancient systems. It is thus plausible (even in the face of evidence that in our own time some societies impose talionic penalties) that interpretations of the code's provisions should assume less violent rather than more violent alternatives. That "an eye for an eye" really means that the tortfeasor had better reach a settlement with the unfortunate victim or his family, is, therefore, a superior interpretation. Of course, the potential of such settlements will serve as a fair deterrent to antisocial activity.

The second contextual comment about the code's treatment of unsuccessful surgeons concerns its balancing of rewards and liabilities. Logic does not necessarily demand any such balance, but the Code of Hammurabi is hardly alone among legal systems in raising the penalty for failure precisely in those circumstances in which it is generous with rewards for success. For example, the surgeon who operates on a severe wound receives 10 shekels of silver—a sum equal to the damages called for when a man strikes a woman and causes her to miscarry and five times the price set for the construction of a typical boat or the employment of a builder for 72 days. Whether or not these fixed fees could be bargained down, it is clear that in return for the risk of liability the surgeon was to be paid handsomely for success. At a minimum, it appears that he earned a 100 percent premium in return for the risk of liability. In an important way the code thus describes the inevitable companion to strict liability—higher prices, or forced insurance. The surgeon's liability for failure and, if the earlier discussion is correct, the compensation of injured patients are financed by the patients who are treated.

The system described above is usefully compared with our modern-day use of negligence in the medical context. Superficially, negligence seems ill-suited to the task of deterring sloppy medical practices. Proof problems abound because the patient finds it extremely difficult to reconstruct the steps taken by his doctor. As such, it is tempting to suggest that negligent behavior will go undetected and undeterred and, therefore, that strict liability would be a more appropriate standard for judging medical services. On the other hand, as every thoughtful modern tort student knows, strict liability in this area is especially unworkable because patients go to doctors when things are already wrong and no fact-finding system could hope to hold doctors strictly liable for their contribution to the harm while separating out preexisting conditions. The same probably can *not* be said for the surgeons of ancient Babylon. While their *negligence* was probably very hard to distinguish *ex post,* it is not difficult to imagine that the drafter of this code section comprehended that the liability his rules imposed would follow a patient's death *during* or, at most, soon after surgery. Separating out preexisting conditions in this inexact way is obviously relatively easy because the patient either survives surgery or does not. In support of this view, it should be noted that the code provides a reward, or fee, for the surgeon who heals a broken bone but does not provide for liability when a bone is improperly set. In this setting, the precondition is not easily separated from that which a surgeon causes.

· · ·

In sum, it is at least plausible that there are as many as three strict liability cells in the code . . . and that these cells can be explained in much the same way as more modern uses of strict liability in the common law can be explained. Hammurabi's law and our own are thus similar, or at least analogous, in their treatment of tortfeasors. One might expect such similarity given the behavioral consequences of negligence and strict liability rules. Indeed, it is possible that the choice of negligence rather than strict liability as the basic rule in Hammurabi's day reflected (as it may in our own system) the practical problems of valuation in a complex society. Moreover, even the exceptions to the general (uniform negligence) rule in the Code of Hammurabi appear on close inspection to be more similar to, than different from, our own exceptions to the negligence rule.

Exodus

. . .

The section of Exodus dealing with torts begins with cases of one person injuring another. Talionic rules are set out, although here it is even clearer that apart from homicide the law describes a system of monetary compensation, rather than strictly in-kind retribution. The text then contains the following verses from chapter 21:

> 33. And if a man shall open a pit, or if a man shall dig a pit and not cover it, and an ox or an ass fall therein,
>
> 34. the owner of the pit shall make it good; he shall give money unto the owner of them, and the dead beast shall be his.
>
> 35. And if one man's ox hurt another's, so that it dieth; then they shall sell the live ox, and divide the price of it; and the dead also they shall divide.
>
> 36. Or if it be known that the ox was wont to gore in time past, and its owner hath not kept it in; he shall surely pay ox for ox, and the dead beast shall be his own.

Chapter 22 begins with the rules providing for multiple restitution by thieves and then returns to tort law:

> 4. If a man cause a field or vineyard to be eaten, and shall let his beast loose, and it feed in another man's field; of the best of his own field, and of the best of his own vineyard, shall he make restitution.
>
> 5. If fire break out, and catch in thorns, so that the shocks of corn, or the standing corn, or the field are consumed; he that kindled the fire shall surely make restitution.

Verses 33, 34, and 36 of chapter 21 appear to announce a negligence rule. They focus on two clear examples of negligent behavior and then carefully provide for simple restitution or single damages. Verses 33 and 34 do not suggest a strict liability rule for all pit diggers, because it is quite clear that if the digger covered the pit and then, say, a storm or passer-by uncovered it, the digger would not be liable. It is possible that even commentators who view these ancient rules as fact-specific rather than illustrative would agree that these two verses contain a generally applicable negligence rule. A commonly used hermeneutic method suggests that when a text lists more than one example of something, it implies the inclusion of unmentioned items with similar characteristics. Although it is not surprising to find a negligence rule in such an ancient system, note that a strict liability rule also would not be surprising.

Both rules deter negligent behavior by forcing the negligent party to feel, or internalize, the hurt he causes his victim. It would be astonishing to find a negligent party paying less than single damages. But it is clear that at least in the . . . Code of Hammurabi and the Book of Exodus we find uniformity on this matter: negligent actors pay.

As mentioned earlier, there is some reason to expect variety in the treatment of nonnegligent actors who cause injuries. There is the possibility of no liability (a negligence rule), full liability (a strict liability rule), or one-half liability (an even "splitting" rule). Indeed, any sort of split liability is possible: a one-third splitting rule was discussed in the context of Hammurabi's Code and a variable splitting rule is found in general average contribution in admiralty law. . . .

Admiralty law contains a similar response to uncertain causality in the mechanics of the rule of general average contribution. If, to lighten a ship during a storm, its master jettisons some cargo, then all owners of property at risk—including the owner of the ship itself—are treated as a community and each shares in the loss in proportion to the value of the goods he had at stake in this community. This is obviously a (somewhat more modern) kind of splitting rule, for the ship captain's behavior during a storm is hardly negligent, and yet the shipowner, like everyone else, shares in the loss. This general average rule is probably more than just an example of variety between negligence and strict liability when behavioral consequences do not dictate a rule. A rule of no liability (negligence) might be exploited by corrupt behavior; shippers could pay the captain to toss someone *else's* goods overboard. General average may thus be a way to ensure fair treatment among shippers and to save them the cost of seamy negotiations. A rule of strict liability— with its costs presumably passed on to the shippers in the form of higher freight rates—would not have been a bad alternative but, arguably, shippers (at least at some points in history) prefer the choice between partial coverage (general average) and "homemade" strict liability through the purchase of what would be, in effect, insurance coverage.

In any event, an important detail of general average is that, in computing each shipper's proportion of the enterprise, money and valuables carried by passengers are excluded from general average calculations. Although it is possible that this exclusion reflects the realities of fact-finding, for one would hardly want to search all survivors after a storm, it is also possible that the rule reflects the fact that the value of money and other possessions, such as gems, will rarely correlate with their role in causing the disaster. It is plausible that the original rulemaker (or parties to a private general average agreement) would have preferred a rule that made parties contribute to a loss according to the proportional weight of their cargo; after all, the point of jettison is generally to lighten the ship. But a rule based on weight, with some mixing in of the value of the ship or of an arbitrary percentage of cargo losses to be paid by the shipowner in order for it to make sense, would have been slightly less workable. . . .

The splitting rules these two legal systems contain can be understood, at least in part, as influenced by the causal uncertainties in the circumstances they address. These rules may be more accurately described as subtle examples of comparative negligence than as compromises between strict liability and negligence. Either way, they suggest that variety among legal systems will be more pronounced when the choice of rule will not much affect the incentive system necessary for a complex society to flourish.

. . .

Exodus also contains rules . . . that may form cells of strict liability. The grazing in verse 4 is, however, probably not a good illustration of such a strict liability rule. First, it is not clear whether grazing on another's land might have reflected negligent herding rather than intentional wrong-doing. We need more information than what is available on the herding and fencing practices of the people first governed by these rules in order to determine whether or not verse 4 does more than describe an intentional tort. . . .

Verse 5 of chapter 22 contains the clearest example of strict liability in ancient laws. By describing the fire as kindled and then as catching in thorns, the drafter contrasts verse 5 with verse 4 in which a beast was let loose and another's field or vineyard caused to be eaten. I think it not unlikely that this contrast was meant to distinguish accidental from negligently caused losses. The fire may have spread accidentally while no attempt was made to constrain the movement of the beast. The verses which precede and follow verse 5 make it clear that this strict liability rule is fact-specific and not illustrative, for the nonnegligent actor generally incurs no more than split liability. Nor is this variety particularly unexpected. Fires, like explosions, must often consume evidence that would show negligence in starting or tending a fire. As between a negligence rule, with underdetection of negligent behavior, and a strict liability rule, the latter may be preferable. This may be so even though a strict liability rule generates numerous valuation proceedings. Early English law also had a strict liability rule for firestarters. Whether the demise of this rule is correctly traced to the desire of lawmakers to discourage urban density, to encourage precautionary behavior by every property owner, or to encourage participation in volunteer firefighting units, it is clear that more than one negligence-based legal system has been influenced by arguments in favor of strict liability for firestarters.

In short, a literal reading of Exodus yields powerful evidence in support of the uniformity-variety thesis. . . . There is uniformity when behavioral consequences require it, and a fair amount of variety otherwise. . . .

Mongolian Customary Law

The earliest Mongolian codes contain many rules regarding class distinctions, taxes, and family law, and few concerning torts and contracts. Later material and traditions collected by conquerors and scholars do, however, contain many tort, contract, and property rules. In many cases, the rules vary little among the various tribes. And inasmuch as virtually all the peoples in the part of the world we now call Siberia and Mongolia have been herdsmen and hunters, rather than farmers and manufacturers, it is not surprising that the various tribal customs deal repeatedly with similar questions. . . .

With few exceptions, Mongolian tribal laws do not let losses lie where they fall. Minors up to the age of eight are free of responsibility in at least one tribe, and a fenceowner is not liable if someone happens to injure himself against the fence. There are other such examples but by and large, as we will see presently, splitting rules prevail, and clearly negligent acts are treated almost as harshly as crimes. Examples of uniformity and variety are thus especially interesting in Mongolian law because splitting—which is a curiosity in other legal systems—dominates the legal terrain of Mongolian law.

These splitting rules impose liability on many nonnegligent and even, perhaps, on some negligent causal agents. The general average contribution rule of admiralty law is perhaps the construct most similar to these splitting rules. At the other end of the tort spectrum, an intentional wrongdoer, such as one who insults, spits, throws mud, robs, or misappropriates, is punished rather severely. Finally, simple acts of negligence incur single-damage liability much as one would expect. It is difficult, however, to generalize about the treatment of the accidental killing or maiming of another person. There is little, if any, of the talionic spirit in these laws. But is it multiple, full, or partial liability when an owner of cattle must pay "nine animals and a valuable thing" when his cattle kill a nobleman? The question defies any answer and mocks our own infatuation with determining the value of lost life and pain and suffering. It is, I suppose, possible that an informed answer to this question would undercut the uniformity-variety thesis. But there is, for instance, no reason to think that intentional wrongdoers are regarded as less liable or blameworthy than unintentional but negligent tortfeasors. The vignettes and rules concerning personal injury are probably best excluded from an investigation of the strict liability versus negligence issue in these tribal laws. While these rules regarding personal injury do display a preference for clear rules that require little litigation, they neither support nor contradict the thesis that splitting, or sharing, is the hallmark of these legal systems.

Perhaps the clearest example of the requirement in Mongolian tribal law that negligent actors pay and nonnegligent ones split, as in Exodus 21:33 and 21:35, is found in the customary laws of the Buriats, an eastern Mongolian tribe. If disease spreads from A's cattle to B's, A pays half the damages. But if A has not previously reported the outbreak of contagious disease in his herd to the authorities, then A is fully liable to B. This rule, that antisocial, negligent actors pay and nonnegligent ones split, is, however, neither neat nor uniform in Mongolian tribal law. Here, the variety, or inconsistency, is found at times within the system itself, but still within the very range in which variety will not be harmful. . . . [W]hile negligent actors must pay, nonnegligent actors must normally pay or split in a variety of ways, depending on the circumstances. The details of these tribal systems are thus very different from our own and from other "primitive" legal systems. But at a different level of understanding, the systems are remarkably uniform; the details vary only within the predictable ranges.

Conclusion

The uniformity-variety thesis is not limited to the question of tort liability and it is surely not limited to primitive law. [This] discussion . . . has primarily focused on primitive legal systems because in these settings there is less need to debate the relevance of borrowing among systems, and because the behavioral importance of legal rules is sometimes clearer when far removed from our own culture. Moreover, it is remarkable, I think, to learn that many of the problems we wrestle with in the law today were considered with subtlety and imagination equal to our own thousands of years ago and in civilizations thought to be quite primitive.

Notes and Questions

1. What version of corrective justice might the rule in section 53 of Hammurabi's Code be said to reflect? What about sections 218 and 219? Can you rationalize these rules?

2. Consider the "splitting" rules found in Exodus 21:35 and in the Mongolian rule regarding the infection of cattle via a herd whose contagious disease had been reported. Such splitting is rare in American law, although many observers seem to find it attractive. Can you think of explanations as to why it is a little-used tool in our legal system?

3. This selection concentrates on presumed behavioral effects of legal rules, with little information as to the historical and cultural contexts in which these rules were found. A very different approach to comparative law emphasizes the contexts of different legal systems. An obvious argument against the sort of analysis undertaken in this reading is that it ignores the likelihood that legal rules are influenced, if not entirely determined, by the culture and unique history of the lawmakers or people governed by the various rules. Moreover, even if we had some knowledge about the cultures of which these rules were a part, we could never really understand the roles they played or the effects they had. A pragmatic counterargument is that we know so little about the context (and manner) in which many rules were applied that an insistence on context is really a decision to ignore the information we have about distant cultures.

Can you think of legal rules that you know to be different in different states in the United States? Is your explanation of these differences functional, cultural, or historical?

Tort Law and the Economy
in Nineteenth-Century America:
A Reinterpretation

GARY T. SCHWARTZ

The prevailing view of American tort history regards nineteenth-century tort doctrine as deliberately structured to accommodate the economic interests of emerging industry. According to this view, the courts jettisoned a potent pre-nineteenth-century rule of strict liability in favor of a lax negligence standard, leniently applied that standard to enterprise defendants, administered a severe defense of contributory negligence, and placed strong controls on negligence law under the name of "duty."

. . .

Frequently, nineteenth-century tort law is described as providing "subsidy" to economic enterprise. A liability rule presumably amounts to a subsidy if it entails a departure from an otherwise appropriate liability standard designed to relieve a class of injurors from the expense of liability. Of course, in the nineteenth century, innovative industry may have been perceived as providing social benefits that could have justified a subsidy.

. . .

Even if a subsidy of enterprise makes economic sense, it seems simply unconscionable to exact that subsidy from the individual victims of serious accidents by depriving them of their right to compensation from the enterprises responsible for their injuries. . . . Evaluation of the negligence rule as a subsidy thus gives rise to a powerful moral objection.

How accurate, then, is the subsidy interpretation? This selection tests that interpretation by scrutinizing nineteenth-century tort law as it developed in two quite different American jurisdictions—California in the new West and New Hampshire in the old Northeast.

. . .

Subsidy scholars [have] . . . tended to assume that early American doctrine had simply followed English models. Not accepting this assumption, [Professor Morton Horwitz undertook a] substantial American law investigation. . . . Horwitz' most inclusive argument is that in 1800 both trespass and case were governed by a firm rule of strict liability, a rule that American courts later overthrew in favor of a negligence standard. To document this sequence, he cites three opinions in New York, Massachusetts, and Pennsylvania, all rendered between 1817 and 1833, and identifies them as "turning point[s]" in the law. The three opinions plainly do espouse a negligence point

Abridged and reprinted without footnotes by permission of The Yale Law Journal Company and Fred B. Rothman & Company from *The Yale Law Journal*, vol. 90, pp. 1717–(1981).

of view. In none of them, however, is there any indication of a prior general rule of strict liability that the courts thought they were abrogating. . . .

There is nothing in the New Hampshire record during this half-century that confirms Horwitz' view that American judges consciously intervened to overthrow a solid, general rule of strict liability. . . . In subsequent New Hampshire opinions in the 1850s and 1860s, "tort" gradually emerged as a coherent body of law, with negligence its guiding liability principle. This process was later ratified by Chief Justice Doe's opinion in *Brown v. Collins.* *Brown*'s holding, in denying liability, was actually quite narrow.* But the Doe opinion . . . contained a critique of certain strict liability ideas, including those discernible in the array of English opinions in *Rylands v. Fletcher.*

In California, the case law did not begin, of course, until the opening of the state court system in 1850. What is impressive is how frequent tort suits were from the outset and how immediately negligence emerged as the almost unquestioned liability standard. Both before and after *Rylands,* the California Supreme Court decided escaping water cases on a negligence basis, never even considering any *Rylands* argument. The court found that early state statutes superseded the English rule of strict liability for trespassing cattle, and deemed negligence to be an ample basis for liability in two blasting cases. Only with respect to the custody of animals known to be dangerous did the court accept strict liability, and even here one judge dissented, finding insufficient reasons to depart from the negligence standard.

The California common-law opinions during this half-century altogether avoided anything resembling a general discussion of industrial or economic policy. This is true of the 1850–1900 New Hampshire opinions as well, with the interesting exception of *Brown v. Collins,* which frequently alludes to "progress and improvement" as providing support for its negligence views. What I understand Doe to be saying is not that enterprise should be relieved of liability as such, but rather that, in light of the nineteenth-century's hardly deniable public interest in economic development, liability should not be imposed when there is no proper "legal principle" or "legal reason" for doing so. As a study of the possible "legal principles" for strict liability, Doe's opinion surely ranks as an impressive document, especially when measured against the then-existing literature. The conclusion he reached was that a rule of strict liability did not seem justified.

This conclusion, it should be noted, was one that was shared by all the leading legal intellectuals of Doe's generation. In *The Common Law,* published eight years after *Brown,* Holmes declared that strict liability "offend[ed] the sense of justice." Wigmore, writing in 1894, praised Doe's *Brown* opinion as "masterly." Ames expressed the view that strict liability asserts an "unmoral standard," and Thayer recorded his agreement with the "fundamental proposi-

---

*In *Brown,* the defendant's horse, frightened by railroad noise, bolted out of the defendant's control and damaged a post on the plaintiff's property: "[W]hatever may be the full legal definitions of necessity, inevitable danger, and unavoidable accident, the occurrence complained of in this case was one for which the defendant is not liable, unless every one is liable for all damage done by superior force overpowering him, and using him or his property as an instrument of violence."

tion of the common law which links liability to fault." Thayer's view rested in good part on his assessment of "[h]ow powerful a weapon the modern law of negligence places in the hands of the injured person. . . ." This assessment is at least roughly congruent with my own observations on the nineteenth-century negligence system, which are set forth in the following sections.

Economic Developments and Tort Responses

A high degree of abstraction afflicts the legal historians' typical references to nineteenth-century "entrepreneurs" and "infant industries." Except for the railroads, the enterprises in question are rarely identified, and the entire nineteenth century is generally treated as a single, undifferentiated economic episode. Yet any effort to ascertain how nineteenth-century tort law related to the American economy should refer to the specific economic developments occurring during that century.

. . .

[T]extile mills constituted the first significant stage of American industrialization. . . . In 1840, the vast majority of all "factories" in the United States were textile mills; as late as 1860, textiles remained the country's largest manufacturing industry. . . . Through 1840, then, textile production was the infant industry that was foremost in America. Yet the textile mills were almost wholly absent from tort law decisions in New Hampshire during the first half of the nineteenth century. Indeed, the tort problems that the New Hampshire Supreme Court addressed simply had nothing to do with modern economic activity; most of the case law looked backward to a more traditional and largely rural society. . . . With respect to the textile industry, therefore, the tort law subsidy thesis is not so much false as irrelevant. . . .

In the late 1840s and 1850s, the first of several major railroad-building booms occurred. The opening of the railroads resulted in new or improved transportation services for persons and freight and spurred growth in industries that provided the railroads with necessary supplies like machinery and rails. . . . The railroads, unlike the textile factories, were highly conducive to accidents and injuries; the case law in New Hampshire from the late 1840s and in California from the mid-1860s is replete with opinions on railroad liability. . . . In California, the first personal injury action against a major railroad was *Kline v. Central Pacific Railroad*, decided in April 1869. Kline, a teenager, illegally boarded a train. After the railroad conductor employed sharp language and put a hand on his shoulder in ordering him off the train, he jumped from a moving car and suffered injury. Although the California court agreed that the plaintiff was a "wrongdoer," it granted him a recovery, finding that his wrongdoing was "remote" and that the railroad was legally required, having discovered his presence, to use reasonable care in removing him from the railroad car. The court thus commemorated the imminent completion of the Central Pacific's transcontinental by requiring that railroad to compensate a mere trespasser.

In New Hampshire and California, later railroad cases fell into recurring categories. Railroads were most likely to cause injuries to passengers, to persons riding in carriages at railroad intersections, to livestock wandering on railroad tracks, and to farmers whose crops were set on fire by sparks. Two basic tort issues persist throughout the courts' opinions: the negligence or wrongdoing of the railroad, and the contributory negligence of the plaintiff-victim. In the passenger cases, only rarely did the railroads escape liability on the grounds that they were not negligent. Because the high speeds of the new railroads created "hazards to life and limb," and because the railroads were "entrusted [with] the lives and safety" of their passengers, the New Hampshire Supreme Court held railroads liable to passengers for "even the smallest neglect." If a railroad car derailed, the California Supreme Court declared a presumption of negligence on the railroad's part; if a car experienced a sudden jerk, both the California and New Hampshire courts agreed that this event was prima facie evidence of negligence. Obviously, the reasoning in these cases consisted of an unlatinized, common-sense version of the doctrine now called res ipsa loquitur.

Railroad passengers were only rarely denied a recovery on account of contributory negligence. If a passenger was injured while boarding or deboarding a moving railroad car, the California court, while hardly regarding such conduct as intelligent, nonetheless ruled that it was not necessarily negligent and hence affirmed jury verdicts against the railroad. As for the passenger who failed to look ahead while deboarding, the New Hampshire court specified that if she was in a "flustered state of mind" because the railroad had overshot its station, a jury finding of no contributory negligence was appropriate.

In every railroad intersection case, the plaintiff was able to present enough evidence on the initial issue of the railroad's negligence to take the case to the jury. Frequently, the plaintiff in a crossing case established negligence by showing that the railroad had failed to comply with a state statute or local ordinance applicable to railroad operations; by inferring negligence from a legislative violation, the courts were effectively developing the doctrine of negligence per se. Of the pure common-law rulings, the most interesting was *Huntress v. Boston & Maine Railroad,* in which the New Hampshire Supreme Court conceded that neither the railroad's engineer nor its fireman was negligent as the train approached the crossing. The court then observed, however, that "railway managers may be presumed to have special knowledge of the dangers of their business, and to be aware of the constant peril arising at level crossings. . . ." Relying on this observation, the court concluded that the jury could properly find the railroad negligent for not having "guard[ed] against accidents by stationing flagmen [at the crossing] or slackening the speed of the trains." . . .

The Late Nineteenth-Century Transformation

The textile factories introduced industrialization into the United States in the early nineteenth century, while the railroads were a primary force after mid-century. It is the period between 1870 and 1910, however, that is generally credited as completing the transformation of the American economy and as accomplishing a transformation of American society as well. By 1900, the United States had emerged as the world's premier economic power.

. . .

Beginning in the 1890s, American cities began to assume their present form; in particular, the use of structural steel girdings and the development of electric elevators made possible the erection of high-rise office buildings in downtown areas. Intraurban transit systems, which previously consisted of horse-drawn vehicles, were converted to electricity. In Los Angeles during the mid-1880s, a number of entrepreneurs began to build transit lines; by the early twentieth century, Los Angeles developed the most extensive rail transit system in the country.

The enterprises that contributed to this late nineteenth-century transformation produced many tort issues, some of them novel. In *Gregg v. Page Belting Co.,* the New Hampshire court approved the right of the passenger-victim of an elevator fall to sue, under a negligence theory, both the owner of the building containing the elevator and the manufacturer and repairer of the elevator itself, the absence of privity notwithstanding. In *Treadwell v. Whittier,* the California Supreme Court affirmed the jury's verdict on behalf of the plaintiff injured when the elevator in the defendant's store inexplicably fell. The court ruled that mere proof of the elevator fall established a presumption of negligence compelling a plaintiff's verdict unless adequately rebutted. Also, by imaginatively classifying elevator operators as "common carriers," the court required them to conduct inspections according to "the best known tests reasonably practicable."

. . .

In *Redfield v. Oakland Consolidated Street Railway,* a streetcar rolled downhill, injuring a passenger, in part because the car was operated by only one employee. The railroad sought to show that one-employee operation was customary within the industry. Historians advise that nineteenth-century tort law frequently regarded a defendant's compliance with industry custom as fatal to any claim of negligence; only with Judge Hand's famous opinion in *The T.J. Hooper,* they suggest, did the modern view solidify that compliance with custom is merely evidence of non-negligence. Yet in *Redfield* itself, the court ruled not only that compliance with industry custom did not require a verdict for the defense, but also that it was not even admissible as evidence on the question of negligence. "[C]ustom may originate in motives of economy, or the stress of pecuniary affairs, or in recklessness, and not from considerations based upon the proper discharge of their duty toward others using their cars." The court added that even if the challenged practice had not resulted in any previous accidents, "that circumstance is only a matter of wonderment, and is an instance of how good luck will sometimes protect carelessness for long periods."

Only rarely did the defense of contributory negligence bar a street railway victim from recovery. While the California court appreciated that pedestrian accidents could be almost entirely avoided if pedestrians exercised "the greatest care and caution," it nevertheless held that the law required no more than the ordinary care of "people in general." In addition, the California court denied that parents were contributorily negligent as a matter of law if they allowed their children to play unattended on the street where they might be injured by a negligently driven streetcar. To find parents at fault in such a case, stated the court, "would be harsh and unreasonable, especially to the poor, in every town and city." . . .

Nineteenth-Century Tort Doctrine

The previous part has questioned the subsidy thesis by considering the nineteenth-century cases on an industry-by-industry basis. This part, cutting across industry boundaries and looking at all the case law, reviews nineteenth-century tort doctrine in a somewhat more general way.

Negligence

Nineteenth-century tort law in New Hampshire and California emphasized negligence as the standard of liability. The negligence principle proceeded to demonstrate its vitality. It could be ambitious in its detection of activities that harbor appreciable risks. The more substantial the New Hampshire and California courts perceived the risks in the defendant's activity to be, the higher the level of care that they required the defendant to exercise. Insisting on more "care" when an activity entails greater risks is, of course, in line with the twentieth-century understanding of the negligence rule; the cases thus suggest that the nineteenth-century rule already possessed a distinctly modern character.

The factor of private profit was seen as a reason for being skeptical, rather than appreciative, of the propriety of risky activity engaged in by enterprise. In general, the New Hampshire and California courts were reluctant to find that economic factors justified a defendant's risktaking. Neither court even once held that mere monetary costs rendered nonnegligent a defendant's failure to adopt a particular safety precaution. The California court seemed quite unimpressed with defendant's claims that their conduct complied with industry customs. On one occasion, the New Hampshire court proved openly hostile to a claim of justification couched in enterprise terms. In *Sewall's Falls Bridge v. Fiske & Norcross,* the plaintiff's bridge had been damaged by the accumulation of logs that the defendants had floated down the river. The court ruled irrelevant and hence inadmissible the defendants' evidence that "a large amount of timber at the head waters of the Merrimack . . . cannot be taken to market, without costing more than its value in market, in any other mode than that which they practiced." According to the court, "the prospective extent of [the defendants'] interests and the contemplated magnitude of their operations, could not give them any special privilege to manage their business in a careless and negligent manner. . . .

The New Hampshire and California courts elaborated on the negligence standard so as to facilitate the plaintiff's ability to prove his case. Both courts were eager to find negligence when the defendant violated a statute or ordinance designed to promote safety. As the California court explained, if an ordinance seeks to prevent placing "in jeopardy the lives of men, women, and children," that ordinance's safety purpose should be furthered by holding the defendant to a "strict legal accountability" when his violation of the ordinance causes private injury. The California court, while rarely using the exact phrase "res ipsa loquitur," still applied the *res ipsa* notion to a range of business-caused injuries; indeed, *res ipsa* logic had taken hold in California in 1859, four years prior to *Byrne v. Boadle,* the English case that is usually credited with having originated the *res ipsa* idea. In addition, the California and New Hampshire courts assigned to the lay jury primary responsibility for both ascertaining the facts of the defendant's conduct and evaluating the negligence significance of those facts.

All of these factors converged to produce a nineteenth-century negligence standard with a highly expansive quality, as the case results indicate. A tort rule that denies liability in the absence of negligence is most open to controversy when the defendant has engaged in risky (but somehow nonnegligent) conduct that has brought about the plaintiff's injury without the concurrence of any negligent (or clearly risk-producing) conduct on the part of any third party or of the victim himself. Among all of the nineteenth-century cases that met these conditions, there were only a very small number in which the courts reached the stark conclusion that the defendant's conduct was free of negligence and hence immune from liability.

. . .

It is true that during the nineteenth century the negligence standard acquired a new prominence and publicity; but to conclude, as subsidy writers suggest, that the emphasis on negligence entailed the dramatic or deliberate overthrow of an ambitious prior rule of strict liability requires a reading of the historical record that is unsubtle at best and inaccurate at worst.

In general, the New Hampshire and California case law resists the claim that the nineteenth-century negligence system can properly be characterized or disparaged as an industrial subsidy. The courts expanded on the negligence standard in ways that rendered it ambitious and demanding, narrowing the gap between negligence and strict liability. Far from erecting a duty prerequisite to every tort claim, the courts easily recognized that everyone owes a duty to everyone else to abstain from negligent conduct. The courts applied the defense of contributory negligence sparingly and sympathetically, and developed a variety of extenuating maxims that virtually excluded from the law the concept of "slight" contributory negligence.

The record in New Hampshire and California reveals no tendency on the part of the judiciary to shelter emerging industries from what would otherwise be their liability in tort. If anything, novel forms of risktaking generated by the profit motive were viewed with enhanced, rather than reduced, suspicion. To this extent, the courts, far from being vulnerable to a populist critique, were themselves operating on the basis of populist impulses. Turnpikes and especially textile mills played an important role in the early nineteenth-century New Hampshire economy; yet that state's court subjected turnpike companies and textile factories to emphatic liabilities. The railroads loomed large during the last half of the century; yet in the New Hampshire and California courts railroad companies suffered defeat on the vast majority of contested issues. Electric power supply and new elevator systems typified the late nineteenth-century's economic and societal transformation. Yet in opinions animated by a concern for safety, the California court spurned a power company's implicit request for a liability rule subsidy and held elevator operations to an exacting liability standard.

The larger lessons of my research project are several. One is that the basic assertions included in the subsidy thesis are either false or misleading. Indeed, the reinterpretation supported by most of the two-state evidence is that the courts, in implementing the negligence system, were solicitous of victim welfare and generally bold in the liability burdens they were willing to impose on corporate defendants. If my conclusions are revisionist, they are also somewhat conservative. They suggest that the overall performance of tort law in the two states studied need not be disowned as offensive or discreditable; in truth, they indicate that there is a surprising degree of continuity between nineteenth-century tort law and the law we now recognize in the late twentieth-century. The subsidy thesis is conducive to the view that judicial decisionmaking is under the direct control of elites wielding effective economic power. Insofar as the evidence fails to support that thesis, the reader can reject the claim of economic determination and continue to adhere to whatever his beliefs might be as to the integrity or distinctiveness of the common law process.

Notes and Questions

1. What sort of evidence in the New Hampshire and California cases might have convinced the author of the subsidy thesis?

2. Do you suppose that a typical business is more often the plaintiff or the defendant in a tort suit? Can you think of some kinds of businesses that would probably prefer strict liability over negligence (or at least be indifferent as to this choice)?

3. Is it your sense that common-law decisions have important social consequences? This question will be explored in part VII.

Some Effects of Uncertainty on Compliance with Legal Standards

JOHN E. CALFEE AND RICHARD CRASWELL

When analyzing legal standards, it is convenient to assume that the parties subject to a standard know exactly what behavior is required of them. In practice, however, such certainty is rarely present. This selection analyzes some ways in which uncertainty about the application of legal standards can give parties economic incentives to "overcomply" or to "undercomply"—that is, to modify their behavior to a greater or lesser extent than a legal rule requires. . . .

Thinking About Uncertainty

The primary focus of this selection is on parties who must choose some course of action from a more or less continuous range of choices. Thus, our analysis might apply to a motorist deciding how fast to drive, a factory owner deciding how much pollution to emit, a seller deciding how soon to deliver ordered merchandise, or a monopolist deciding how far to reduce prices to meet the threat of a new competitor. [We are] less concerned with simple "either/or" situations where the actor has only two possible choices (e.g., to murder or not to murder).

Our focus is also limited to situations where there is some level of behavior that would be "optimal" from society's viewpoint, and where that optimal behavior does not lie at the extreme end of the available range of choices. Thus, we do not consider situations where society has no interest in the defendant's behavior, or where the social ideal would involve reducing the level of activity to zero. This last condition eliminates from our analysis activities such as murder or rape, but virtually any activity whose desirability depends on a balance among competing factors remains for consideration. We will speak most frequently of the balance between economic costs and benefits, but most of our conclusions would apply with only slight modifications to a balance that included other social values.

In the most general terms, uncertainty occurs whenever people cannot be sure what legal consequences will attach to each of their possible courses of action. Such uncertainty arises from a number of sources. Perhaps the most common source (and the easiest to think about) is that people may not know in advance just where the legal standard will be set. For example, it is difficult

Abridged and reprinted without footnotes by permission of Virginia Law Review Association and Fred B. Rothman & Company from 70 *Virginia Law Review* 965 (1984).

to predict where a negligence jury will draw the line between "reasonable" and "unreasonable" speeds, or how an antitrust court will distinguish between "predatory" and "competitive" price cuts.

Although much of our analysis can be interpreted in terms of the difficulty in gauging the location of the legal standard, uncertainty may arise from other aspects of the legal process. For example, a 55-mile-per-hour speed limit is a clear legal standard known in advance, but potential defendants cannot be sure that they will be able to convince a court that they complied with that standard. Indeed, there is always some chance of error in the legal system, and at the time defendants must choose their behavior it will usually be hard to predict the kinds of evidence that will be available when they are brought to trial, the persuasiveness of the witnesses and advocates who will participate in that trial, or the temperament of the judge or jury at the time of their decision. Defendants may also be uncertain whether a private plaintiff or a public prosecutor will bring them to trial at all. Finally, defendants may be uncertain about the size of the damages or fine they might have to pay if found liable.

These uncertainties, taken either singly or together, imply that defendants do not face a simple choice between actions certain to lead to liability and actions bearing no risk of liability at all. Instead, each possible action is accompanied by an associated probability that a defendant will be tried, found liable, and made to pay damages or a fine. There may be regions at either extreme where that probability approaches zero or one. For example, a polluter may know that if he puts out enough smoke to poison an entire city, he is sure to be liable, and that if he emits no smoke at all he is sure not to be liable. But in the more interesting (and, usually, the more relevant) region, the defendant can only attempt to estimate the likelihood of liability at each level of pollution that he has the capacity to emit.

These estimates of the likelihood of liability attached to each course of action can be referred to as the defendant's "distribution of probabilities." To the extent that defendants are influenced by the fear of liability, their behavior will be influenced by this distribution of probabilities, rather than simply by the nominal legal standard. Indeed, from the defendant's point of view the rule of law is that distribution of probabilities. Even if the legal system aspires to a more definite standard—such as a pollution standard of X parts per million or a negligence standard defined as the cost-effective level of care— liability-conscious defendants will consider the probability that they will or will not be held liable at a greater or lesser level of compliance.

Unfortunately, very little is known about the actual distribution of probabilities for any legal rule. It is plausible, however, to assume that the likelihood of liability increases as a defendant's behavior becomes more dangerous (that is, as its social costs increase). The motorist who causes an accident while driving at eighty miles per hour should, at least, be more likely to be sued and found negligent than a motorist who causes a similar accident while traveling at 40 miles per hour. Indeed, a positive (or, at least, non negative) correlation between the likelihood of liability and the social costs of a defendant's behavior could be the definition of a minimally rational legal system.

Uncertainty Under a Negligence Standard

The law attempts to control behavior through many different devices: negligence standards, rules of strict liability, fines or imprisonment, and even bounties and other affirmative incentives. To some extent, the uncertainties discussed in the last section will affect each of these approaches. The effects of uncertainty are more easily illustrated, however, in connection with a traditional negligence standard.

Imagine an activity involving a risk of accidents, such as operating a railroad in an urban area with frequent automobile and pedestrian crossings. The railroad can reduce the number of accidents by posting guards at crossings, or by running its trains at slower speeds. Each of these steps is costly, and the railroad would prefer not to take them. Indeed, from a social point of view we would not want the railroad to take *all* possible precautions, because some of them (such as slowing trains to a speed of 10 feet per hour) would cost far more than the accidents they would prevent.

A negligence standard establishes a legally required level of precautions, usually defined as "reasonable care." Defendants who violate this requirement are held liable for the costs of all resulting accidents, and defendants who satisfy the requirement are not held liable at all (even though some accidents may still take place after all "reasonable" precautions have been taken). Such a rule is depicted by the heavy kinked line in Figure 1. If the straight line marked L gives the expected losses from accidents at each possible level of speed (to focus on just one precaution), and x^* is the speed that

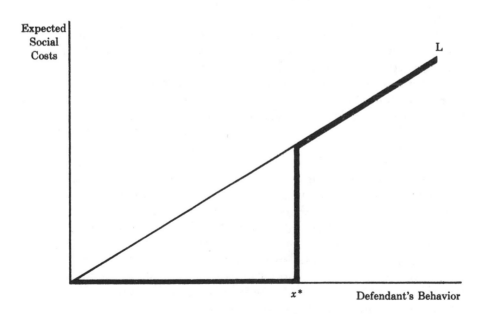

Figure 1

the legal system deems "reasonable," then railroads that operate at lower speeds will pay nothing in damages, and railroads [that] operate at higher speeds must pay for all resulting accidents.

When defendants know the exact location of the legal standard (x^*), the sudden jump in liability for damages at x^* gives them a powerful incentive to comply with that standard. Indeed, if x^* is set at the cost-effective level of care under the "Learned Hand Rule" and defendants know this precise level of care, defendants' incentives will be to comply exactly with that standard and act at the socially optimal level.

As discussed earlier, however, defendants will often be uncertain about the legal standard or its application. They may not know how the "reasonable care" standard will be interpreted, whether they will be able to convince a jury that they were actually driving at a lower speed, or whether they will be sued at all. The result is that the railroad faces only a probability that it will have to pay damages if it drives at speeds above x^*. Similarly, it also faces some probability that it might still have to pay damages even if it drives at speeds below x^*.

This situation is depicted in Figure 2, where the curve marked D represents the defendant's expected damage payments—the expected accident costs discounted by the chance that the defendant will not be found liable and therefore will not have to pay damages. Presumably, at very low speeds there is a very small chance that the railroad will be sued and found liable, so on the left side of the graph D is only a small fraction of L. Conversely, at very high speeds the probability of liability should become much higher, so on the right side of the graph D and L are closer together. If there is some outrageously high speed at which suit and liability are a virtual certainty, the D and L curves would then coincide.

One effect of this uncertainty is to eliminate the abrupt jump in expected damage payments at x^*. More generally, the effect is to change the defendant's *marginal* incentives—that is, the marginal benefits the railroad can expect, in the form of lower expected damage payments, from reducing the speed of its trains. The social benefits from reductions in speed are given by the slope of the L curve: at this rate expected accident costs decline as speed is reduced. The socially optimal speed is the point at which the marginal benefits from reduced speed equal the marginal costs of reduced speed (e.g., slower delivery times, higher operating costs). This is also the speed the railroad would voluntarily choose if it knew that it would always have to pay the full accident costs (L), as it would under a true strict liability system. The railroad would compare the reduction in social costs (the slope of the L function) with the increase in its own costs, and choose the speed that minimized the sum of private and social costs.

Under the system depicted in Figure 2, however, the defendant need not be concerned with the true social costs (L). Instead, a profit-maximizing railroad will look to the reduction in its expected damage payments (D) that would result from reducing the speed of its trains, and compare this to the private costs of reduced speed. If the D function is steeper than the L func-

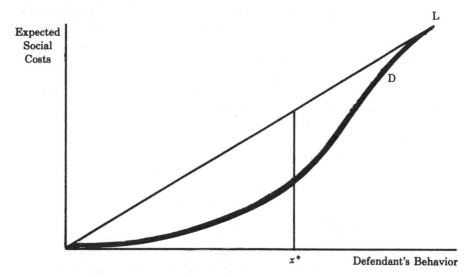

Figure 2

tion, the reduction in expected damage payments from a given reduction in speed will exceed the reduction in expected social costs (i.e, the reduction in expected accident losses) from the same reduction in speed. That is, the change in expected damage payments will overstate the social benefits of a reduction in speed, and give the railroad too great an incentive to reduce its speed. In short, a D function that is steeper than the L function will give defendants an incentive to overcomply.

The important point to notice is that, as drawn, the D function in Figure 2 *is* steeper than the L function over much of its range. This is not due to arbitrary drafting; it is because over much of its range the D function has a greater distance to rise. It must begin on the left (where the likelihood of liability is small) as a tiny fraction of L, but must rise until it almost equals L on the right side of the graph (where the probability of conviction is high). Thus, its average slope over this range must be steeper than that of the L function. This does not mean that it will necessarily be steeper at every point, but does mean that there will at least be a substantial range of points at which the legal system will overstate the benefits of increased care to defendants. Thus, even though the D function is always below the L function, indicating that defendants can expect to pay less than the true social costs of their activity, the steepness of the D function may give defendants an incentive to take more than the socially optimal level of care.

This analysis has not assumed anything about the nature of the uncertainty giving rise to the D curve, except that the probability of being held liable increases as the speed of the train increases. The incentive toward over-compliance depicted in Figure 2 could arise even when the distribution of probabilities was perfectly symmetric and centered on the socially optimal speed of x^*. This result seems paradoxical. If the distribution of probabilities is such that the "average" legal standard (or the defendant's "best guess" about the legal standard) is x^*, why would defendants not choose x^* as the standard with which to comply? Why would they ever have an incentive to overcomply, by reducing their speed further?

The answer turns on the balance between two conflicting tendencies. First, even at a speed like x^*, there is still a positive probability that the railroad "won't get caught"—i.e., that it won't be sued, or that if it is sued it won't be found liable. This factor will lead the railroad to discount (or underestimate) the increased accident risks from excessive speeds. If this factor were the only one present, it would lead defendants to undercomply by taking too little care. For example, if the probability of not being held liable were 50 percent regardless of the speed at which the defendant drove, the D function in Figure 2 would always be exactly half of the L function, and would always have a flatter slope. Economists who have analyzed criminal deterrence note that the possibility of not getting caught reduces the deterrent effect of criminal penalties, and usually recommend that penalties be increased to compensate for that reduction.

But these writers have almost universally ignored the second factor that is also present here. In most situations where the defendant can choose from a range of possible actions, the probability of being held liable varies with the egregiousness of the defendant's conduct. A grossly negligent driver (or flagrant polluter) is surely more likely to be found liable than one driving at a speed that might have been reasonable (or one who is marginally polluting). Whenever this is the case, the reduction in the likelihood of liability—e.g., the ability to reduce from 50 percent to 30 percent the probability of having to pay any damages—provides the extra incentive to take additional care. In other words, additional care not only produces a marginal reduction in total accident costs, thereby reducing the amount the defendant can expect to pay *if* he is found liable, but also decreases the probability that he will have to pay for *any* of those accident costs.

Although the attractiveness of reducing the risk of exposure to any liability will vary from case to case, we can identify those factors that make the incentive to overcomply strongest. If a defendant is currently exercising the socially optimal level of care x^* and is deciding whether to exercise even more care, the attractiveness of reduced exposure to *any* liability will most obviously depend on the amount he would expect to pay if he is found liable. The defendant's incentive to overcomply will therefore be greatest in activities where the socially optimal level of care still involves a significant risk of costly accidents, and smallest in activities where the optimal level of care is one that reduces expected accident costs almost to zero. In the graph in Figure 1, the incentive to overcomply will tend to increase with the size of the vertical jump at x^*, or with the height of the L curve at that point.

Conversely, the incentive to undercomply will be strongest when: (a) the amount the defendant can save in private costs by taking less care than the optimum is relatively large; and (b) the likelihood of not being found liable, or "not getting caught," is quite high even at levels of care slightly below the socially optimal level. If the distribution of probabilities is symmetric around the optimal level of care x^*, the likelihood of escaping liability at lower levels of care will never be more than 50 percent, and the strength of the incentive to undercomply will be limited. If the private gains from engaging in less care are sufficiently large, however, and if the total expected accident costs at the efficient level of care are not very large (so that the incentive to overcomply is weak), the incentive to undercomply can still dominate. . . .

Policy Implications

. . .

It is theoretically possible to make adjustments in the legal system that would correct for any incentives to over- or undercomply. It is difficult to recommend such adjustments with confidence. As mentioned earlier, very little is known about the shape of the probability distribution for any actual legal rule, or about how various changes in the legal system would affect those distributions. Despite this cautionary note, though, we now examine some possible policy implications of our analysis.

Damage Multipliers

One possible adjustment for uncertainty—the idea that penalties should be increased by some multiple of the true social costs whenever there is some chance that the offender will not get caught—has already gained wide acceptance in the economic literature on deterrence. Jeremy Bentham may have been the first to observe that the chance of not being punished will reduce the deterrent effect of any fine or punishment, and to conlude from this that "the more deficient in certainty a punishment is, the severer it should be." Subsequent writers have developed formal economic models that seem to confirm this intuition. As a result, this principle has been invoked not only to justify larger criminal penalties for difficult-to-detect offenses, but also to support related concepts such as punitive damages for certain kinds of torts, the treble damage rule of antitrust law, and proposals to increase the damages available for breach of contract.

This principle does not apply, however, to a number of important cases. Even when the probability of punishment is less than one, if that probability declines as defendants take more care, then defendants may tend to overcomply. In such a case, increasing the expected fine or damage award would only increase overdeterrence, exacerbating the problem rather than curing it.

The Benthamite argument that damages should always be increased to reflect the average probability of not being caught appears to be valid only in three types of situations: (1) when the conduct in question involves a pure either/or choice rather than a continuous range of choices with the optimum falling somewhere in the middle, so that overcompliance is meaningless; (2)

when the conduct can vary over a continuous range, but the probability of being sued and found liable is the same regardless of the conduct actually chosen by the defendant; or (3) when the conduct in question falls into one of the categories analyzed in this selection where undercompliance rather than overcompliance is more likely. As discussed [earlier], undercompliance may be more likely when the damages for even the slightest violation (that the plaintiff would like to avoid having to pay) are extremely small. Undercompliance would be particularly likely under any form of "incremental damage rule," under which the fines or damage awards increased gradually from zero, because in that case there would be *no* expected damages for slight departures from the optimal level of care.

In any of these situations, then, some augmentation of damages may be economically justified. Yet many regulatory and private law subjects do not fit these special cases. Antitrust violations, speeding, pollution, and most negligent behavior, for example, all involve choices along a continuous spectrum, with an optimal level of behavior that is determined by a balancing of desirable and undesirable effects, so that overcompliance is both possible and socially costly. In these situations, arguments that penalties equal to the social costs of defendants' behavior will always lead to underdeterrence and therefore should be increased must be taken with a good deal of salt. A better rule in many cases might be to reduce the damage awards to compensate for what would otherwise be an incentive to overcomply. For example, the routine exclusion from damage awards of such items as attorney's fees or subjective losses may be less harmful than is often supposed, and could even be beneficial.

Such a recommendation cannot be unqualified, because it would be difficult, if not impossible, to calculate the size of the optimal adjustment. Moreover, the entire analysis of this selection has been limited to cases where the nominal legal standard, or the mean of the distribution, was otherwise "correctly" centered. If the legal standard were for some reason far below the optimal level of care, for instance, defendants could overcomply with respect to that standard without necessarily more than the optimal level of care. However, both of these difficulties—the difficulty of calculating the optimal adjustment and the possibility that the substantive legal standard may not have been set correctly—apply with equal force to the Benthamite proposal to increase damages. The suggestion that damages ought to be reduced, instead thus rests on at least as strong a ground as the Benthamite position, and in many cases its support may be a good deal stronger.

Cost/Benefit Standards

Another position widely accepted in the law and economics literature is that liability standards (such as the level of "reasonable care" under a negligence system) ought to be set at the level that equates marginal costs and benefits. Of course, when uncertainty is recognized defendants will be faced not with a sharp line but with a distribution of probabilities, and it becomes difficult even to speak in terms of "the" legal standard. If all errors or variations in the application of the nominal standard are assumed to be unbiased, however, then the nominal legal standard will equal the mean of the resulting distribution. On this assumption, the recommendation that the nominal standard always be set at the cost-effective level of care can no longer be supported on economic grounds.

This conclusion follows directly from our earlier analysis. If defendants have an incentive to overcomply, setting the nominal standard at the optimal level of care will lead defendants to take more than optimal care, which by definition is undesirable. Society would be better off if juries were told to try for a standard somewhere below the cost-effective level of care, so that defendants who overcomplied relative to this standard would end up exercising the optimal amount of care. Of course, in those cases where the net incentives favored under- rather than overcompliance, the nominal standard would need to be set *above* the optimal level of care.

The difficulties with this recommendation are similar to those associated with trying to adjust damage awards: it would be very difficult to calculate the size of the optimal adjustment, and it is not clear that actual distribution of probabilities are centered at the nominal legal standard. Yet the same objections can be raised against the traditional argument for an exact cost-benefit standard. It is often difficult to calculate the cost-effective level of care, and when courts or juries apply a cost-benefit standard there is no guarantee that the resulting distribution of probabilities will actually be centered at the cost-effective level.

This analysis would cast doubt on the idea that the common law is usually efficient if common law courts have been trying to set the negligence standard at the cost-effective level of care. However, the evidence for this last proposition appears to consist of statements of judges such as Learned Hand, and the observations of Richard Posner and others, that negligence decisions vary in a way that at least reflects variations in costs and benefits. The difficulty with this evidence is that courts would respond to changes in the costs and benefits of care even if they were not trying to set the standard exactly at the cost-effective level of care (i.e., exactly at x^*). A system where courts tried to set the standard slightly below the optimal level of care would reflect the same pattern of changes, so this evidence is consistent with either theory. Furthermore, because judges' statements of their grounds for decision are rarely given much weight when they seem to oppose various economic interpretations of legal rules, it is dangerous to rely too heavily on them in this context as support for one theory or another.

Those who believe that the common law tends to be efficient should suspect that in some cases courts do aim for a standard of care below the cost-effective level and should thus begin to question judicial statements, such as Learned Hand's, to the contrary. Courts might be applying something closer to a gross negligence standard, or a simple negligence standard that includes some room for error, under which the reasonable man is only required to come "reasonably close" to the cost-effective level of care. Although either of these lower standards would lead to too little care in a world of absolute certainty, they may well be superior to a strict cost/benefit rule if there is any uncertainty about legal outcomes.

Reducing Uncertainty

Finally, there is a third approach to correcting the distortions caused by uncertainty: reducing the uncertainty itself. In legal terms, this could involve improving the fact-finding process to reduce the chance of random error, promulgating enforcement guidelines to make enforcement decisions more predictable, or changing the substantive legal rule from a vague but flexible standard to a bright-line test. Reducing the uncertainty will not eliminate any incentives to overcomply, but it should reduce the extent of such overcompliance.

Reducing uncertainty is not costless, however. For example, administrative expenses would increase if special masters were hired to improve the fact-finding process. Reducing uncertainty may also impose other, more subtle costs. When increased certainty is obtained by adopting strict enforcement or bright-line legal standards, these rules must be stated in terms of easily observable factors that are usually only imperfect proxies for the variables that we would prefer to measure. A 55-mile-per-hour per se negligence standard is certainly more predictable than a "reasonable speed" standard, but there will surely be some situations where that speed limit is far higher than the socially optimal speed, and others where it is too low.

As a result, there will often be a conflict between the goal of reducing the variance of the distribution of probabilities and the goal of keeping the mean of that distribution correctly centered. A bright-line standard may produce fewer distortions due to uncertainty, but more distortions due to imperfect location of the mean. Discovering which effect will be stronger in any given case presents a very difficult empirical question.

Conclusion

In some respects, our analysis of uncertainty is simply a more formal treatment of an issue that has often interested legal scholars. Discussions of uncertainty frequently appear, for example, in debates over the choice between strict liability and negligence. Critics of the negligence standard have emphasized the difficulties facing courts attempting to determine the efficient level of care, and consequently, the small likelihood that the negligence system will induce efficient behavior. Our analysis suggests that even an uncertain rule can be adjusted, in theory, to produce the optimal level of compliance by those subject to the rule. However, the information needed to calculate and implement the proper adjustment appears to be at least as complex as that required to calculate the cost-effective level of care. Thus, the need to account for uncertain legal outcomes makes an efficient negligence system even more difficult to implement than otherwise might have been supposed.

Legal debates have focused more particularly on the relationship between uncertainty and overcompliance. Not surprisingly, these concerns have been expressed most strongly in contexts where overcompliance by defendants is seen as particularly costly to society. In a tautological sense, overcompliance is always costly because we have defined overcompliance to mean a level of care that is above the socially optimal level, but these costs may be greater in some contexts than in others.

. . .

One purpose of this article is to demonstrate formally that uncertainty does indeed affect people's incentives to comply. Propositions that seem intuitively obvious to writers accustomed to thinking in terms of simpler models— that liability standards should be set at the cost-effective level of care, that equal chances of error in either direction will have no net effect on the incentives of risk-neutral defendants, and that penalties should generally be increased by a factor reflecting the probability of not being punished—turn out to be of doubtful validity once legal uncertainty is admitted into the analysis. Other intuitive judgments concerning, for example, the effect of excessive damage awards, or the role of contributory negligence, begin to make sense only when uncertainty is introduced.

Notes and Questions

1. If we applied the death penalty to motorists who exceed the 65-mile-per-hour (mph) speed limit, drivers would in all likelihood proceed not at 65 or 64 mph but rather at 50 or so. What kind of uncertainty is assumed in this hypothetical example? Do drivers have a similar incentive when the penalty for speeding is much less? If not, why is the analysis put forth in this reading not relevant? If so, what could be done to avoid this effect?

2. Imagine that a speeding motorist travels at 75 mph and causes $100,000 in damage. The fact finder determines that even at a nonnegligent speed of 55 mph there would have been an accident, but no more than $30,000 of damage would have resulted at the safer speed. Should the negligent driver pay $100,000 or $70,000? Can you produce some case authority for either result? Which result seems more consistent with a corrective-justice approach? Which result is assumed by the authors?

III

The Requirement of Causation

It is plain that if A drives recklessly while B is the victim of a hit-and-run driver many miles away, A does not pay B. A only pays when A is more closely related to an injury. This requirement of causation has many facets. There is the question of liability when a plaintiff knows that one of several parties caused her harm, but can not precisely identify the cause of the injury. There are also questions of remoteness in time and of intervening actors. Christopher Columbus is not held responsible for today's torts even though, in a sense, if not for Columbus the world might be sufficiently different that the torts we are familiar with today would never occur. Yet a party may be held liable for the acts of another if the tort were a consequence of the first party's actions. This part begins by addressing why it matters in compensating an innocent victim whose negligence caused a harm.

Remarks on Causation and Liability

JUDITH JARVIS THOMSON

Plaintiff Summers had gone quail hunting with the two defendants, Tice and Simonson. A quail was flushed, and the defendants fired negligently in the plaintiff's direction; one shot struck the plaintiff in the eye. The defendants were equally distant from the plaintiff, and both had an unobstructed view of him. Both were using the same kind of gun and the same kind of birdshot; and it was not possible to determine which gun the pellet in the plaintiff's eye had come from. The trial court found in the plaintiff's favor, and held both defendants "jointly and severally liable." That is, it declared the plaintiff entitled to collect damages from whichever defendant he chose. The defendants appealed, and their appeals were consolidated. The California Supreme Court affirmed the judgment.

Was the court's decision in *Summers* fair? There are two questions to be addressed. First, why should either defendant be held liable for any of the costs? And second, why should each defendant be held liable for all of the costs—that is, why should the plaintiff be entitled to collect all of the costs from either?

Why should either defendant be held liable for any of the costs? The facts suggest that in the case of each defendant, it was only .5 probable that he caused the injury; normally, however, a plaintiff must show that it is more likely than not, and thus more than .5 probable, that the defendant caused the harm complained of if he is to win his case.

. . .

The court's argument seems . . . to go as follows. The plaintiff cannot determine which defendant caused the harm. If the plaintiff has the burden of determining which defendant caused the harm, he will therefore be without remedy. But both defendants acted negligently "toward plaintiff," and the negligence of one of them caused the harm. Therefore the plaintiff should not be without remedy. Therefore it is manifest that the burden should shift to each defendant to show that he did not cause the injury; and, if neither can carry that burden, then both should be held liable.

The argument does not say merely that both defendants are wrong-doers, or that both defendants acted negligently; it says that both defendants acted negligently "toward plaintiff"—that is, both were in breach of a duty of care that they owed to the plaintiff. Suppose, for example, that the plaintiff had brought suit, not against the two hunters who were out quail hunting with him, but against three people: the two hunters and . . . Smith, who was

Abridged from Judith Jarvis Thomson, Remarks on Causation and Liability. Copyright © 1984 by Princeton University Press. Reprinted without footnotes by permission of Princeton University Press from 13 *Philosophy and Public Affairs* 101 (1984).

driving negligently in California that day and . . . nearly ran the plaintiff down as the plaintiff was on his way to go quail hunting. . . . All three were wrongdoers, all three acted negligently, and indeed negligently toward the plaintiff, and one of the three caused the harm, though it is not possible to tell which. But it could hardly be thought fair for all of them, and so a fortiori for Smith, to have to carry the burden of showing that *his* negligence did not cause the harm. As it stands, the argument does not exclude Smith, for he *was* negligent toward the plaintiff. So we must suppose that the court had in mind not merely that all the defendants were negligent toward the plaintiff, but also that their negligent acts were in a measure likely to have caused the harm for which the plaintiff sought compensation.

There lurks behind these considerations what I take to be a deep and difficult question, namely: Why does it matter to us whose negligent act caused the harm in deciding who is to compensate the victim?

It will help to focus on a hypothetical, variant of the case, which I shall call *Summers II*. Same plaintiff, same defendants, same negligence, same injury as in *Summers;* but *Summers II* differs in that during the course of the trial, evidence suddenly becomes available which makes it as certain as empirical matters ever get to be, that the pellet lodged in the plaintiff Summers's eye came from defendant Tice's gun. Tort law being what it is, defendant Simonson is straightway dismissed from the case. And isn't that the right outcome? Don't we feel that Tice alone should be held liable in *Summers II?* We do not feel that Simonson should be dismissed with a blessing; he acted very badly indeed. So did Tice act badly. Tice also caused the harm, and (other things being equal) fairness requires that he pay for it. But why? After all, both defendants acted equally negligently toward Summers in shooting as they did; and it was simple good luck for Simonson that, as things turned out, he did not cause the harm to Summers.

· · ·

What we are concerned with here is not blame, but only who is to be out of pocket for the costs. More precisely, why it is Tice who is to be out of pocket for the costs. It pays to take note of what lies on the other side of this coin. You and your neighbor work equally hard, and equally imaginatively, on a cure for the common cold. Nature then smiles on you: a sudden gust of wind blows your test tubes together, and rattles your chemicals, and lo, there you have it. Both of you acted well; but who is to be in pocket for the profits? You are. Why? That is as deep and difficult a question as the one we are attending to. I think that the considerations I shall appeal to for an answer to our question could also be helpfully appealed to for an answer to this one, but I shall not try to show how.

There is something quite general at work here. "B is responsible for the damage to A's fence; so B should repair it." "The mess on A's floor is B's fault; so B should clean it up." Or anyway, B should have the fence repaired, the mess cleaned up. The step is common, familiar, entirely natural. But what warrants taking it?

I hazard a guess that the, or anyway an, answer may be found in the value we place on freedom of action, by which I mean to include freedom to plan on action in the future, for such ends as one chooses for oneself. We take it that people are entitled to a certain "moral space" in which to assess possible ends, make choices, and then work for the means to reach those ends. Freedom of action is obviously not the only thing we value; but let us attend only to considerations of freedom of action, and bring out how they bear on the question in hand.

If A is injured, his planning is disrupted; he will have to take assets he meant to devote to such and such chosen purpose and use them to pay the costs of his injury. Or that is so unless he is entitled to call on the assets of another, or others, to pay the costs for him. His moral space wold be considerably larger if he were entitled to have such costs paid for him.

But who is to pay A's costs? On whose assets is it to be thought he is entitled to call? Whose plans may *he* disrupt?

A might say to the rest of us, "Look, you share my costs with me now, and I'll share with you when you are injured later." And we might then agree to adopt a cost-spreading arrangement under which the costs of all (or some) of our injuries are shared; indeed, we might the better secure freedom of action for all of us if we did agree to such an arrangement. The question which needs answering, however, is whether A may call on this or that person's assets in the absence of agreement.

One thing A is not entitled to do is to choose a person X at random, and call on X's assets to pay his costs. That seems right; but I think it is not easy to say exactly why. That is, it will not suffice to say that if all we know about X is that X is a person chosen at random, then we know of no reason to think that a world in which X pays A's costs is better than a world in which A pays A's costs. That is surely true. But by the same token, if all we know about X is that X is a person chosen at random, then we know of no reason to think that a world in which A pays A's costs is better than a world in which X pays A's costs. So far, it looks as if flipping a coin would be in order.

· · ·

A is injured. Let us supply his injury with a certain history. Suppose, first, that A himself caused it—freely and wittingly, for purposes of his own. And suppose, second, that it is not also true of any other person X that X caused it, or even that X in any way causally contributed to it. Thus:

(1) A caused A's injury, freely, wittingly, for purposes of his own; and no one other than A caused it, or even causally contributed to it.

· · ·

For example, A might have broken up one of his chairs to use as kindling to light a fire to get the pleasure of looking at a fire. It is harder to construct examples of injuries which consist in physical harm which have histories of this kind. But it is possible—for example, A might have cut off a gangrenous toe to save his life.

· · ·

It is a plausible first idea that the answer lies in the concept "enrichment." Suppose I steal your coffee mug. I am thereby enriched, and at your expense. Fairness calls for return of the good: I must return the coffee mug.

. . .

Well, fairness needn't call for the return of the very coffee mug I took, and surely can't call for this if I have now smashed it. Replacement costs might do just as well. Or perhaps something more than replacement costs, to cover your misery while thinking you'd lost your mug. In any case, anyone can pay those costs. But I must pay them to you because I was the person enriched by the theft of the mug, and at your expense. So similarly, perhaps we can say that B must pay the costs of having the fence repaired because B was the person who enriched himself, and at A's expense, by the doing of whatever it was he did by the doing of which he damaged the fence.

Enrichment? Perhaps so. B might literally have made a profit by doing whatever it was he did by the doing of which he made a mess on A's floor (e.g., mudpie-making for profit). Or, anyway, he might have greatly enjoyed himself (e.g., mudpie-making for fun). Perhaps he made the mess out of negligence? Then he at least made a saving; he saved the expense in time or effort or whatever he would have had to expend to take due care. And he made that saving at A's expense.

But this cannot really be the answer—it certainly cannot be the whole answer. For consider Tice and Simonson again. They fired their guns negligently in Summers' direction, and Tice's bullet hit Summers. Why should Tice pay Summers's costs? Are we to say that is because Tice enriched himself at Summers's expense? Or anyway, that Tice made a saving at Summers's expense—a saving in time or effort or whatever he would have had to expend to take due care? Well, Simonson saved the same as Tice did, for they acted equally negligently. It would have to be said "Ah, but Tice's saving was a saving *at Summers's expense*—and Simonson's was not." But what made Tice's saving *be* a saving at Summers's expense? Plainly not the fact that his negligence was negligence "toward" Summers, for as the court said, Tice and Simonson were both "negligent toward plaintiff." If it is said that what made Tice's saving be a saving at Summers's expense is the fact that it was Tice's negligence that caused Summers's injury, then we are back where we were: for what we began with was why that fact should make the difference.

. . .

Perhaps it pays to set aside the concept "enrichment" and attend, instead, to what we have in mind when we characterize a person as "responsible." Consider again: "B is responsible for the damage to A's fence; so B should repair it." Doesn't the responsible *person* pay the costs of damage he or she is responsible *for?* And don't we place a high value on being a responsible person?

Similarly, the responsible person pays the costs of damage which is his or her fault.

This is surely right; but what lies behind it? *Why* do we think it a good trait in a man that he pays the costs of damage he is responsible for? Why do we expect him to?

Suppose now that having caused himself the injury, A wants for one or another reason to be made whole again. That will cost him something. Here is B. Since (1) is true of A's injury, B's freedom of action protects him against A: A is not entitled to call on B's assets for the purpose—A is not entitled to disrupt B's planning to reverse an outcome wholly of his own planning which he now finds unsatisfactory.

That seems right. And it seems right whatever we imagine true of B. B may be vicious or virtuous, fat or thin, tall or short; none of this gives A a right to call on B's assets. Again, B might have been acting very badly indeed contemporaneously with A's taking the steps he took to cause his own injury: B might even have been imposing risks of very serious injuries on A concurrently with A's act—for example, B might have been playing Russian roulette on A, or throwing bricks at him. No matter: if A's injury has the history I described in (1), then B's freedom of action protects him against the costs of it.

If that is right, then the answer to our question falls out easily enough. Let us suppose that A is injured, and that B did not cause the injury, indeed, that he in no way causally contributed to A's injury. Then whatever did in fact cause A's injury—whether it was A himself who caused his injury, or whether his injury was due entirely to natural causes, or whether C or D caused it— there is nothing true of B which rules out that A's injury had the history described in (1), and therefore nothing true of B which rules out that A should bear his own costs. Everything true of B is compatible with its being the case that A's costs should lie where they fell. So there is no feature of B which marks his pockets as open to A—A is no more entitled to call on B than he is entitled to call on any person X chosen at random.

Causality matters to us, then, because if B did not cause (or even causally contribute to) A's injury, then B's freedom of action protects him against liability for A's costs. And in particular, it is Simonson's freedom of action which protects him against liability for Summers's costs in *Summers II*, for in that case it was discovered that Tice had caused the injury.

I have been saying that freedom of action is not the only thing we value, and that is certainly true. But if I am right that it is freedom of action which lies behind our inclination to think causality matters—and, in particular, our inclination to think it right that Simonson be dismissed once it has been discovered that he did not cause Summers's injury—then these considerations by themselves show we place a very high value on it, for those inclinations are very strong.

Since the question we began with was the question why causality matters to us, we could acceptably stop here. But I think it pays to press on, to see how far attention to freedom of action will carry us.

For as we know, however much causality matters to us in assessing liability, it is on no plausible view sufficient for liability. The fact that Tice caused

Summers's injury does not by itself yield the conclusion that he is properly to be held liable for it; what yields this conclusion is the conjunction of the fact that Tice caused the injury *and* the relevant facts about Tice's fault—that he was negligent, that the injury was of a kind such that Tice's act was negligent in that he did not exercise the care which is called for precisely in order to avoid causing an injury of that kind, and so on. Suppose A's injury was caused by B as in

> (2) B caused A's injury by some freak accident—by doing something which he took all due care in the doing of, and which he could not have been expected to foresee would lead to harm.

Then alas for A, it seems right that A's costs lie where they fell; the fact that B caused the injury does not suffice for imposing liability on him. When fault is added to causality, however, things look very different to us. If A's injury was caused by B as in

> (3) B caused A's injury wrongfully—by intention, or out of negligence

then B must plainly pay.

Why this difference between (2) and (3)? It might be thought we could say this. In (3), B caused A's injury by doing what it was wrong [of] him to do, and freedom of action has its limits: one is not free to act wrongly. By contrast, in (2) there was nothing B did which it was . . . wrong [of] him to do, no constraint of morality that he violated; so it is his freedom of action that protects him against liability for A's costs in (2).

But I think this account of (2) and (3) is oversimple. In the first place, I think we are free to act wrongly—so long, that is, as we cause no harm to others (more generally, infringe no right of theirs) in doing so. It is not the fact that B acted wrongly in (3) that makes him liable for A's costs in (3): B can have been acting as wrongfully as you like concurrently with the coming about of A's injury and is all the same not liable for A's costs if A caused his own injury—as in (1). What makes B liable for A's costs in (3) is rather that in (3) he wrongfully caused A's injury. It is *that* which fixes that his freedom of action does not protect him against liability for A's costs in (3).

Second, it is not the fact that B did act wrongly in (2) that protects him against liability for A's costs in (2). For what if A's injury had a history of the following kind:

> (4) B caused A's injury, and did so freely and wittingly, but did so to save himself from a very much greater injury, and was justified in so acting.

B did not act wrongly in (4), and is all the same properly held liable for A's costs in (4). A case of the kind I have in mind, which comes from the legal literature, is *Vincent v. Lake Erie Transportation Co.,* in which a ship's captain tied his ship to a dock to protect it from the risk of being sunk in a storm. The dockowner's dock was damaged by the ship's banging against it in the storm, and he sued the shipowner. The court declared that the ship's captain had acted properly and well; but it (surely rightly) awarded damages to the plaintiff. A second case of the kind I have in mind, which comes from the literature of moral theory, is Joel Feinberg's story of a hiker, lost in a sudden mountain storm, who broke into an empty cabin and burned the furniture to keep warm; the hiker was plainly justified in so acting, but he owes the cabin owner compensation for the damage he did.

Why is B protected against liability for A's costs in (2), but not in (4)? B acts wrongly in neither case, and it is not at all easy to see the source of the difference.

Richard A. Epstein offers the following justification for the imposition of liability on the defendant in *Vincent:*

> Had the Lake Erie Transportation Company owned both the dock and the ship, there could have been no lawsuit as a result of the incident. The Transportation Company, now the sole party involved, would, when faced with the storm, apply some form of cost-benefit analysis in order to decide whether to sacrifice its ship or its dock to the elements. Regardless of the choice made, it would bear the consequences and would have no recourse against anyone else. There is no reason why the company as a defendant in a lawsuit should be able to shift the loss in question because the dock belonged to someone else. The action in tort in effect enables the injured party to require the defendant to treat the loss he has inflicted on another as though it were his own. If the Transportation Company must bear all the costs in those cases in which it damages its own property, then it should bear those costs when it damages the property of another.

These seem to me to be very helpful remarks. Suppose the name of the man who actually owns the dock is Jones. What Epstein points to is this: If the dock had belonged, not to Jones, but to the Lake Erie Transportation Company, then the company would not have been entitled to call on Jones's assets for funds to repair it. Why not? Consider again

(1) A caused A's injury, freely, wittingly, for purposes of his own; and no one other than A caused it, or even causally contributed to it.

If the dock had belonged to the Lake Erie Transportation Company, then the company would have caused itself an injury (by causing an injury to its own dock); so the history of its injury would have been as described in (1), and the company would not have been entitled to call on Jones's assets for the costs.

So far so good. But all of that is counterfactual. The dock in fact belongs to Jones, not to the Lake Erie Transportation Company, and how do we get from the counterfactual remarks about what would have been the case if the company had owned the dock to what Jones is entitled to, given Jones does own the dock? Perhaps Epstein's thought is that the step is warranted by what is said in the final sentence of the passage . . . . : "If the Transportation Company must bear all the costs in those cases in which it damages its own property, then it should bear those costs when it damages the property of another." . . .

But that is unfortunately overstrong. If B has to pay the costs of any injury of A's which B causes (as B has to pay the costs of any injury of his own which he causes), then B may properly be held liable, not merely in

> (4) B caused A's injury, and did so freely and wittingly, but did so to save himself from a very much greater injury, and was justified in so acting.

but also in

> (2) B caused A's injury by some freak accident—by doing something which he took all due care in the doing of, and which he could not have been expected to foresee would lead to harm.

Doesn't B have to pay the costs of any injury of his own which he causes himself by accident? But it really does seem wrong to hold B liable in (2).

I suppose it is arguable that what makes it seem wrong to hold B liable in (2) is not any considerations of fairness to B, and, in particular, that it is not B's own freedom of action that protects him against liability in (2). For it is arguable that what blocks shifting A's costs to B in (2) is a rule-utilitarian argument issuing from our concern for freedom of action for all of us—that is, from our desire to be able to count on being free of costs for harms which we cause others, but which we could not, or anyway, morally speaking need not have foreseen and planned for. (Such an argument would have to make out, more strongly, that we prefer being free of costs for harms which we cause others in this way to being free of costs for harms which we are caused by others in this way.) If that is the ground for leaving A's costs to lie where they fell in (2), then Epstein's point could be restated as follows: *in general* B must pay the costs of any injury of A's which B causes—but that is not so where utility is maximized by the adoption of a rule which relieves B of liability, and utility *is* maximized by the adoption of a rule which relieves B of liability in (2), but not so in (4).

But I fancy that there is more to be said about (4) than Epstein says. Let us look, not at what B's actions caused, but at the content of B's planning before he acts. In a case that will later be describable by (4), B has an end in view that he wants to reach, and he figures he will be able to reach it if he does something which he is aware will cause A a harm, and thereby impose costs on A. I stress: B is aware of the fact that his acting will cause A a harm. In a case that will later be describable by (2), B is not aware of the fact that his acting will cause A a harm, and has no moral duty to find out whether it will. Considerations of freedom of action (namely, A's), however, suggest that if B is aware that his acting will cause A a harm, then—other things being equal— B must buy from A the right to cause A that harm, and must do this before acting. In *Vincent,* the dockowner was not there to be bargained with. (So also was the cabinowner not there to be bargained with in Feinberg's story of the hiker.) So other things were not equal in *Vincent*. But surely the fact that a right-holder is not there to be bargained with for possession of the right cannot be thought to entitle the one who wants it to have it free.

These remarks are far too brief; the differences between (2) and (4) call for far closer attention than they can be given here. But I have in any case wanted only to suggest that considerations of freedom of action will take us a long way—not merely into the question why causality matters, but also into the question when and where it does.

. . .

Let us return . . . to *Summers v. Tice*. I said earlier that we must suppose the court had in mind not merely that all defendants were negligent toward the plaintiff, but also that their negligent acts were in a measure likely to have caused the harm for which the plaintiff sought compensation. Only those who were likely to have caused the harm should be among the candidates for liability, [on the assumption] that liability should ideally only be imposed on those who did cause the harm, whatever the degree of fault in others.

Ideally. But here was a case in which there was no way of knowing who caused the harm.

Isn't it unfair to the defendants not to dismiss them both?

Why might one feel that fairness to them requires dismissing them both?

Perhaps simply the fact that it is not known about either that he caused the harm.

. . .

One might have a stronger ground for thinking it unfair to the defendants in *Summers* to hold them liable than the fact that it is not known about either that he caused the harm, namely, the fact that it is only .5 probable in the case of each that he caused the harm, and thus not even more probable than not in the case of each that he caused the harm.

It seems to me that there really is a measure of unfairness to the defendants issuing from this consideration. And doesn't it increase as we consider a hypothetical . . . in which there are 10 shooters, and only a .1 probability in respect of each that he caused the harm?

I suspect that the *Summers* court itself felt this unfairness. The court said of the defendants: "They are both wrongdoers—both negligent toward the plaintiff." So far so good. But then it went on to say: "They brought about a situation where the negligence of one of them injured the plaintiff, hence. . . ." The implication of that sentence is one I did not mention in my summary of the court's argument . . .: . . . the defendants *jointly* brought about a situation in which the plaintiff would be injured. . . . [It] thus hints that they were acting in concert. Stronger still is the implication of the court's next sentence: "The injured party has been placed by defendants in the unfair position of pointing to which defendant caused the harm." That implies, not merely that they acted in concert, but also that they jointly acted in such a way as to make the plaintiff be unable to point to which defendant caused the harm. (Compare a possible case in which a pair of defendants jointly destroy evidence which would have made it possible for the plaintiff to identify which is the actual harm-causer.) If we thought that even the weaker of these two implications were true, then we would feel no unfairness at all in the court's holding both defendants liable. If they had been acting in concert—if, for example, they had had a plot, Simonson to shoot at Summers first and Tice to shoot immediately thereafter if he thinks Simonson missed—then it would not matter from the point of view of liability *whose* shot in fact caused the injury: both should be held liable.

. . .

On the other hand, whether or not one feels that holding both defendants liable is in a measure unfair to them, that feeling is certainly swamped by the feeling of unfairness which is generated by the thought of the plaintiff's being without remedy from either of them. That is the court's point; and that is why, at a minimum *on balance* fairly, it affirmed that liability should be imposed on both. . . .

Was the court's decision in *Summers* fair? . . . Why should the plaintiff be entitled to collect all of his costs from either?

Fairness does seem to allow of, indeed require, holding both defendants liable; but doesn't it require, if both are to be held liable, that they pay in proportion to the probability that they caused the harm, and thus that the plaintiff be entitled to collect only half of the costs from each?

Or perhaps fairness requires that, since each was .5 probably the harm-causer, each should be given a .5 probability of paying all of the costs. But if the defendants are risk-neutral, it must be all the same to them whether each pays half of the costs, or each is given a .5 probability of paying all, and I shall therefore ignore this idea.

So why did the court decide as it did?

The court's only real argument for thinking that the plaintiff should be entitled to collect all of the costs from either defendant, as he chooses, . . . is this: there is a rule to the effect that if two or more people are acting in concert (joint tortfeasors) and between them cause a harm, then each is responsible for the whole damage—that is, the plaintiff may collect his entire costs from any one of them. What is the reason for this rule? It would be unfair to deny the injured redress simply because he cannot prove how much damage each of the defendants did, when it is certain that between them they did it all; therefore they should have the burden of apportioning the costs among themselves. The premise of that argument does not advert to the fact that the defendants are acting in concert, and since the argument succeeds when the defendants are acting in concert, it succeeds also when they are not acting in concert but are both independent causes of the injury ("plural causes")—as, for instance, where two motorcyclists both make loud noises, frightening the plaintiff's horse, the sound made by each being sufficient to frighten the horse.

Now the premise of the argument no more adverts to the defendants' being, both of them, causes of the injury than it adverts to their acting in concert; and it would be no surprise if the *Summers* court therefore thought the argument succeeds when the defendants are not "plural causes" but "alternative causes"—that is, where one or the other of them (though not both) caused the injury. So, the *Summers* court concludes: "The wrongdoers should be left to work out between themselves any apportionment."

Collecting from a defendant may be hard. If it is open to one defendant to sue another for a contribution to the costs—and in most states this is a possible proceeding—then isn't it fairer to the plaintiff that he not have the double burden of collecting half of his costs from each? Shouldn't each of them have the following conditional burden: If the plaintiff fastens on me for the whole of his costs, then I collect half from the other defendant?

In any case, if the events take place in a state that allows suits for contribution, the outcome of holding both defendants jointly and severally liable is likely to be roughly equivalent to what strikes one intuitively as the fair outcome, namely, holding each defendant liable for that fraction of the costs which is the probability of his having caused the harm. Only "likely," because the suit for contribution might not succeed, and only "roughly" because of the costs of such a suit, but perhaps close enough so as to warrant no concern on the ground of fairness. After all, both of the defendants acted negligently, and one of them caused the plaintiff's injury.

. . .

[In] *Sindell v. Abbott Laboratories,* which was decided by the California Supreme Court in 1980, plaintiff Sindell had brought an action against 11 drug companies that had manufactured, promoted, and marketed diethylstilbesterol (DES) between 1941 and 1971. The plaintiff's mother took DES to prevent miscarriage. The plaintiff alleged that the defendants knew or should have known that DES was ineffective as a miscarriage-preventive, and that it would cause cancer in the daughters of the mothers who took it, and that they nevertheless continued to market the drug as a miscarriage-preventive. The plaintiff also alleged that she developed cancer as a result of the DES taken by her mother. Due to the passage of time, and to the fact that the drug was often sold under its generic name, the plaintiff was unable to identify the particular company which had manufactured the DES taken by her mother; and the trial court therefore dismissed the case. The California Supreme Court reversed. It held that if the plaintiff "joins in the action the manufacturers of a substantial share of the DES which her mother might have taken," then she need not carry the burden of showing which manufactured the quantity of DES that her mother took; rather the burden shifts to them to show they could not have manufactured it. And it held also that if damages are awarded here, they should be apportioned among the defendants who cannot make such a showing in accordance with their percentage of "the appropriate market" in DES.

In short, then, the plaintiff need not show about any defendant company that it caused the harm in order to win her suit.

Was the court's decision in *Sindell* fair? I think most people will be inclined to think it was. On the other hand, it is not easy to give principled reasons why it should be thought fair, for some strong moral intuitions get in the way of quick generalization. What I want to do is to bring out some of the sources of worry.

. . .

The *Sindell* court in fact rejected the plaintiff's claim that she should prevail on the rationale which generated the decision in *Summers,* and we should first ask about its grounds for doing so. The court said: "There [i.e., in *Summers*], all the parties who were or could have been responsible for the harm to the plaintiff were joined as defendants. Here, by contrast, there are approximately 200 drug companies which made DES, any one of which might have manufactured the injury-producing drug." Thus in *Summers* there were two who could have caused the harm, and both were defendants, whereas by contrast in *Sindell* there were 200 who could have caused the harm, but only ten of them were among the defendants.

Why did she not join all 200 in her action? Presumably because they were not all reachable. . . .

In any case, the court said she could prevail on an "adaptation" or "modification" of the rule in *Summers.* In particular, it said that if she "joins in the action the manufacturers of a substantial share of the DES which her mother might have taken," then the burden should shift to them to show they could not have manufactured the DES which her mother took. . . .

Consider a hypothetical. . . . Ten shooters all fired negligently, and all are equally likely to have caused the harm; one has quit the country, so Summers sues the remaining nine. The trial court dismisses. The appeals court says that if Summers joins in his action those who imposed a substantial share of the risk— for example, 90 percent of it—then the burden should shift to the defendants to show that they could not have caused the harm. Would that be fair?

I should think that anyone who takes it to be fair for liability to have been imposed on Tice and Simonson in *Summers* will take it to be fair for liability to be imposed—in some way and in some measure—on the defendant shooters. . . . But in what way and what measure?

Here are two possibilities. First:

> (a) Each defendant x is to pay $.n$ of Summers's costs, where $.n =$ the probability that x caused the harm.

Thus if Summers sues all nine reachable defendants, each pays .1 of Summers's costs (since each was .1 probably the harm-causer); and Summers collects only .9—thus 90 percent—of his costs. The second possibility is more complex:

> (b) Each defendant x is to pay $.n/.m$ of Summers's costs, where $.n =$ the probability that x caused the harm, and $.m =$ the probability that the harm-causer is among the defendants.

Thus if Summers sues all nine reachable defendants, each pays .1/.9 of Summers's costs (since each was .1 probably the harm-causer, and the probability that the harm-causer was among the nine was .9); and Summers collects .9/.9—thus 100 percent—of his costs.

I think that intuitively (a) does strike one as fairer, for under (a) each defendant pays only that fraction of the costs of the harm which is the probability that his negligence caused it.

But if the hypothetical appeals court . . . had had (a) in mind, why would it have placed the following condition on the action: that Summers must join in his action those who imposed a substantial share of the risk? If no shooter would be held liable for more than that fraction of the costs of the harm which is the probability that his negligence caused it, why not leave it to Summers to sue as many as he can reach, *or* as few as he wishes?

A number of commentators on *Sindell* have asked why it should matter whether or not Sindell's defendants manufactured a substantial share of the DES which her mother might have taken—for didn't the court go on to say that damages should be apportioned in accordance with percentage of "the appropriate market" in DES? Suppose that Sindell sues nine drug companies, who among them sold 90% of the DES which her mother might have taken; if no drug company is to pay a larger share of her costs than its percentage of the market, then surely the court must have had in mind

(a') Each defendant x is to pay $.n$ of Sindell's costs, where $.n = x$'s share of the DES her mother might have taken,

under which she recovers less than her costs, rather than

(b') Each defendant x is to pay $.n/.m$ of Sindell's costs, where $.n = x$'s share of the DES her mother might have taken, and $.m =$ the defendant's joint share of the DES her mother might have taken,

under which she recovers 100 percent of her costs. But if the court did have (a') in mind, why require that Sindell sue a defendant group which manufactured a substantial share of the DES that her mother might have taken? She would be a fool to sue fewer than she could reach; but why impose the requirement?

The question is not merely theoretical, that is, not merely a question as to the rationale of the decision. Suppose a person is harmed by a drug manufactured by 200 companies, and 195 of them have now gone out of business, and the remaining five manufactured only 40 percent of the drug then sold. (Or is 40 percent a substantial share? If so, choose an appropriate smaller percentage.) It might be worth suing for 40 percent of one's costs if one's costs were heavy; but the rule in *Sindell* would not allow of success in such a suit.

The court seemed to think it would not be fair for the plaintiff to win if her defendants did not, among themselves, manufacture a substantial share of the then available supply of the drug which caused her harm, and this because if she does win, and they do not manufacture a substantial share of the drug, then there will be a substantial likelihood that "the responsible manufacturer, not named in the action, will escape liability." But if *any* manufacturer is excluded from the class of defendants, then there is some likelihood that the responsible manufacturer will escape liability; the smaller the market share of the excluded manufacturer, the smaller the likelihood that the responsible manufacturer will escape liability, but some likelihood all the same. Why does it matter whether the likelihood of escape is small or large, given that no defendant will be held liable for more than that fraction which is its share of the market?

Perhaps what was at work in the court was simply the familiar fact that it should be the causer of the harm who pays the costs of the harm; and the less likely it is that the defendants caused the harm, the more likely it is that the causer of the harm will escape liability.

Perhaps what was at work in the court was a concern about fairness in distribution of liability. If the plaintiff sues only those who manufactured (as it might be) 40 percent of the drug, and they lose, then admittedly they pay only 40 percent of the plaintiff's costs; but the others who manufactured the remaining 60 percent of the drug have no liability imposed on them at all. All imposed a risk; is it fair that only some be required to pay their share, while a great many are not required to pay theirs? (Compare bringing criminal charges against Jones, and not bringing them against Smith, though Smith and Jones did exactly the same thing.)

But I think that what was at work in the court was something different and simpler. For the fact is that I have merely been pretending that it is clear how the court wishes the plaintiff's damages to be apportioned among defendants. What exactly did the court have in mind by "the appropriate market" in DES? All of the DES which was then on the market, and which Sindell's mother might have taken? Or: all of the DES which was marketed by the defendants against whom she brings suit? The decision is ambiguous. If the former, the Court had (a') in mind, if the latter, the court had (b') in mind; and it does not explicitly say which.

What is nowadays commonly called "the *Sindell* rule" is (a'); but I hazard a guess that it was (b') that the court had in mind. My first reason is textual. The court says: "Once plaintiff has met her burden of joining the required defendants [i.e., manufacturers of a substantial share of the DES which her mother might have taken], they in turn may cross-complaint against other DES manufacturers, not joined in the action, which they can allege might have supplied the injury-causing product." What would be the point of a manufacturer's doing that, if he were going to have to pay at most that fraction of her costs which is his share of the DES which her mother might have taken?

More interesting, second, it is only if the court had (b') in mind that fairness really would require that the companies joined in the action have manufactured a substantial share of the DES Sindell's mother might have taken. If all of her costs are to be paid by a group which does not include all who might have caused her harm, then fairness would require (at a minimum) that they, among themselves, have imposed a substantial share of the risk.

In short, it is because the court had (b') in mind that it imposed the requirement that Sindell join in her action the manufacturers of a substantial share of the DES which her mother might have taken.

Notes and Questions

1. Imagine that a plaintiff suffers $100,000 of damage in an automobile crash and is able to convince a jury that there is a 75 percent chance that A was negligent and that A caused the plaintiff's troubles. Would Thomson (and you) think it fairer for A to pay $100,000 because A is likely to have been the cause of the harm, or for A to pay $75,000 as a way of reflecting the jury's degree of certainty?

2. In *Summers v. Tice,* suppose that the defendants had been shooting from different angles such that the probability of hitting Summers in the eye was not the same for both defendants but two-thirds and one-third, respectively. What result do you think the court would have reached? What would Thomson's essay suggest?

3. Consider the relationship between the "freedom of action" that figures strongly in Thomson's analysis and the choice between a negligence rule and a strict liability rule. Is freedom of action violated by a rule requiring tortfeasors to pay for the harms they cause, regardless of whether they are negligent? Even when negligence is the primary basis for liability, there are often exceptions for hazardous activities, such as blasting, or for classes of tortfeasors, such as common carriers. Do such pockets of strict liability violate freedom of action, or is blasting with explosives (or serving as a common carrier) different from shooting quail?

Property, Wrongfulness, and the Duty to Compensate

JULES L. COLEMAN

Causation and Wrongdoing

Suppose *A* harms *B*. Thomson's view is that *A*'s causing *B*'s loss provides *B* with a morally relevant reason for calling upon *A*'s assets to rectify whatever losses he may have incurred. Not every fact about *A* can provide *B* with a morally relevant reason for such a claim. First, not every fact about A *particularizes* him, that is, distinguishes *A* from the rest of the world. Second, not every distinguishing fact about *A* is *morally* relevant, that is, provides *B* with the basis of a moral claim against him. For Thomson, *A*'s causing *B*'s harm serves both to particularize *A* and to provide a moral basis for claims *B* makes against *A*'s resources. It is *A,* after all, not *C, D,* or *E* who causes *B*'s damage. Moreover, *A*'s harming *B* is normatively significant in the light of a complex moral theory that emphasizes free action. *B*'s claim against *A*'s resources is consistent with the value we place on agents acting freely, since liability imposes a hardship on *A* not for free action as such but only for harmful action.

This article first appeared in 63 *Chicago–Kent Law Review* 451 (1987) and is abridged and reprinted without footnotes by special permission of IIT Chicago-Kent College of Law.

There is a good deal that is puzzling about this view. . . . First, negligence (when present) particularizes injurers just as well as causation does. If *A* unreasonably puts *B* at risk, then this is a fact about *A* that is not true of everyone. Moreover, it is a fact about *A* that is morally relevant to *B*'s claims against *A*'s resources. For it is consistent with the value we place upon freedom of action that individuals are encouraged not *unjustifiably* to impose risks on others. One response to Thomson, then, is that both causation and negligence can particularize injurers and do so in morally relevant ways.

Professor Weinrib agrees that causation is normatively significant, but not because it particularizes injurers; after all, it does not. Rather causation particularizes *victims:*

> The difficulty with Thomson's explanation is that it concentrates on the wrongdoer, the moral quality of whose act is unaffected by whether the potential for harm that it releases actually comes to pass. Accordingly, the tort requirement of causation makes no sense if we conceive of the law as passing judgment on this moral quality as such. Causation becomes pertinent only when we focus on the plaintiff's receipt from the defendant of an amount of money representing the harm suffered. This compensatory transfer shows that tort law is not concerned solely with the defendant's emission of a harmful possibility but with that possibility's coming to rest on a particular plaintiff. *Inasmuch as cause particularizes, it does so with reference to the plaintiff rather than the defendant.*

Causation particularizes the victim in the *analytic* sense that a victim, by definition, is someone who suffers harm. Thus, the fact that *A* causes *B* harm is normatively significant because it demonstrates that *B*, not someone else, was harmed, by *A*. So if *A* must pay someone, it must be *B*, not *C*, *D*, or *E*, none of whom were harmed by *A*.

To this point in the argument Weinrib has not claimed that *A* should pay damages to anyone, only that if he is to pay, it must be *B* he pays. It was *B* who *A* injured. For Weinrib, Thomson is right to find normative significance in causation, but wrong to identify that significance with the *liability* of the injurer.

The natural question for Weinrib is: what grounds the injurer's liability? Why and when must injurers pay? One answer is: an injurer should pay damages whenever he or she causes another harm. Were this true, we would have a full theory of liability *and* recovery based entirely on causation. *A* pays whenever he causes harm, and he pays whomever he harms.

Weinrib correctly points out that, of contemporary tort theorists, only Richard Epstein holds the view that *A*'s causing harm is (prima facie) sufficient to warrant liability. Epstein, however, cannot (and does not) avail himself of Thomson's argument for the causal condition. He cannot because Thomson's argument establishes neither the moral necessity nor the sufficiency of causation, only its moral relevance. Thus, Thomson's argument is too weak for Epstein's purposes. Instead, on behalf of the claim that causation is sufficient for prima facie, liability, Epstein relies upon one essentially conceptual and three normative arguments. The conceptual argument is as follows. Suppose *A* harms *B*. If *A* is made to bear *B*'s loss, then *A* is treated by law in the same way he would be were he to harm himself. By holding *A* liable whenever he harms *B,* we treat him as if he had harmed himself. Had he harmed himself, moreover, he would have no grounds for objecting to his having to cover his own losses. Therefore, he can have no greater reason for objecting to his having to cover the expenses he causes others. This point, as Weinrib eloquently points out, cuts absolutely no normative ice:

> Epstein concludes that because the defendant would have borne the loss if he were identical with the plaintiff, the defendant should therefore bear the loss when the litigants are restored to their separate existences. One can equally argue, however, that because the plaintiff would have no cause of action if he were identical with the defendant, so no cause of action should be available when their individual identities are restored. Epstein assumes that the relevant feature of his hypothetical is that the superperson suffers an irrecoverable *loss* that should remain the actor's loss in the two-party situation. But the significant feature may be the superperson's *irrecoverable* loss, that should remain irrecoverable when transposed into the actuality of litigation. This reading allows no liability for any losses.

The normative arguments for strict liability are also unpersuasive. At one time Epstein argued that *A* should be liable for the damage he causes on the grounds that *A* is *responsible* for what he causes, and a just theory of liability must be based on a theory of moral responsibility. The problem here as I have argued—and I have no reason to believe Weinrib would disagree—is that *causation* is neither necessary nor sufficient for moral responsibility. I am *not* justly liable for all sorts of harms I cause you—those resulting, for example, from fair competition; whereas, I *am* sometimes morally responsible for harms I ought to have prevented but did not in fact cause.

Epstein has also argued that *A* should be liable for the harms he causes *B* on grounds of *corrective justice.* A person's harming another upsets the equilibrium that existed between the parties prior to *A*'s action. Therefore, *A* is responsible for setting matters right by reestablishing the previous equilibrium. But not every departure from the status quo ought to be annulled or rectified. Only if in reducing *B*'s welfare, *A* does something wrong, ought *B*'s loss be rectified. In that case, corrective justice would require negligence or wrongdoing, not merely causation, as a condition of liability. Thus, corrective justice does not support a theory of strict liability.

Epstein appears to have finally settled on a defense of strict liability in which the principle of liability falls out of a theory of property. Judging from Weinrib's paper, he too appears to maintain that a theory of liability is presupposed by the concept of property. The difference between them is that whereas Epstein believes that strict liability is presupposed by property, Weinrib claims negligence is. In fact, no substantive theory of liability is entailed by the concept of property. Let us see where both Epstein and Weinrib go astray.

Weinrib puts the question this way: "what is the liability regime correlative to the idea of property?" Epstein's argument that strict liability derives from a theory of property is the following:

> [I]f you deny the plaintiff the prima facie right to recover against a stranger without proof of negligence, then you have *taken* a limited property interest. . . . By definition, every liability rule is tied to a correlative property interest that the law protects; to alter the one is necessarily to change the other. The linkage is not empirical, it is analytical, *a function of the way in which we do use, and must use, all legal language.*

For Epstein, my property rights mark the boundaries of my moral space. Any intrusion of my space is action contrary to my right. If your intrusion results in harm, you owe me, and it does not matter whether your intrusion was wrongful or innocent. Your liability is part of what it *means* for me to have a right; it is part of the concept of a property right. If you should harm me without compensating me, then you fail to understand the concept of property.

To understand Epstein's view, we must first define the notion of "harm." On some views, a necessary condition of harm is an invasion of a right. . . . So when *A* harms *B, A* invades a right of his. If one has an extensive view of property, like Epstein, then the rights *A* invades are *B*'s property rights. But what does it *mean* to have a property right? For *B* and *A* it means that the latter cannot act contrary to the former's wishes without his (*B*'s) consent. If he does, then in doing so he takes *B*'s property. He does what he has no right to do, and it does not matter whether his action was innocent, justifiable or wrongful. To show that he understands the concept of property and to respect *B*'s property right, *A* must make amends. . . .

According to Weinrib, the concept of property tells us he should pay whenever his conduct is *negligent.* But he cannot be required to pay even if he is negligent if there is no one his negligence harms. Only victims can be compensated. The concept of property tells us that only negligent parties can compensate. It does not tell us that *all* negligent parties *must* compensate. Since they can only compensate people who are victims, negligent actors must compensate all and only their victims.

Can Weinrib make good on his claim that property yields negligence liability? Imagine *A* and *B* again and the concept of property. Weinrib invites us to consider three cases. In each case, *B*, not *A*, has an alleged property right. In one case, *A* asserts the right to use what is in fact *B*'s as he, *A*, sees fit. In the second, *A* asserts the right to act as he sees fit, knowing that occasionally his doing so will impose on *B* an unreasonable risk of harm. *A* does not, however, claim the right to use *B*'s property as he sees fit. Instead, *A* claims the right to act without restrictions provided he does the best he can. If and when he fails to do as best he can, he may be subject to liability for negligence. His negligence, in other words, is measured by a *subjective* standard. In the third case, *A* claims the right to act as he sees fit provided in doing so his conduct does not fall below an *objective* standard of negligence. Weinrib argues that the first claim is logically inconsistent with the concept of property, and that the second is inconsistent with the concept of equality entailed by property. Only the objective negligence standard embodied in the third claim is consistent both with the concepts of property and equality entailed by it. Thus, property mandates negligence.

There are three separate arguments here. Common to each, however, is the alleged connection between the concepts of property and equality. "Implicit in the notion of property is the equal standing of all property owners. . . . [A]ll property holders are as property holders equal to each other." Weinrib claims that this tautology has important normative consequences. To see how Weinrib is led to this position, let us consider how this premise figures in the first case Weinrib invites us to consider: the case in which A asserts the right to use B's property as he sees fit.

Weinrib argues first that if A really can assert such a claim against B, then the property could not really be said to be B's at all. For it is A, not B, who has control over its use. To have property, on this view, is to have a domain of authority or control. Second, he argues that "[i]nasmuch as all property holders are equal," A's claim to the free use of B's property could "equally be made by everyone with respect to everything." Indeed, B could make the same claim against A with respect to the same property. "Such a network of crosscutting claims would not be a regime of property for all but the impossibility of property for anyone." Thus, A's assertion of a right to use B's property is "inconsistent with the notion of property."

There are two arguments here, only one of which explicitly relies on the concept of equality. The other relies on treating property as specifying a domain of autonomy. Neither argument is sound, however. Focusing first on the autonomy argument, suppose B owns a house. A claims to use B's house at his discretion. Is recognizing the legitimacy of A's claim inconsistent with the claim that B owns the house—i.e., that it is B's, not A's? Suppose A is free to use B's house at his will, and B never is, but each time A uses B's house A must compensate B, that is, pay him a rent. Imagine another case. A asserts no right to use B's house. B can exclude others, but cannot himself use his house as he sees fit. In neither case does B have any freedom to use his house as he sees fit. In the first case, A has that freedom; in the second case, no one does. In neither case is it obviously false that B has or owns property. In the first case, for example, B's property right may not entail control over his property, though it may be the basis of his claim to compensation for A's use of it. Without a property right, B may be unable even to claim relief for A's use. More generally, as I have argued elsewhere, the concept of a right need not entail any specific claims to alienate or to exclude. Such claims are not part of the *meaning* of property. Rather they follow from particular *normative* conceptions or theories of property. No doubt, property without alienation or autonomy may be morally unattractive, but it is hardly incoherent. The house remains B's even in the face of A's assertions. It is just that property in such a regime may not be worth all that much.

Weinrib's argument from the concept of equality is even more troubling. He contends that A's claim to use B's property makes property impossible.

The basic argument is supposed to show that if we admit that everyone is equal as a property holder, then as soon as *A* claims to use *B*'s property, everyone is equally entitled to make similar claims against everyone else. Recognizing the legitimacy of these claims simultaneously makes property impossible.

Suppose we grant that it would be impossible to sustain a property regime in which each person's claim to use everyone else's property was sustainable all the time. But that is not the case Weinrib presents. In his case, *A* claims a right unqualifiedly to use *B*'s property. How can that claim when conjoined with a principle of formal equality make property impossible? Recognizing *A*'s claim to the free use of *B*'s property means that as a matter of formal equality we are committed to recognizing as valid all claims made by others that are *relevantly similar* to *A*'s. It does not mean recognizing as valid *all* claims regardless of their similarity to *A*'s. It is not *A*'s making an assertion to use *B*'s property that is problematic. What counts is the basis of his assertion and the similarity of those grounds to the grounds others present. What is the basis of *A*'s claim to use *B*'s property? He may claim no basis at all; or he may claim his status of being equal as a property holder to *B;* or he may have some other reason, for example, need. If *A*'s claim is groundless, then, if we recognize it, we are committed to recognizing all similar—i.e., groundless—claims. Property may then be unimaginable. Also, if *A*'s claim is based merely on his status as a property holder, we are again committed to recognizing all such claims. In that case, the concept of property which forms the basis of *A*'s claim unravels. On the other hand, if *A* has a substantial basis for his claim to use *B*'s property, then, while the principle of formal equality may commit us to recognizing as valid all similar claims, it does not commit us to a set of crosscutting claims that render property impossible. No one, merely by virtue of their status as a property holder, has the same claim to *B*'s property that *A* does.

Weinrib is mistaken in claiming that property is impossible whenever we recognize a single claim to the discretionary use of another's property. The question remains whether he is correct in claiming that the concepts of equality and property yield negligence. For Weinrib, reasonable care is specified in cost-benefit terms. *A*'s conduct is negligent if and only if the costs of prevention are less than the costs of harm to *B* discounted by the probability of its occurrence. "The virtue of the negligence standard is that it regulates the relationship between the property holders on the basis of equality." The theory of strict liability violates this principle of equality because it entails the judgment that the victim's property is always more valuable than the injurer's free action. The theory of absolute victim liability—i.e., the principle of *non*recovery—violates the principle of equality because it entails the judgment that the injurer's freedom is always more valuable than the victim's property. The *subjective* theory of negligence violates the principle of equality because it gives a special status or preference to the injurer's capacities. Only negligence counts the relevant interests of injurers and victims equally. Only it is required by the concept of property.

There are two problematic facets of this argument. The first is that the actual argument for negligence never invokes the concept of property, and so it can hardly be said to be entailed by it. Second, the principle of formal equality is compatible with *every* scheme of liability—from negligence to strict liability to no liability. Let us examine these problems in turn.

Weinrib's argument for negligence never invokes, nor need it invoke, the concept of property. Instead, it is an instance of the familiar argument that standard utilitarianism (cost-benefit analysis or efficiency) is not only compatible with but, in fact, embodies an ideal of equality. According to this conception of equality, each person must count for one and no more than one; this is the idea of equal standing. Utilitarianism satisfies or embodies this ideal because in determining right conduct each person's interests, preferences or desires count for one. The costs to the potential victim count no more nor less than the benefits to the potential injurer, and right conduct is determined by an objective balancing of the two. There is no need to invoke the idea of a property right in defending the negligence standard. Indeed, in ordinary normative discourse, arguments from rights are invoked precisely to *counter* the utilitarian argument, not to serve as a premise in its derivation. For the very point of appealing to property rights is to establish that considerations of utility maximization—in which each person's interests are counted equally— are inadequate to overcome certain claims. Claims of right implicitly deny that normative conclusions are to follow exclusively from balancing based on an equal consideration of interests. When *B* claims a property right against *A*, part of what he means to assert is that it does not matter whether *A*'s taking property from him can be shown to have desirable utilitarian consequences; *A* simply has no right to take. The whole point of appealing to property rights is to deny that *A*'s interests are to count equally with *B*'s.

It is possible to make property *rights* yield a utilitarian theory of negligence by advocating a utilitarian theory of rights. Those rights we have and the claims to which they give rise are those that maximize utility. In this case, the cost-benefit theory of negligence rests on a *normative* theory of property—not on the idea or concept of property itself. Moreover, while the argument for negligence based on a utilitarian theory of rights is compatible with the principle of formal equality, simply because utilitarianism is, the argument itself in no way relies upon the principle of equality. In other words, it is possible to derive a negligence standard from property directly without recourse to a principle of equality. Moreover, the negligence principle derives from a contestable *normative* theory of property, not from the concept of property itself, or from its alleged corollary, the principle of equality.

In summary, the principle of formal equality can be spelled out in such a way as to require no more than the equal consideration of interests. From the principle that everyone is entitled to an equal consideration of interests, it may be possible to derive a principle of negligence. Doing so invokes the principle of equality but not the concept of property. On the other hand, one can invoke a particular theory of property rights in order to derive the principle of negligence liability. But the theory of property that's needed is a utilitar-

ian one. In that case, the theory of negligence requires a contestable normative theory of property. Once again negligence does not derive from the idea
of property itself, but from a theory of property that is in any case incompatible with various other *conceptions* of property—notably the Lockean one.
The problem with both Epstein and Weinrib, then, is that one cannot derive a
substantive theory of liability from the concept of property.

As far as I can judge from Weinrib's brief discussion of it, the principle of
equality requires that we treat property-right bearers equally as property-right
bearers. It should be obvious, then, that this principle is satisfied no matter
what the liability rule is, provided it applies to all property right holders. For
example, a rule of strict liability does not favor one set of property holders over
another. All property owners, as property owners, will be entitled to repair
whenever they are injured by the conduct of others; and all property owners
(and others) who injure property owners will be required to make repair. The
same can be said for a rule of no liability. (Remember property owners are
given equal respect even when they are all treated with no respect at all.) In the
same way that one cannot derive substantive normative conclusions from the
premises allegedly elucidating conceptual connections, one cannot derive substantive claims about liability rules from purely formal principles.

Notes and Questions

1. If negligence derives from a specific theory of property, is it somehow inconsistent for one legal system to use negligence in some settings but strict liability in others?

2. How does the "principle of formal equality" compare with Fletcher's reciprocal
risks? (See part II.)

Probabilistic Recoveries, Restitution, and Recurring Wrongs
SAUL LEVMORE

. . .

Is it sensible for our civil law system to be centered on a preponderance-of-
the-evidence standard of proof, when it might alternatively employ a more
flexible, or sliding, rule under which defendants' payments would increase as
the evidence against them becomes more persuasive? . . .

Abridged and reprinted without footnotes by permission from 19 *Journal of Legal Studies* 691
(1990). Copyright © 1990 by The University of Chicago Press.

Error-Minimizing Rules of Decision

The Attraction of the Preponderance-of-the-Evidence Rule

Imagine that a fact finder is convinced that A and B behaved negligently, that one of these parties caused a victim, V's, injury, and that circumstantial evidence suggests that there is a two-thirds chance that A caused the harm. If V's losses are $99 but the law denies V recovery because proof was not certain, then the expected, or average, error of that "no-recovery" rule is $99. Either A or B did, in fact, wrongfully cause V's injury, and one party is, therefore, underpaying and is underdeterred by $99. In contrast to such a no-recovery rule, a preponderance-of-the-evidence rule permits V to recover $99 from A but nothing from B. That rule generates an expected error of only $66 because there is a two-thirds chance that A in fact caused V's harm, but a one-third chance that the rule has A paying $99 for something A did not do and has B paying nothing when B should pay $99. The expected, or average, error must, therefore, add A's expected overpayment (one-third of $99) to B's expected underpayment, for a total of $66.

Some commentators have suggested that the preponderance rule might be improved on by using a proportional shares, or "purely probabilistic," rule under which A would pay $66 and B would pay $33 because those sums reflect our information about the relative likelihood that A and B caused V's injury. Although this probabilistic rule may at first appear scientific, it generates greater error than the preponderance-of-the-evidence rule. . . . There is a two-thirds chance that A was truly the tortfeasor, in which case the probabilistic rule has A paying $33 too little and B $33 too much. But there is also a one-third chance that B was the wrongdoer. The rule therefore has A overpaying $66 and B underpaying $66. The expected error is thus $\frac{2}{3}(33 + 33) + \frac{1}{3}(66 + 66) = \$88$.

Alternatively, the law might try to minimize error by using a probabilistic approach only after some threshold suggested by the preponderance rule is reached. Defendant A could be charged with two-thirds of V's losses, while B, who is more likely than not to have been uninvolved, would pay nothing at all. The rule disadvantages A more than it does B and may, therefore, contradict another aim of the legal system. Moreover, the error generated by this rule again exceeds that of the preponderance rule. When A truly caused V's injury, A will underpay $33; when B was really the responsible party, A will overpay $66 and B will underpay $99. The error of this "partially probabilistic" rule is thus $\frac{2}{3}(33) + \frac{1}{3}(66 + 99) = \$77$. The preponderance rule, in short, may be the dominant rule in the legal system precisely because it minimizes error—at least for a single case.

The matter can be made more realistic and more interesting by noting that legal rules can themselves affect uncertainty and, therefore, error computations. Faced with a denial of recovery, plaintiffs may work harder to locate the evidence that assures recovery. The no-recovery rule—which generated the highest error of the four rules considered above—could actually minimize error by encouraging plaintiffs to develop further evidence and to identify tortfeasors with great certainty. If, for example, there were more than a one-third chance that V could, with some legwork, positively determine whether A or B was the true tortfeasor, then error would be reduced to zero, because $99 would be transferred from the culpable party to V, and there would be no underpayments or overpayments. Conversely, it is possible that under this rule, which denies recovery unless plaintiff's case is a certain one, the actual tortfeasor will invest in further efforts to conceal his wrongful behavior so that the rule may be neither error minimizing nor encouraging of socially useful behavior.

Such consideration of the behavior that is encouraged by the various recovery rules explains several well-known cases where recovery is denied even when the preponderance rule supports plaintiffs' claims. A rodeo promoter who has collected an admission fee from 499 patrons but finds 1,000 persons in the audience (and, perhaps, a hole cut in the fence that encloses the rodeo grounds) will almost surely not be allowed to collect under the preponderance rule (from even one member of the audience) because the easiest way to minimize error is for promoters to issue tickets or other receipts to those who pay admission. The no-recovery rule encourages the issuance of tickets. A promoter who issued tickets and warned patrons that production of these tickets might be required, and then found that attendance exceeded paid admissions, would, I dare say, be able to collect from (or do worse to) spectators without tickets even if they constituted less than 50 percent of the audience.

Taking incentives into account, however, does not revive the case for the purely probabilistic rule. Under that rule, V is entitled to $66 from A and to $33 from B, and V has no incentive to make any further investigation into the true source of his injuries. It is simply the case that the purely probabilistic rule neither creates desirable incentives nor minimizes error.

In contrast, the partially probabilistic rule may have socially desirable properties because it compromises between immediate error reduction and the creation of fact-finding incentives. The opportunity to increase recovery from $66 to $99 offers V an incentive to discover the true tortfeasor. If this incentive works more than one in seven times, the expected error will drop from $77 to something less than $66. Inasmuch as the partially probabilistic rule succeeds in reducing error only by encouraging further investigation, no conclusive comparison of the various rules can be made without some assessment of the true social cost of additional investigation, but it is likely that in some settings the partially probabilistic rule is superior to the preponderance rule. In similar error-minimization terms, the partially probabilistic rule can be thought of as superior to the no-recovery rule when fact-finding is very sensitive to marginal recovery dollars. In the illustration involving A, B, and V, for example, the no-recovery rule "offered" $99 for (the hope of) a one-in-three chance of positive evidence, while the partially probabilistic rule offered $33 for a one-in-seven chance.

These incentive effects must be kept in perspective. As the probability that A or B was the actual tortfeasor increases, the expected error of the preponderance rule shrinks while that of the no-recovery rule grows. The incentive effects of denying recovery must, therefore, be quite powerful to offset the error-minimization advantages of the preponderance rule. It is in extreme cases like the rodeo example, where incentive effects are palpable, that the legal system is most likely to refuse "naked statistical evidence" and apply the no-recovery rule. But both the basic rule and its exception are explained by the error-minimization principle.

Recurring Defendants and the Probabilistic Rule

If V had 999 interactions with A and B, then the (purely) probabilistic rule would emerge as the error-minimizing rule. An omniscient fact finder would, after all, require A to pay for the 666 accidents caused by A and would require B to pay in the 333 cases that actually resulted from B's wrongful behavior. In all cases, V would collect. When such large numbers are involved, the probabilistic rule reaches the same result. By requiring A to pay $66 and B to pay $33 every time, the rule ensures that A and B will pay the exact amount they would have paid with flawless fact-finding. The scale of repetition allows us to believe that an omniscient fact finder *would* have found that A caused 666 cases and that B caused 333 (and not many more or many fewer). The expected error of the probabilistic rule is thus very close to zero. In contrast, the preponderance rule has A paying $99 each time and so, over 999 cases, overcollects that amount from A and undercollects from B 333 times for a total error of 333 (99 + 99); in expected-error terms (per incident) we already know this to be $66.

Only one element changes when the repetition, or averaging, is not always with V, but for other victims with similar injuries. An omniscient, error-free fact finder might require, for example, that A pay V and W and that B pay another victim, X, because this fact finder knows which tortfeasors injured which victims, while the best the probabilistic rule can do is require A and B to pay two-thirds and one-third, respectively, of the damages sustained by all these victims. The tortfeasors pay and the victims collect in the end exactly as they would have in the hands of the flawless fact finder. It is no longer controversial, I think, to argue that there is no moral or practical reason to insist that victims be denied recovery or that some wrongdoers escape liability simply because we are unable to match wrongdoers with their victims. This point has been made with great force in the torts literature, especially in light of the use of market-share liability for diethylstilbesterol in *Sindell v. Abbott Laboratories*. The argument is also appropriate in other settings where similar claims are likely to be brought many times against the same defendant. The precise degree of repetition that should be required for such a switch from the preponderance to the probabilistic rule for such "recurring cases" is a question that can be left open. The important point is that it is hard to see why errors should be measured in a case-by-case, rather than an aggregate, manner when a probabilistic rule accomplishes virtually all that an omniscient fact finder would do. The other rules are inferior. A no-recovery rule grossly undercollects from and underdeters A and B; a preponderance-of-the-evidence rule overcollects from A and underdeters B; and a partially probabilistic rule underdeters B.

The Problem of Recurring Misses

The discussion in [the first] section used the idea of error minimization both to explain the traditional preponderance-of-the-evidence rule and to account for the exceptional use of market-share liability in mass tort cases. In this section, I consider a second challenge to the preponderance rule, the problem of recurring misses, and show how it, too, may be met with a probabilistic solution.

Consider the celebrated case of *Haft v. Lone Palm Hotel,* where a motel operator failed to meet a statutory obligation requiring it either to provide a lifeguard at its swimming pool or to post a notice that no lifeguard was present. A father and his young son drowned in the pool and there were no eyewitnesses. The California Supreme Court reversed a lower court and held that, since defendant's failure to provide a lifeguard eliminated a likely eyewitness, it was wrong to require the family of the deceased to bear the burden of proving that defendant's statutory violation more likely than not caused the deaths.

The decision is easily ridiculed. Even if a lifeguard would have reduced the Hafts' chances of death by more than 50 percent, the statute explicitly allowed a pool operator simply to post a notice. This notice might have reminded swimmers (and nonswimmers, like the Hafts) of the dangers of deep waters, it might have disabused Mr. Haft of any belief that a lifeguard was on duty and only momentarily out of sight, and it might have suggested to him that horseplay in the pool would be dangerous. But these possibilities surely do not add up to much; it is safe to guess that the causal connection between drownings and the absence of notice alone was quite small, say 10 percent, and thus well below the preponderance rule's dividing line, given that very few law-abiding pool operators employ lifeguards rather than signs.

Conversely, a directed verdict for the Lone Palm Hotel plainly underdeters wrongful behavior. Most wrongdoers, such as speeding motorists, cannot ignore the possibility that they will cause harm and then lose before a jury under a preponderance-of-the-evidence rule. They may sometimes win when they should lose and lose when they should win, but the deterrence function of tort law will be satisfied at least in a rough way, even if there is some chance in who collects compensation. The full damages paid in some cases offset the complete absence of damages in other cases. In contrast, if pool operators know that the absence of both lifeguards and signs always produces a causal connection on the order of 10 percent, then they will never be liable under the preponderance rule and they can violate the statute with impunity. Cases like *Lone Palm* fall into the category of "recurring misses" for which the preponderance rule is not well suited. The background statistics on such matters as drowning and the efficacy of lifeguards and signs are sufficiently stable to ensure that the preponderance rule will systematically "miss" ongoing instances of antisocial behavior that it should deter.

Unfortunately, this problem of recurring misses appears with some frequency in tort law. Many "failure-to-warn" and informed-consent cases fit this pattern because disclosure would change the behavior of some small fraction but not of most of those who are duly informed. Similarly, many cases in which a cost-justified medical or other precautionary procedure might have been taken, but was not, may be fairly described as dealing with the danger of recurring misses. . . . The identifying feature of these cases is, once again, that there is a wrongful party who is more than 0 percent but possibly never more than 50 percent likely to have caused an injury.

The two most obvious strategies for solving the underdeterrence (and undercompensation) problem presented by recurring misses are, first, to increase the use of probabilistic rules and, second, to move away from ex post, conventional tort rules to ex ante, regulatory rules. These two strategies can be mixed and matched to create a number of solutions to the problem of recurring misses. The first is simply to use regulatory or administrative sanctions to overcome the deterrence gaps in the tort system created by recurring misses. Unresponsive pool operators, like speeding motorists, could be influenced with fines or other penalties rather than with the threat of tort liability. The California statute in *Lone Palm Hotel* did, in fact, provide for a fine and for possible imprisonment for each day that a pool operator provided neither lifeguard nor notice. This first "solution," of course, calls for legislative intervention. It offers no help to a judge confronting a recurring miss, frustrated, perhaps, that low levels of risk systematically elude tort sanctions.

The second solution is internal to the tort system and is to employ a probabilistic, or expected value, rule for recurring misses. A 51 percent "connection" would not lead to 100 percent liability as under the preponderance rule, but to only 51 percent liability; similarly, a 37 percent connection would lead not to no liability, but to liability for 37 percent of the plaintiff's losses. There might be some correction or forgiveness for very low probabilities where the administrative or litigation costs are large relative to the risks that would be controlled. And this probabilistic rule could be confined to recurring misses, by judicial "certification" suspending the preponderance rule. Judges might recoil at so revolutionary an innovation, but they might nonetheless be willing to smuggle in such a probabilistic rule as part of a comparative negligence regime by insisting that there was enough of a causal connection, with plaintiff perhaps nine times more responsible than defendant. In many cases, however, it will not be possible to camouflage a probabilistic rule in this fashion.

The third solution exploits the principles of proximate cause to mimic a probabilistic rule. A judge considering *Lone Palm Hotel* might have thought it plausible that Mr. Haft would have kept his son (but not himself) out of the pool if notice had been posted, so that the jury could find that defendant "caused" the son's death but not the father's. This Solomonic allocation of injuries is unlikely to match exactly a formal probabilistic rule, but its spirit and effect can be quite similar. A fact finder can make a defendant pay only a fraction of a set of harms by finding that the defendant caused some but not all of the various separate injuries. This mechanism, however, is more usefully thought of as a precursor to comparative negligence than as a solution to the problem of recurring misses.

The fourth solution is the one that was adopted in the case: the court switched the burden of proof to the defendant to show the absence of a causal connection between its statutory violation and the deaths by drowning. I can imagine the thinking behind this decision to be as follows:

> If I decide that the plantiff need not show a causal connection and I direct a verdict for the plaintiff, then I substitute a serious problem of overdeterrence for an annoying problem of underdeterrence, for if 10 percent of the time the presence of a sign might have made a difference, then surely in 90 percent of the cases it would not. I could hope, of course, that my style of statutory construction will not be followed by all other judges in the future, and ideally if 90 percent of judges stick to the older rule then we will have achieved optimal deterrence through backhanded means. All this is too much to hope for.

> But there is another approach I might try that does not require other judges to stand up later and disagree with me. I could switch the burden of proof to the defendant so that the plaintiff escapes a directed-verdict motion. This may not do the poor plaintiff much good because the jury may be persuaded that the father knew full well what the situation was when he took the fatal swim. But if this case, and others like it, settle for about 10 percent of their face value, or if 10 percent of such juries are persuaded otherwise, then switching the burden of proof will do the job. But I am surely taking a chance, for if settlements or juries are more generous than what I have in mind, then I would have done more good with a simple no-recovery rule. But I shall take my chances, keep my own counsel, and hope for the best.

I like to think that the decision in *Lone Palm Hotel* was reached in this manner, notwithstanding the clear requirements of the preponderance rule. That a later swimming pool decision in California labored to distinguish itself from *Lone Palm* supports this view. It is noteworthy, however, that even a less ambitious judge, frustrated by misbehaving defendants or especially compassionate toward a particular plaintiff, has the capacity, however unconscious, to imitate a probabilistic rule in loose fashion.

Finally, a fifth solution to the problem of recurring misses goes outside the boundaries of tort law into restitution law, with its different principles of recovery. [Courts can try to combat antisocial behavior by extracting from the defendant that gain which accrues from bad behavior. In most settings, however, this gain is quite small and it is difficult to deter antisocial behavior with the restitution remedy alone.]

Recoveries for "Lost Chance"

The "lost chance" cases illustrate both the different attributes of tort and restitution systems and the importance of the problem of recurring misses.

Consider, for example, *DeBurkarte v. Louvar,* where a physician negligently delayed ordering a biopsy for a patient who later suffered terribly from cancer. The probability of successfully treating the patient and arresting the cancer was undoubtedly decreased by the substantial delay before correct diagnosis, but the causal connection between the late treatment and the subsequent harms could probably not have been established sufficiently to satisfy the preponderance-of-the-evidence rule. Delay in diagnosis is likely to increase the chance of a loss, but by leading inexorably to a verdict for the defendant, the preponderance rule ignores the reduction in odds that the plaintiff has been forced to bear. The physician who blunders in the midst of a procedure may often be caught by the preponderance rule, but one who fails to make a timely diagnosis will rarely, if ever, be held liable under the rule. The recurring-miss problem of *Lone Palm Hotel* cannot be uncommon in health care settings.

Plaintiffs who have brought such medical malpractice claims have been met with four kinds of judicial responses:

1. Some courts have simply required that the raw preponderance rule be met and have thus found in favor of defendants. Perhaps these decisions were rendered with an understanding that the particular kind of negligent behavior engaged in by the defendant would sometimes cause a harm by the preponderance of the evidence, so that these cases do not address the question of whether wrongful behavior of the sort in question will regularly escape the reach of liability rules. It is, therefore, impossible to know whether all these decisions simply follow the traditional preponderance rule or whether some actually reflect a view that there is no recurring-miss problem.

2. Some courts have allowed complete recovery even when the evidence clearly indicates that plaintiff cannot prove causation to the degree required by the preponderance rule. Again, these decisions may be following the *Lone Palm* model, by compassionately or cleverly assigning liability, without seeking to announce a universal rule. One frequently quoted decision is instructive in this regard:

> When a defendant's negligent action or inaction has effectively terminated a person's chance of survival, it does not lie in the defendant's mouth to raise conjectures as to the measure of the chances that he has put beyond the possibility of realization. If there was any substantial possibility of survival and the defendant has destroyed it, he is answerable. Rarely is it possible to demonstrate to an absolute certainty what would have happened in circumstances that the wrongdoer did not allow to come to pass: The law does not . . . require the plaintiff to show to a *certainty* that the patient would have lived had she been hospitalized and operated on promptly.

It goes almost without saying that the passage slides over the real question. The law never requires certainty—it asks only for the preponderance of the evidence—and the problem here is that the uncertainty is so great as to make it less likely than not that the defendant's wrongdoing caused the plaintiff's entire loss. Of course, this sort of evasion may be the best indication that the court recognizes the problem of recurring misses and that it intends to work only rough justice.

3. Some courts march rather plainly toward a probabilistic rule which, it will be recalled, is the most straightforward solution to the problem of recurring misses. This was, in fact, what the court in *DeBurkarte* did, allowing recovery, but "*only* for the lost chance of survival." It is unclear whether this court, or indeed any other court, has grasped the systematic implications of fractional recovery. Thus, the court does not say that it believes that the lost-chance cases will cluster below the 50 percent level. And no court has indicated whether it will limit the plaintiff to a probabilistic recovery when causation can be established by the preponderance of the evidence, as is necessary to prevent overdeterrence.

4. Finally, it will come as no surprise that some courts create a gray area between responses 2 and 3, allowing some recovery, without specifying whether plaintiff should collect full damages or damages reduced by the probability that the defendant did not cause the harm. There is, again, no indication whether any of these courts believes that the defendant's wrongdoing would ever be linked to a victim's loss by the preponderance of the evidence.

These developments in medical malpractice "lost-chance-of-survival" cases are both reassuring and suggestive. They reassure us that our discussion of possible solutions to the recurring-miss problem was not entirely radical, and indeed may represent an emerging rule, at least in medical malpractice cases and perhaps, in time, elsewhere. There are, however, two reasons why probabilistic innovation has begun in the medical malpractice area. The first has to do with the availability of information, such as that contained in mortality tables, and the second returns us to the possibility that the unjust-enrichment principle could be used as a solution to the problem of recurring misses.

One great advantage of an all-or-nothing, rather than a probabilistic, rule is that at times the fact finder will see relatively easily that the truth lies far away from the dividing line set out by the all-or-nothing rule. In these cases there is no need for further inquiry. For example, cases in the 0–20 percent or 80–100 percent range are easy. In contrast, under a probabilistic rule the fact finder must develop precise estimates in order to discount properly the plaintiff's total damages, which, in situations like *Lone Palm Hotel,* are likely to be enormously demanding and wildly inexact. One's intuition is that a lifeguard provides added safety and that a warning sign provides perhaps some, but not very much, safety, but it would be a difficult task to attach precise numbers to these sentiments. There is, however, in the cancer cases a respectable body of statistical information about the likely consequences of a delayed or missed diagnosis that provides the factual information for using a probabilistic approach. Mortality tables, after all, teach the habit of thinking about lost chances, and it can be no surprise that it is in the area of tort law most closely related to the regular use of these tables that lost chances make their appearance in deciding the question of liability. Put somewhat sadly, death is the loss of chances, and legal decisions involving death have been able to make use of this reality because of the availability of information. The exact consequences of using these tables depends, of course, on the choice of decision rules under uncertainty, so it is not surprising to see courts vacillate between total and partial recovery in this area.

The second explanation for the use of lost-chance arguments in the medical malpractice area brings us back to the practical problems of restitutionary rules. Recovery for unjust enrichment is a refreshing but mostly unusable tool in the medical malpractice area because the negligent party's wrongful behavior so often involves not an extra, unnecessary, profitable procedure but, rather, a delay, a moment of forgetfulness, or a plain omission. As such, a recovery grounded in restitution principles would be so small that it would neither compensate nor deter anyone. Most victims would not bother to pursue their claims. It is, therefore, possible that a probabilistic solution emerged in the medical malpractice area because there is both the problem of recurring misses and no expectation that a restitutionary solution to this problem could emerge.

Recurring Misses and Restitution

The analysis thus far leads to the conclusion that recurring misses be treated with one of three tools: (1) a probabilistic tort rule, (2) a restitution rule, or (3) occasional radical decisons, such at that found in *Lone Palm Hotel,* that violate the preponderance rule. Fortunately, the third and most dangerous of these tools can be supplanted with some creative use of restitution principles.

While the first of these tools is an attractive option in the "lost chance" claims in the medical malpractice area, it is much less attractive in a case like *Lone Palm Hotel,* where there is a dearth of statistical information about the effectiveness of statutory warnings and other factors. The problem with the restitution alternative in *Lone Palm,* however, is that the gain to the defendant from not posting a sign was so small that restitution appears to provide insufficient deterrence. But what if plaintiff had argued that since there was some chance, say X percent, that a law-abiding pool operator would have employed a lifeguard (rather than posted a sign) at a substantial cost over the last several years, recovery in restitution should now include X percent of that cost, or savings. The claim is obviously not terribly precise and does not provide full compensation, but it has the ring of law to it, and combined, if necessary, with some recovery of litigation costs, it might form the foundation of a workable, efficient system of social control where other tools are far more problematic. Perhaps the courts in *Lone Palm* and in other recurring-miss cases would have welcomed this aggressive, restitution-based argument.

In other settings, the restitution solution may be especially attractive because it is easier to get tolerable measures of the defendant's enrichment. For example, a drug manufacturer that fails to provide a low-cost warning about some side effect might be sued not only for the unjust enrichment derived from not printing the warnings but also for the sales that might have been lost if such a warning had been provided. It is, after all, likely that disclosure was withheld precisely because the manufacturer feared that too many risk-averse customers would decline to purchase drugs that came with negative information. In contrast, in *Lone Palm* it is quite difficult to assess how much business was gained by the motel because it did not post the required sign, and, indeed, inasmuch as few potential patrons examine safety equipment and signs around swimming pools before registering at motel desks, it is possible that no business was wrongfully gained. In the drug case, however, the extra sales enjoyed as the result of suboptimal warnings provide the most convenient means of fueling a restitution-based solution to the recurring-miss problem. To be sure, the magnitude of these marginal drug sales is quite speculative and might be enormously exaggerted (even to equal gross revenues) in the hands of some fact finders who were swayed by particular circumstances and unconcerned about chilling future research and production. But restitution law has not been filled with such wild recoveries, and my instinct is that the creative use of restitutionary (and of probabilistic) recoveries is preferable to the norm of no recoveries in these recurring misses and is more constrained than the solution found in *Lone Palm Hotel.*

Notes and Questions

1. Consider again *Summers v. Tice,* in which each defendant was negligent and each had a .5 chance of having caused the plaintiff's most serious injury. Does the rule of joint and several liability minimize error, as defined in this reading?

2. The reading refers to one of the most popular hypotheticals dealing with probabilistic analysis, namely, the case of a rodeo promoter who has collected an admission fee from 499 patrons but finds 1,000 persons in the audience and a hole cut in the surrounding fence. What are the arguments against alllowing recovery by the promoter who sues a randomly chosen member of the audience, in light of the fact that it is more likely than not that this potential defendant did not pay admission? See David Kaye, The Paradox of the Gatecrasher and Other Stories, 1979 *Arizona State Law Journal* 101.

Imagine now that there are 1,000 people in the audience, that the promoter has gone to the trouble of issuing tickets and announcing that stubs must be retained, that 998 admissions have been collected, and that three members of the audience are unable to produce stubs. One of the three has apparently lost the stub, while two have yet to pay the admission fee. From whom (and how much) should the promoter be able to collect? The problem is discussed in the unedited version of this reading in 19 *Journal of Legal Studies* 695, n.6.

Economic Loss in Tort

WILLIAM BISHOP

The law of torts severely restricts recovery by a plaintiff for financial losses suffered in consequence of the negligent conduct of the defendant. In American law this is usually, and perhaps misleadingly, summarized by saying that there is no recovery for negligent interference with a contract. . . . In other common-law jurisdictions this is referred to . . . as the problem of economic loss. Here the term may mislead because many financial losses may not be losses at all in an economic sense, that is from the point of view of net social welfare.

The principle theoretical idea used here is this. In a range of cases private economic loss caused by a tortious act is not a cost to society. Therefore the exclusion of economic loss from the liability which tortfeasors bear will be efficient in such cases. In this section this proposition is illustrated and some of its general implications discussed in succeeding sections.

Two examples will help to clarify the main idea. They are based on a railway accident that happened in Canada in November 1979.

Example 1: Imagine two contiguous towns called Mississauga and Etobicoke. A railway runs through them. The railway company is considering installing special equipment to minimize the danger of derailments of trains carrying dangerous chemicals. For simplicity assume that there is no danger to life if people are evacuated in time. Then the main direct effect of any derailment is to induce mass evacuation from the town in which it occurs to the other town, for about a week. The cost of this equipment is $10 million. It has a ten-year life. The cost to residents of evacuation is $4 million. About one derailment per decade will be prevented by the equipment. If these were the only relevant costs we would not want the railway to install the equipment. It costs more than it saves.

Now suppose each town is served by a butcher, a baker, a candlestick maker and numerous other tradesmen and merchants. Each businessman has a normal weekly volume of trade. Suppose further, and this is crucial, each can accommodate more than his normal volume of trade, for a short time at least, at no extra cost beyond cost of raw materials. Assume initially that each businessman is risk neutral.

The effect on businessmen of any derailment and evacuation in Mississauga is this. The Mississauga businessmen make no sales for a week and at the end of the year find their bank balances smaller than they might otherwise have been. The Etobicoke businessmen double sales for a week and have larger bank balances at the end of the year. The effect of the derailment in this respect is to transfer wealth from one set of people to another.

Abridged and reprinted without footnotes from 2 *Oxford Journal of Legal Studies* 1 (1982) by permission of Oxford University Press.

Suppose the Mississauga businessmen have a right of action to recover "pure economic loss." The suffer loss (accounting) profits of $8 million during the week. If the railway must pay out this sum as well as direct losses to residents it faces total expected derailment costs of $12 million. This sum is greater than the cost ($10 million) of the special equipment. So the railway will decide to install the equipment. But this would be the wrong decision from the society's point of view. Social-welfare calculation still reveals a net loss of only $4 million, not $12 million. The $8 million economic loss is not a social cost but rather a transfer payment. It is a loss to one group but a gain to another group.

Example 2: The facts are identical to *Example* 1, except that businessmen cannot easily expand services for a short time to accommodate an upsurge in demand. Each Etobicoke butcher, baker, and candlestick maker must hire extra help who will undertake the work only if paid for the trouble. Further, existing staff are annoyed and overworked and demand extra pay for their trouble. Prices for meat, bread, and candlesticks rise for all consumers, both for those who normally buy in Etobicoke as well as for refugees from Mississauga.

These facts describe a rise in the real costs of production in the short run. They are captured in the usual diagramatic analysis by an upward-sloping marginal cost curve for each firm. The facts of *Example* 1 imply flat marginal-cost curves in both towns at least for temporary fluctuations in demand.

On these facts, "economic loss" to Mississauga business is *not* merely a transfer to those in Etobicoke. The Etobicoke businessmen incur higher real social costs. Further, some consumers are not now served, because prices have risen too high for them to buy. Their losses are real social costs insofar as the price rise pays for the Etobicoke businessmen's extra costs. Any extra profit to Etobicoke businessmen is only a transfer payment to them. The exact sum of real social loss from the accident depends on the facts. A superhuman court with large resources and with great insight and acumen, in principle, would calculate it accurately, as could a similarly endowed potential tortfeasor railway firm.

Numerous plausible stories could be told about the Mississauga railway accident. For example if Etobicoke consumers have inventories of goods they may react to the arrival of refugees not by buying more but by running down inventories. Then these are built up gradually after the refugees have gone home. Here there is no sudden sharp rise in demand and the likelihood of cost increases in Etobicoke so much the less.

In Mississauga many firms may find the lost work compensated for by higher sales in later weeks. This typically will be so for durable goods (e.g., radios) or some services. Here purchase often simply will be delayed rather than lost.

It might be objected on theoretical grounds that the losses to Mississauga businessmen must be real costs. It might be said that if purchases can be accommodated costlessly in Etobicoke then we have no explanation for why the firms exist in Mississauga in the first place. But this argument ignores the

$4 million compensation paid to the Mississauga refugees. This is enough to leave them as well off as if they had stayed at home. Any payment to third parties whose market falls in consequence will be a case of *double counting* so long as business costs are not raised in Etobicoke. For example if I open a circus in Etobicoke and persuade all the Mississauga residents to move voluntarily to Etobicoke then the businessmen in Mississauga whose market falls have no complaint and ought to have none.

. . .

Next consider a variation on the facts in *Example* 1. Extra demand is accommodated costlessly so that the only cost induced by the tort is increased risk. Suppose there are two butchers and two bakers in each town. One of each is independent. The other is owned by a chain. Call the chain Loblaws. If we allow recovery to the independent for lost profit, should we allow recovery to Loblaws as well? That would be very strange since Loblaws has suffered no private loss at all. It merely records in one week larger sales that usual in Etobicoke and smaller sales than usual in Mississauga. In effect, Loblaws is a self-insurer against this type of loss.

If there are only independent butchers then, since risks are for them real costs, part of the loss suffered prima facie should be included. If there arc only chain butchers then no part of such loss is a real social cost and prima facie none should be included. But, which rule should be adopted if there are both chain and independent butchers? Here the criterion of "least-cost avoider" or "efficient-risk bearer," as between injurer and victim, does not yield a unique result. Awarding such damages to one and not the other would affect incentives in retailing. In effect, it could operate as a subsidy to small-scale distribution. At the margin it would lead to some inefficient retailing patterns. The possibility of efficicnt self-insurance is one more reason for caution in awarding damages for economic loss. . . .

Suppose next that butchers and bakers have opportunities to avoid losses arising from accidents. They can invest in plant and equipment which can be converted temporarily to some other use when the regular market fails—for example, sharpening tools (for butchers) or baking pies for the frozen-food market (for bakers). If such action can be taken in the case of all types of equipment then it is covered in the present law by the requirement of reasonable mitigation. But if this course is open only to some bakers or butchers, those with appropriate equipment, it is not so covered. If no compensation is offered for economic loss, then each butcher or baker, acting in isolation, has an incentive to make the appropriate investment. But if compensation is offered this incentive is absent. Such capital switching, if it can be invested in cheaply enough, will be advantageous to society. In the long run, it will reduce to a socially optimal level the total capital needed by society to produce sharp tools, pies, baked goods and meat cutting.

If all capital were perfectly and instantly malleable and mobile, then there would never by any case for awarding lost profit. Or more precisely, there would be no lost profit arising from idle resources. All capital is not malleable and cannot cheaply be made so. But refusal to award economic loss gives the right incentives for firms to undertake such efficient investments as can potentially be made.

. . .

Return to *Example* 2 and add the following facts. Jack and Jill work in Mississauga—Jack in the bakery and Jill in the candlestick shop. They are evacuated to Etobicoke. They are compensated for the expense and upset of the move. Jack takes a job in Etobicoke during the week of evacuation. The temporary job pays the same wage but Jack misses his old surroundings and workmates. Jill does not take a temporary job. She very much enjoys her week off despite the lost wages, and she uses the time to knit and to read *War and Peace*.

. . .

Jack would pay something to keep his old job despite the fact that the money wages are equal between it and the new one. This is a real cost to Jack and to society. It should enter into the cost-benefit analysis of people who decide about accident-avoidance expenditure.

Conversely Jill loses her full money wage but has gained leisure. This leisure is valuable. If the tortplanner has to pay this as compensation then he will overestimate the social cost of an accident and spend too much on avoidance. The cost should include only the difference between Jill's wage and the sum she would accept if offered a week off without wages or work.

. . .

It is time to draw some tentative conclusions from this theoretical inquiry. The private cost of an accident as seen by plaintiffs need not correspond to the social cost and, in general, it will not. But what exactly is the social cost of an accident is a very complex matter. The answer depends upon innumerable particular facts of interacting markets. Further, there are many complex incentive effects of liability or nonliability that any decision-maker needs to know before he can decide whether imposing liability is efficient or not. Now, amid all this complication and resultant unclarity, one fact stands out as emphatically clear. Courts are not equipped to make the requisite technical judgment in each individual case. Further, even if they were so equiped, the exercise would be pointless unless each potential tortfeasor in the economy could predict in advance the impact on him of likely court decisions. Probably the best that can be hoped for is a series of rules or presumptions which achieve approximately efficient results on the average and which yield tolerably clear guidance to decision-makers.

The system of rules and presumptions could start from a presumption of recovery and then build in exceptions for types of cases in which recovery should be denied. This is the solution of civilian legal systems where there is no general bar to recovery for economic loss. These systems make liberal use of a doctrine of remoteness to exclude many types of economic loss. On the other hand, the system of rules and presumptions could start from the other direction with a presumption of nonrecovery and then build in exceptions, allowing recovery for certain types of cases.

From the point of view of economic efficiency, it is probably a matter of indifference which system is chosen. The common law has chosen the latter course. . . .

In general, there is some reason to exclude liability for economic loss since it may induce too much avoidance activity by potential tortfeasors. However, there are many exceptions. In the following sections, a number of well-known cases are analysed. They are classified into . . . groups, each carrying a general assumption of nonrecovery . . . or recovery . . . with exceptions or qualifications.

. . .

In *Cattle* v *Stockton Water Works*, where the rule that economic loss cannot be recovered in tort was first stated, Blackburn J justified it by reference to what he regarded as a clear case. He referred to *Rylands* v *Fletcher*, where a mine was flooded and, inter alia, workmen lost their jobs when the mine shut down permanently. To Blackburn J social policy clearly required that the workers should not recover compensation in tort law for their lost jobs. On lost jobs, cost-benefit criteria suggest the following points.

First, where the plant shutdown is permanent there is in general little case for compensation of workmen. Here, the factors of production should be encouraged to move speedily to other uses. Normally the only real social costs generated by the tort are the job-search expenditure of workmen. These are suitable items for compensation.

Second, if the plant shutdown is temporary then workers laid off will probably not be able to find alternative employment. Here, labour is like capital. It is quasi-fixed. Failure to compensate it will in theory lead to workers in risky industries demanding a premium above the wage demanded in nonrisky industries. These costs are real ones caused solely by the risks of plant closure due to tortious acts. Pro tanto these costs should be built into the calculations tortfeasors perform.

Third, in some situations there may be a case for compensating a workman for lost rent. Suppose a man moves from the closed mine to a job as a farm labourer at the same wage. He might still have suffered losses if he had been earning a "rent" in his mining job.

Fourth, even in cases where awarding compensation to employees is efficient, we must be careful to distinguish lost wages from workers' real losses. The remarks made [earlier] about leisure may apply. The value of leisure gained must be offset against wages lost. Where the worker finds another job at lower pay his loss is the reduced income and not his full wage in the former job. Both of these cases are part of, or easily assimilated with, the familiar legal category of mitigation.

Practical problems of evidence and calculations will normally preclude any adjustment for rent, leisure, search cost and risk—the adjudication cost will likely exceed the efficiency gain. In general, Blackburn J was probably right to conclude that a rule against damages awards for lost jobs and wages is the proper rule.

. . .

If all cases should be subsumed neatly under the general principles set out above, then economic loss would be a quiet backwater of tort law. In fact, it is a difficult and contentious area. The reason is that there are many cases in which economic loss *should* be recovered by a plaintiff.

. . .

[A] case in which the awarding of economic loss is required for efficient protection of public resources is illustrated by *Union Oil v. Oppen*. In that case the oil company had negligently spilled oil into the ocean off California, destroying a fishery. The owners of vessels which prosecuted the fishery sued. They were awarded judgment. . . . The California court attempted to apply Professor Calabresi's criteria to the case. It decided that Union Oil was the best briber as it could have bought out the fisherman. . . . Here above all deterrence was called for and ought to have formed the basis of the judgment. . . .

The economic case for recovery of economic loss is overwhelming in such cases. The reason is this. Most resources in our society are owned by someone. They are the subject of property rights. That is normally a necessary condition for their efficient use. A fishery is the classic textbook example of a resource in which private-ownership rights are impractical being too expensive to police. But this detail ought not to blind us to the fact that the valuable real resources are likely to be destroyed if potential injurers have the wrong incentives.

Two caveats are in order. First we might hesitate to award the plaintiff's private losses as damages if the resource is being overexploited—the bane of common-property resources. The ideal solution is a tort remedy plus some other scheme for inducing optimal exploitation of the resource. Second we might hesitate over the identity of the proper plaintiff. Perhaps the government of the state . . . should have the right of action since the proper measure of damages is the whole value of the resource over the period during which it is rendered unavailable for exploitation. There is no reason to award this to the fishermen—save that they may be more reliable enforcers than politicians who may be subject to political pressures. . . .

In this selection a theoretical framework has been developed to analyse the problem of economic loss in tort. The relevant economic considerations are complex and depend upon facts that are usually unknown. It is not surprising that this topic has so long proved so intractable: it is a very complicated subject. . . .

In general the courts, in deciding cases, have reached results that seem broadly efficient. This may seem surprising since explicit judicial reasoning has largely ignored the matters relevant to economic efficiency. But if we remember that an important part of economic theory is at bottom only common sense made systematic then it is far less surprising that judges, feeling their way by vague intuition, have developed rules that seem fundamentally sound from the point of view of economic theory.

Notes and Questions

1. How might Bishop explain the fact that if A injures B, A pays for B's medical expenses and lost earnings but not for the cost endured by B's employer in hiring C to fill in for B while the latter recuperates? Could A possibly collect from C some portion of the gain enjoyed by C as a result of A's injuring B? Do we ever allow such recoveries by a wrongdoer?

2. Consider Bishop's example in which a negligently operated railroad derails, a town must be evacuated, and lost profits in the affected town are offset by increased profits in a neighboring town where the evacuees are temporarily housed. If there is no recovery for "economic loss," perhaps because there is little social loss, what deters the railroad from negligent operation? Are there likely to be other claims against the railroad?

3. The discussion of *Union Oil v. Oppen* implies that recovery is necessary only for the protection of public resources. Is there a more general conclusion that can be drawn by comparing the railroad-evacuation case with *Union Oil?*

4. There is a difficult doctrine in tort law, known as "intentional interference with contract," under which X may sometimes be able to recover from Y when the latter has caused Z to break a contract with X. Why do you suppose recovery can be triggered by intentional action and not by negligence? Might Bishop's explanation of *Union Oil* be seen as falling in this category? What about the railroad example?

Women, Mothers, and the Law of Fright: A History

MARTHA CHAMALLAS WITH LINDA K. KERBER

. . .

Fright, of course, is not gender-specific. Men have suffered heart attacks and other serious injuries from fright, arising both from concern for themselves and for their loved ones. The relevant tort rules have never been explicitly gender-based. Historically, however, women have tended to bring claims for fright-based injuries far more often than [have] men.

Until quite recently, courts and commentators indicated an awareness of this gender disparity, although they lacked a feminist framework in which to assess its significance in shaping the law. The present state of legal scholarship, however, has lost sight of the gendered aspect of fright law.

. . .

In *Spade v. Lynn & Boston Railroad* [an 1897 case] . . . Margaret Spade was a passenger on the defendant's railway when the defendant's employees forcibly ejected two drunken passengers standing beside her. Although plain-

Abridged and reprinted without footnotes from 88 *Michigan Law Review* 814 (1990) by permission.

tiff was actually grazed by one of the ejected passengers as he "lurched" off the car, Margaret Spade admitted that it was the fear caused by the incident that produced her subsequent hysterical paralysis and other ailments.

The *Spade* opinion is notable for its candor. The court did not rely on the pegs of remoteness or absence of proximate cause to deny recovery, but instead focused squarely on the relevant qualities of the plaintiff and defendant. The court believed that only "a timid or sensitive person" might suffer a physical injury because of fright. The court's unwillingness to believe that a normal person would suffer physical harm from fright is much like that of its predecessors. But the court also went beyond prior cases to reach the normative conclusion that the law ought not to be structured to protect the interests of this group of unusually sensitive persons. The Massachusetts court openly expressed its concern that a ruling for the plaintiff would jeopardize "[n]ot only the transportation of passengers and the running of trains, but the general conduct of business and of the ordinary affairs of life." In its view, the "logical vindication" of the impact rule* was that it would be unreasonable to hold a "merely negligent" defendant to pay for even the real consequences of fright.

The image that emerges from *Spade* is one of a hypersensitive plaintiff whose claims pose a threat to business-as-usual. Even though ordinary tort principles allowed recovery only when the defendant was negligent and the plaintiff was free from contributory negligence, the Massachusetts court feared that the application of these ordinary principles in cases of fright-based injuries would impose a disproportionate liability on defendants. This sense of disproportion arose both from the court's view of plaintiff's injury as marginal and from its belief in the social importance of defendant's activity. . . .

Finding Gender in the Impact Rule:
Pre-realist Commentary, 1902–1921

In general, traditionalist legal scholars in the first two decades of the twentieth century were critical of the reasoning and most of the results of the early restrictive cases. Many were unconvinced that there were persuasive reasons for a wholesale denial of recovery for fright-based physical injuries. The liberal attitudes expressed reflected a faith in the ability of medical science to aid the law in this area, with little conscious focus on the gender-related nature of the tort beyond a recitation of the facts of the major cases.

The first widely cited American law review article on fright was written by Francis Bohlen, a University of Pennsylvania law professor who would later become the first Reporter for the Restatement of Torts. The major thesis of his 1902 article, which would be taken up by the courts when they wished to dismantle the impact rule, was that a fright-based physical injury case should

---

*[The traditional "impact rule" allows the plaintiff to collect for pain, suffering, humiliation, nervousness, and so forth only when the defendant has caused an actual impact with plaintiff's body resulting in some physical injury.—Ed.]

be viewed as just another kind of physical injury case, rather than as an attempt to expand recovery to encompass purely mental suffering. By viewing fright as the mechanism by which legally cognizable harm is produced, Bohlen undercut the argument for viewing these cases as conceptually distinct. Most importantly, he predicted no special difficulty with proof of causation, because of his belief that medical testimony could prove the causal connection "quite as accurately" in nonimpact cases as in impact cases. He did not display open mistrust of the credibility of pregnant women and apparently was not worried that this class of plaintiffs would falsely or mistakenly attribute the cause of their miscarriages or other birth-related traumas to the actions of defendants. Bohlen's response to the standard remoteness argument was equally straightforward: it was problematic to view pregnancy as extraordinary or unusual because there was always a certain number of pregnant women in every community. Bohlen's liberal position thus stemmed from an assessment that the logic of the early cases was deficient, coupled with a confidence that medical science could be relied upon to delimit the scope of legal liability. His argument was ultimately reflected in the position of the *First Restatement of Torts,* which repudiated the impact rule. For the next three decades, Bohlen's article would serve as the foundation for many of the traditional critiques of the impact rule.

Another leader of the Restatement movement—Herbert Goodrich—employed science, rather than legal argument, as his principal weapon against the impact rule. Goodrich's 1921 article was unusual because it went beyond traditional case analysis to value scientific thinking over legal precedent in arguing for a broader scope of recovery. Goodrich cast his argument in Darwinian terms. Relying principally on two medical treatises for empirical support, his basic assumption was that fear, properly viewed, is not a "purely emotional thing." He variously described the physical manifestations of fear as putting the body in shape "for fight or flight," as "clearing the decks for action," and as putting the "whole human body on a war basis." He speculated that harmful physical effects from fear are produced when there is no opportunity "to put strong feeling into action." As for the right of legal recovery, Goodrich argued simply that as long as we "can trace and can see" the physical effects of fright, the law ought to afford recovery.

From a feminist perspective, the fascinating aspect of Goodrich's article is the masculine rhetoric in which he generally described human response to frightening situations. From his account of the physiological changes that occur in the face of fear, Goodrich encouraged the reader to believe that fear is something that happens to men. Goodrich compared "[t]onight's passenger in the luxurious Pullman car, awakened from slumber by the . . . cries of the injured in a railroad wreck" to "his prehistoric ancestor, battling bare-handed for his very life with his enemies of the jungle." For examples to support his thesis, he referred to the reactions of soldiers, a urine test done on the players of the Harvard football team, and the responses of businessmen following a decline in the New York stock market. Finally, to make the point that some kinds of emotional stimuli are "reinvigorating" rather than debilitating, Goodrich claimed that "the tired businessman guesses right when he chooses the excitement of a whirly-girly show, instead of going home to bed."

These images of combative, worldly men exposed to frightening or emotionally stimulating events contrast sharply with the image that emerged from the case law of the fragile, pregnant woman who suffers a miscarriage from nervous shock. The only instance in which Goodrich referred explicitly to a female subject was in a hypothetical case based loosely on the facts of *Spade*. He posited that "*A,* in a condition of partial intoxication which slightly interferes with his locomotion, meets a woman on the sidewalk. He does nothing. But she becomes frightened at him, suffers a shock, subsequent miscarriage, nervous prostration, other ills ad lib." In analyzing his own hypothetical case, Goodrich assured the reader that because *A* had not acted negligently, no recovery would be forthcoming regardless of the injuries sustained. He then generalized that given the requirement of proof of negligence, "there seems slight danger that defendants will be placed at the mercy of the hysterical, the morbid, or the emotionally unbalanced."

Goodrich's association of women with unreasonable emotional response is reminiscent of the arguments advanced in support of the impact rule. The interesting feature of the article, however, is that Goodrich generally supported recovery for fright-based injury, at least when the injury was conceived of as happening to normal men. The distrust of claimants surfaced when the injury was feminized and did not fit as well into Goodrich's masculinized account of the human response to fear.

Frightened Mothers: The Legal Origins of the Bystander Rule, 1923–1935

In jurisdictions where the impact rule was in force, the question of whether recovery could be granted to witnesses to accidents did not need to be addressed. With the lifting or softening of the impact limitation, however, plaintiffs other than the primary accident victims were able to bring suit and the courts were forced to consider whether physical injury produced by fright at witnessing the fate of another should be actionable. Like the near-miss cases before them, the early bystander cases involved claims of women. The near-miss cases, however, frequently involved pregnancy-related injuries and thus were linked to women in a biological sense. The relationship of women to the bystander cases was more obviously social in nature, rather than biological. Mothers tended to care for young children and to suffer nervous shock at the injury of their children. The recurring claims of mothers suing as witnesses to the injury of their children may also be a function of an assessment by attorneys that mothers would make particularly sympathetic plaintiffs and would increase the chances of prevailing in this legally difficult cause of action.

The early precedents generally denied recovery if the fright causing the injury were produced by concern for another. This bright-line limitation had the advantage of limiting the class of plaintiffs to those who were close enough to the dangerous instrumentality to be physically endangered themselves. It had the significant disadvantage, however, of discounting intimate family relationships, because the only legally actionable fright was individually based. . . .

The maternal relationship emerged as a decisive factor in the 1925 English case of *Hambrook v. Stokes Brothers,* a rare case permitting recovery to a mother whose child was severely injured in a traffic accident. Mrs. Hambrook had just parted from her children on their route to school when she saw a runaway lorry careening down the hill in the direction of her children. When she learned from bystanders that a little girl had been struck and taken away, she sustained a nervous shock and a severe hemorrhage. Two and a half months later, she suffered a stillbirth and died shortly thereafter.

The court . . . relied instead on its perception of the quality of the mother-child relationship as a justification for recovery. The requirement that plaintiff prove fear for herself as opposed to fear for another struck the court as unacceptable when the "bystander" was the victim's mother. To demonstrate the injustice of the bystander limitation, one judge hypothesized the following scenario:

> Assume two mothers crossing the street at the same time when this lorry comes thundering down, each holding a small child by the hand. One mother is courageous and devoted to her child. She is terrified, but thinks only of the damage to the child, and not at all about herself. The other woman is timid and lacking in the motherly instinct. She also is terrified, but thinks only of the damage to herself and not at all about her child. The health of both mothers is seriously affected by the mental shock occasioned by the fright. Can any real distinction be drawn between the two cases? Will the law recognize a cause of action in the case of the less deserving mother, and none in the case of the more deserving one? Does the law say that the defendant ought reasonably to have anticipated the nonnatural feeling of the timid mother, and not the natural feeling of the courageous mother? I think not.

By thus associating fear for another with maternal instinct, the court explicitly feminized this species of tort claim. The court restated the question as a choice between validating the maternal instinct by permitting recovery or following a "discreditable" system of jurisprudence that rewarded nonmaternal mothers who valued themselves above their children.

Today *Hambrook* may appear sexist because of its acceptance of unselfishness as the marker of natural motherhood. It is no longer assumed that mothers are naturally or biologically courageous. Nor does it seem particularly courageous or desirable to forget about one's own personal safety in a dangerous situation. But what is striking about the *Hambrook* analysis is that it fashioned the rule of the case with women, and the particular plaintiff, uppermost in mind. In contrast to the cases denying recovery, the court did not characterize the injury as unusual or remote, in part because of its view that mothers would naturally worry about their children and suffer fright at their peril. The court expanded the objectivity of the law to encompass the reality of a familial relationship. By placing a high value on motherhood, the court created an adequate counterweight to the fear of disproportionate liability. *Hambrook* indicates that courts may sometimes be gender-conscious in favor of women. It would be almost 50 years, however, before the *Hambrook* arguments would prove powerful enough to turn the tide in favor of recovery in the United States. . . .

From Bystander to Mother:
The *Amaya* and *Dillon* Cases, 1962–1968

Amaya v. Home Ice, Fuel & Supply Co.

The breakthrough case that first permitted recovery for mothers who witnessed their children's injury was a California appellate decision written by Justice Matthew O. Tobriner in 1962. At the time *Amaya v. Home Ice, Fuel & Supply Co.* was decided, the courts seemed to be hardening their stance against recovery in "bystander" cases, except in the rare "danger zone" situation in which the plaintiff was also physically imperiled. . . .

· · ·

Amaya had all the ingredients of the paradigm fright-based injury case. Lillian Amaya watched from a distance of 80 feet as her 17-month-old son was crushed beneath the wheels of a negligently driven ice truck in the driveway of their home. Lillian Amaya was seven months pregnant at the time of the accident and claimed that the trauma of the event produced physical shock and other bodily injuries. She refused the opportunity to amend her complaint to allege that she also feared for her own physical safety and insisted that her only fear was for the safety of her child.

· · ·

Justice Tobriner's ruling for the intermediate appellate court in favor of Amaya was self-consciously reformist. Although he attempted to discount the significance of some of the adverse precedents, Tobriner's basic strategy was to declare that the prevailing view was fundamentally unjust and simply "not consonant with the reactions, or the mores, of the society of today." In this first round of the debate in the California courts on bystander recovery, Tobriner did not elaborate on the contemporary conditions that compelled the conclusion that mothers who witnessed the negligent injuring of their children deserved compensation. He made only a passing reference to the "teachings of psychiatry" to undercut the familiar argument that emotional disturbance is too hard to measure or too easy to fake. For the most part, Tobriner's public policy argument for recovery in *Amaya* was unstated. He proceeded from the assumption that it was "obvious" that a denial of recovery to mothers is unfair, and sought to expose contrary legal precedent as "anachronistic" without stating exactly what had changed.

Tobriner's ruling for the plaintiff in *Amaya* was reversed by the California Supreme Court in a 4–3 decision. In fact, the result was even closer than the split indicated. By the time of the supreme court ruling, Justice Tobriner had himself been elevated to the supreme court. Unfortunately for the plaintiff, however, Tobriner was required to recuse himself from the *Amaya* litigation in the supreme court because he had decided the case in the lower court. The judge sitting in for Tobriner supplied the critical vote against the plaintiff to produce the very shakiest of precedents against recovery.

. . .

By using the term "spectator injury" to describe Lillian Amaya's harm, the majority cast her injuries in an unsympathetic light. Like the term "by-stander," the "spectator" image likens a mother's injury to the emotional response of any witness and places no special weight on her relationship to her child. This aspect of the majority opinion infuriated dissenting Justice Raymond E. Peters. Peters' opinion emphasized that "[t]he plaintiff here is not just anyone. She is a *mother* of a 17-month-old *infant child.*" The split in the court in *Amaya* was between the majority's view of the plaintiff as a bystander and the dissent's focus on the plaintiff as a mother.

Dillon v. Legg

Only five years after deciding *Amaya,* the California Supreme Court over-ruled it in *Dillon v. Legg.* Justice Tobriner wrote for the new majority in 1968. He stressed that the plaintiff was a mother, not an ordinary bystander, who witnessed the negligent killing of her child. For Tobriner, this type of case was "the most egregious case of them all: the mother's emotional trauma at the witnessed death of her child."

The facts of *Dillon* closely resembled *Amaya.* On September 27, 1964, Margery Dillon was sitting on the porch of her home in Sacramento watching her four-year-old daughter Erin cross the street. At the time of the accident, Erin's older sister, Cheryl, was also on the curb nearby. Both Margery Dillon and Cheryl saw Erin struck and killed by a car negligently driven by David Legg. Margery Dillon's complaint alleged that she suffered both physical and emotional injuries from witnessing the death of her child.

The only arguably relevant factual distinction between *Amaya* and *Dillon* was that, in the latter case, Cheryl, but not her mother, was in the "danger zone"—Cheryl was located sufficiently near the accident to have feared for her own safety, as well as the safety of her sister. Under prevailing precedent that allowed recovery only to witnesses who were within the danger zone, Cheryl stood a chance of recovering for her fright-based trauma while her mother did not.

Tobriner would not allow the ruling in *Dillon* to depend on this narrow factual distinction. Instead he flatly rejected the danger zone approach as a "hopeless artificiality" that would produce the anomalous result of granting "relief to the sister . . . and yet deny it to the mother merely because of a happenstance that the sister was some few yards closer to the accident." The new rule announced in *Dillon* was that liability would depend on the more flexible test of whether the accident and the harm were reasonably foresee-able. Tobriner expressly noted that the closeness of the relationship between the plaintiff and the victim should be a key determinant of foreseeability. The old physical danger zone was transformed by Tobriner into a larger zone of emotional danger.

In *Dillon,* Tobriner made it clear that he was an unapologetic judicial activist, shaping the law to accommodate social change. But it was clear also that his was a cautious activism, that his vision of appropriate change emerged from a long historical perspective and from a liberal's skepticism of letting traditional categories get in the way of equitable solutions. It was "indefensible orthodoxy" that Tobriner opposed.

Dillon set a new legal course based on what Tobriner described as the "natural justice" of the mother's claim. "*Dillon* is unique," Tobriner asserted in a reflective essay two years after the decision, "in its emphasis upon the *rights of the victim* as opposed to a historical concern with the violation of a duty of due care." As he worked out his justification for extending the rights of the victim and the scope of the defendant's responsibilities, Tobriner envisioned the victim not simply as Margery Dillon the person, but as Margery Dillon the *mother,* whose presence on the scene of the accident was inextricably linked to her role as parent. The view of the world embedded in Tobriner's opinion is one in which fathers go to work and mothers care for their children, a world in which children are central to their mothers' emotional lives. Tobriner was only narrowly reversed on appeal in *Amaya;* his decision there is not very different from the one that "stuck" in *Dillon.* As a liberal judge writing in the 1960s, he was already deeply committed to expanding the rights of those who experienced loss, and he was willing to open up the categories of legal harm and duty.

Tobriner was conscious of and made use of the gender of the plaintiff in his opinion in *Dillon.* Much of Tobriner's opinion is written in gender-specific language (*mother,* rather than parent) although he ultimately articulated the criteria for recovery in future cases in gender-neutral terms. Tobriner deployed the gender of the plaintiff both to guarantee the genuineness of the injury suffered and to argue for the foreseeability of the injury. His use of gender and maternity, however, was not woven explicitly into his legal rationale for a new doctrine. For example, Tobriner did not explain why it took until 1968 to recognize this obviously just claim for maternal injury. He pointed to no specific changed social circumstances, beyond an oblique reference to "modern medical knowledge" as support for treating fright on a par with physical impact. He drew no connection between the "indefensible orthodoxy" underlying the more restrictive doctrines and societal bias against women. Nor did the fact that Margery Dillon was a divorced parent of limited financial resources who may have had a particularly compelling need for legal recognition of her interests find its way into his opinion.

Rediscovering *Dillon:* A Feminist Perspective, 1990

Neither Margery Dillon nor her lawyers, so far as we can tell, understood their victory in terms of women's rights. It is also unlikely that the judges who dissented in *Amaya* or joined the majority in *Dillon* understood themselves to be responding directly to the gendered nature of the issues in these cases. Subsequent legal commentators have also ignored the role that gender played in *Dillon*.

The recognition of the cause of action in *Dillon,* however, has an ideological dimension that, like the fright precedents before it, is tied to gender. *Dillon* should also be mined for what it can tell us about cultural attitudes in the 1960s, especially about the relationships between women and their children and how those relationships bear on the physical well-being of women. Timing is crucial. "Mother love" surely was not a creation of America in the 1960s. Why then did it take so long for the courts to produce *Dillon?*

. . .

The years between 1960, when an ice truck injured James Amaya, and 1968, when Tobriner handed down the California Supreme Court's decision in *Dillon,* were transitional years in American social history. They marked the end stages of an era that we remember as "the cold war," a period usually dated from 1945 to 1965 or so, in which an overarching demand for security in foreign affairs filtered into a "domestic revival" that offered marriage and motherhood as the only appropriate roles for women. "The cold war consensus and the pervasive atmosphere of anti-communism made personal experimentation, as well as political resistance, risky endeavors with dim prospects for significant positive results," writes historian Elaine Tyler May. In this atmosphere, "[p]rocreation . . . took on almost mythic proportions. . . . [P]arenthood [was] the key not only to responsible citizenship and a secure future but to a personally fulfilling life. . . . Along with the baby boom came an intense and widespread endorsement of pronatalism—the belief in the positive value of having several children." The family, particularly the mother, became a symbol of stability and security.

This ideology was sustained despite the fact that mothers increasingly worked outside the home; in 1960, 40 percent of women over 16 held a job, and one-third of working women were mothers of children under 18. In the face of changing economic reality, many elements in American society sought to dissuade mothers from thinking of themselves as independent from their children and thus to preserve the cult of domesticity. In newspaper accounts, even women of substantial professional accomplishment "were carefully described as mothers first." In a social context in which women's identities were understood to be motherly and domestic, even women's political action had to find a domestic language in which to express itself. As Amy Swerdlow's study of Women Strike for Peace has shown, when thousands of women wanted to protest atomic bomb testing in 1961, they found that the most effective strategy—indeed, the only effective strategy—available to them was to organize as mothers who feared for the health of their children and who worried particularly about strontium-90 in milk. Any other stance was vulnerable to being attacked as unwomanly, un-American, and pro-communist.

The valuing of children and the privileging of the maternal relationship that pervaded the domestic ideology of the cold war period might well have reinforced Lillian Amaya's and Margery Dillon's sense that they personally had been wronged by harms to their children and strengthened their decisions to seek compensation for their own suffering. Clearly the lawyers for Amaya and Dillon emphasized the special "maternal" nature of the injuries their clients suffered. Lillian Amaya's complaint, for example, underscored her role as mother and mother-to-be: she alleged that at the time of the accident, she was "in a state of pregnancy that had advanced to approximately seven months; that the natural maternal bonds of affection existed between plaintiff and her infant son." At the oral argument before the California Supreme Court, Margery Dillon's lawyer spoke of the "traditional classic concept of motherhood . . . which we all admire and look up to." It was thus in their roles as *mothers* that Lillian Amaya and Margery Dillon came before the courts.

The *Dillon* cause of action signifies that the law regards a mother's anguish at witnessing the death or injury of her child as a harm that qualifies for legal protection. This recognition of the claims of mothers was congruent with the deeply domestic political ideology of cold war society.

The *Dillon* opinion was also congruent with a traditional view of women as the normal caretakers of their young children. The court refuted the defendant's contention that a mother/witness was an unforeseeable plaintiff in child pedestrian cases by stating what it thought to be the obvious: mothers were often likely to be found in the vicinity of their young children. The opinion also revealed a sentimental view of maternity. In criticizing the requirement that recovery must be predicated on the plaintiff's fear for her own personal safety, Tobriner commented on the example in the *Hambrook* case of the selfish mother who fears only for her own safety and the selfless mother who cares only for the safety of her child. Tobriner thought that it would be "incongruous and somewhat revolting" to let the selfish mother recover and deny recovery to the selfless mother. The normative judgment that mothers should place the well-being of their children ahead of their own comes out clearly. The image of the "good mother" emerges; it is the good mother who is afforded legal relief.

It is also clear that the *Dillon* judges were writing in the intensely political year 1968, toward the end of a decade in which both the civil rights and anti-war movements had been conducting a national teach-in, as it were, on the topic "the personal is political." The Presidential Commission on the Status of Women, chaired by Eleanor Roosevelt, had been established in 1961. Betty Friedan's manifesto, *The Feminine Mystique,* had been published in 1963, one year after the *Amaya* decision was handed down. When *Dillon* was decided, a two-year-old National Organization of Women had articulated a call for an equal rights amendment, for access to safe, legal abortions, and for the enforcement of antidiscrimination legislation. In 1968, feminists picketed the Miss America pageant. If the rhetoric of the *Dillon* opinion looks back to the privileging of motherhood and the superficial separation of spheres that characterized the 1950s, the result in *Dillon* is also congruent with the feminist movement as it would grow in the five years following 1968, culminating in the

Roe v. Wade decision of 1973. Despite the sentimentality for motherhood it encompassed, *Dillon* pushed against the marginalization of recurring injuries in the lives of women and gave women a claim of legal right. A decision that had the concrete effect of putting money into the hands of women through the redistributive mechanism of tort liability may be understood to serve women's interests.

From this perspective, *Dillon* deserves to be placed with other rights-affirming precedents of the last two decades that have been particularly important to women. The most familiar of these legal developments are changes in public law, particularly expansion of the constitutional notion of equality that has been employed since 1971 to restrict explicit sex-based discrimination. Tort law has also been part of this movement. For example, the first recognition of a cause of action in tort for "wrongful birth" came in 1967, a year, before *Dillon* was decided. This claim allows women to hold their obstetricians accountable for unwanted pregnancies that were caused in part by the physician's negligent treatment or diagnosis. The wrongful birth cause of action complements the *Dillon* claim in that it gives legal recognition to the interest women have in their relationship to their unborn children. The wrongful birth cause of action pressures health professionals to make sure that a pregnant woman (or a woman contemplating pregnancy) is advised of any condition that might affect her own well-being or that of her unborn child. By enlarging the physician's duty, this cause of action tends to expand the notion of women's health to include reproductive health and health of offspring.

This account of *Dillon* also tends to validate women's expanded role outside the domestic sphere. The legal/market valuation placed on women's fright-based injury has the effect of mainstreaming the injury. *Dillon*'s willingness to protect pregnancy and maternal interests outside the narrow confines of the home contrasts with the Victorian view that these activities had their proper site *inside* the home.

As the fright-based injury is deprivatized, the law regards it as less unusual, less remote, and more deserving of protection. The "reasonable man" of tort law is redefined to encompass the mother. When a mother's fear for her child is acknowledged as a cause of her own physical harm, we can glimpse the beginnings of a feminization of tort law. Relational interests become a constituent feature of one's own physical integrity. This expanded notion of physical harm resonates in women's experience: in the physical and social experience of pregnancy and in the socially constructed experience of motherhood.

· · ·

That recognizing difference may lead to marginalization, while ignoring difference may lead to inequitable results, has long been the Scylla and Charybdis of feminist theory. A major goal of feminist theory is to find a route past these monsters: first, by being skeptical of conceptual dualisms enshrined in familiar cultural and legal practice, and second, by unmasking claims of difference to reveal unstated norms against which difference is judged. Dualisms are by their nature restrictive. Indeed, the history of women has been confined by dualisms—home versus market, household versus state—as has the history of the relationship *between* men and women—public versus private, culture versus nature, defender versus protected. These dualisms have traditionally been expressed in the language of "separate spheres," a series of metaphors that impose dichotomy on understandings of relationships and interactions. To speak of "women's culture" and "men's culture" or, as de Tocqueville did, "two clearly distinct lines of action" for the two sexes is to leave unexamined the interrelationship of the activities included in each collective noun, constructing a vague and unspecified array of allegedly fundamental differences between women and men. Men and women have been enmeshed in a system of meaning and law that, while aiming at equity, discounted and failed to recognize claims associated with women and their interests. To be locked into a series of dualisms that pose men against women and equality against difference inhibits our ability to establish context and to perceive hidden relationships of dominance and subordination.

Notes and Questions

1. Some courts have continued to abide by the impact rule and have rejected the decision in *Dillon v. Legg*. One reason given for favoring the rule is that it screens out many false claims. Remarkably, a number of courts that reject the *Dillon* innovation do permit plaintiffs to recover for the anguish caused by witnessing property damage. The impact requirement may thus be satisfied by impact with person or property. Does this extension screen out false claims? Is there another explanation for these decisions?

2. The final argument in this selection emphasizes a tension that characterizes thinking about biases in the law. If the law ignores an injury that is likely to be borne disproportionately by women (or another group), then one is inclined to celebrate evolution of the law in the direction of legal recognition of (and recovery for) this injury. On the other hand, if the law had long recognized this injury, it could be said that it demeaned women by insisting that they were more easily injured or in greater need of protection than men. In such a case one might be inclined to celebrate reform in the opposite direction. Are these (hypothesized) reactions necessarily inconsistent?

3. What other doctrines in tort law might be described as reflecting gender bias? The concept of the "reasonable man" is considered in part VI, but there are other tort concepts, such as assault and battery and lost earnings, that impact men and women differently. If the tort doctrines we know reflect a male-dominated world, can you imagine a world with different doctrines?

4. Do you expect the doctrinal evolution described in the reading to extend to fathers who claim emotional injury? Can you separate the mother-father distinction from the claim of gender bias?

IV

The Lawyer's Role

Tort law is not self-enforcing. A properly functioning, fault-based system requires actors to expect that their behavior will be judged impartially and fairly accurately after the fact of injury. The well-known Learned Hand test, arising out of *United States v. Carroll Towing Co.,* 159 F.2d 169 (2d Cir. 1947), is generally thought to call for a cost-benefit analysis in which defendant's actions are evaluated in light of the probability that injury should have been expected, the likely severity of the injury, and the cost of precautions. The economic nature of this test is emphasized in a classic article by Richard Posner, A Theory of Negligence, 1 *Journal of Legal Studies* 29 (1972). These probabilities and costs are often difficult to assess (even after the fact) and a fair amount of tort law is concerned with the lawyer's role in the fact-finding process.

One of the many thorny problems associated with the idea that defendants are liable if they have not taken cost-justified precautions is that the proper scope of the inquiry is often unclear. Thus a plaintiff injured by a handgun might try to show that the weapon could have been better made in some specific way (at low cost to the defendant) or instead might try to show that the aggregate costs imposed by handguns (or a brand of gun) exceed the benefits derived from their manufacture and use. The latter strategy is by no means novel; something like it must explain why the person who tries to manufacture a bomb or airplane in a backyard would quickly be held liable when an injured neighbor sues after an errant explosion or surprise landing. On the other hand, one who is injured in an automobile accident is unlikely to prevail in court with a strategy designed to show that the defendant was not cost-justified in purchasing a car or in driving on the day of the accident. It is apparent that this plaintiff needs to point more specifically to the defendant's negligence. The question of a defendant's "untaken precaution" is the subject of the first reading.

At a more general level, potential negligence will be discouraged (and most victims of negligence will be compensated) only if a large proportion of deserving plaintiffs (under the prevailing liability and damage rules) bring suit. In this, as in other areas of law, the question arises whether plaintiffs find it worthwhile to secure adequate legal representation. In the second reading, Professor Donohue casts new light on the comparison between a regime in which each party pays its own legal fees and one in which the winner of a lawsuit emerges whole because the loser is required to pay the winner's legal costs. The former regime is the American norm and includes situations in which plaintiffs secure representation by agreeing to pay their attorneys a predetermined fraction of any recovery. Such contingency fees, as they are called, are somewhat constrained by rules which do not permit lawyers to take too great an interest in their clients' claims. It is generally impermissible for a lawyer to accept a 100 percent (or even a greater than 50 percent) contingency fee, and there are separate bans on payments from lawyer to client. Thus it is generally forbidden for a lawyer to pay some amount to a client before trial (perhaps on the condition that the client testify in the lawsuit regarding injuries suffered at the hands of the defendant) in return for the right to everything awarded in court. The final reading questions the wisdom of the prevailing regime.

Untaken Precautions

MARK F. GRADY

Introduction

"Untaken precaution" has not yet become a term of art in negligence law. While the idea lurks about the edges of negligence theory, most of the modern scholarship on the subject has been devoted, not to the untaken precaution, but to an extended analysis of the general Learned Hand formula stated in *United States v. Carroll Towing Co.* By this orthodox theory of negligence, the critical inquiry is whether the defendant has chosen a level of precaution that globally minimizes social cost. Yet framing the inquiry in this fashion does implicit violence to the way lawyers and judges, in fact, try and decide negligence cases. Of course, the concern with cost and benefits necessarily remains. Nonetheless, . . . the emphasis in practice shifts from the global to the specific. The key question that courts ask is what particular precautions the defendant could have taken but did not.

My objective in this selection is not to demonstrate that the untaken-precaution approach is economically superior, only that it is more descriptive of legal doctrine than the competing idealization. . . .

Breach of Duty

In a specific negligence case, it is the plaintiff's burden to suggest untaken precautions that serve as the basis of his case. For instance, in *Cooley v. Public Service Co.*, the plaintiff suggested two untaken precautions that would have prevented the harm she sustained from a loud noise that came over her telephone wire: (1) placing a wire mesh basket beneath the defendant power company's wires where they crossed the telephone company's wires, and (2) insulating the wires. The court did not think that either would be a breach of duty, mainly because both precautions would have increased the risk of electrocution to those walking on the street below. The court noted that some other untaken precaution might satisfy the breach of duty requirement, but added that it was the plaintiff's burden to suggest it.

The practice of the plaintiff relying on particular untaken precautions arose very early in American law, and it is still the basic feature of negligence analysis. Events do not define what negligence analysis will be the case, and the court does not define it either. Instead, by selecting an untaken precaution

Abridged and reprinted without footnotes by permission from 18 *Journal of Legal Studies* 139 (1989). Copyright © 1989 by The University of Chicago Press.

on which to rely, the plaintiff defines the analysis that everyone else will use—including the defendant, the court that will try the case, and any court that may hear an appeal. When the defendant has been using less than due care, typically several untaken precautions, just as other inputs have substitutes. The plaintiff makes his choice from among a group of possible contenders and frequently alleges several untaken precautions in the alternative, just as in *Cooley.*

. . .

The orthodox economic theory assumes that a precaution is required or not, based on its potential to reduce the probability of the type of accident that actually occurred. Nothing could be farther from the truth. Piggy-backing on a risk to someone other than the plaintiff can be a highly effective strategy in negligence litigation, especially when the plaintiff's accident has been unusual.

The general rule is that any risk can be included in the calculations as long as it would be reduced by the untaken precaution in question and as long as it was foreseeable. *Polemis* is a famous example of judicial recognition of the risk-adding principle. In that case, the justices ultimately agreed that the risk in question (fire) was unforeseeable from the vantage point of the untaken precaution, but they also agreed that there was a breach of duty. The orthodox theory gives no account of how there can be a duty to take precaution against an unforeseeable risk. (The answer is that courts add risks.) Lord Justice Scrutton said that the duty to take precaution arose not from the possibility of what *did* happen (fire), but from several other risks, all different. "In the present case it was negligent in discharging cargo to knock down the planks of the temporary staging, for they might easily cause some damage either to workmen, or cargo, or the ship [by denting it]."

It is not difficult to see why the orthodox economic theorists have been reluctant to accept the idea that courts add different, reasonably foreseeable risks that are all reduced by the same untaken precaution. For courts to determine the combination of precautions that minimizes the social cost of one risk would be a large task (if there is no presumption that customary precautions do so). It would often be Herculean for courts to determine what combination of different precautions minimizes a social cost that consists of a constellation of different foreseeable risks. Frequently, . . . courts would not have enough information for the task. Yet it is quite conceivable that courts are adding different risk reductions when looking at the benefits from a single untaken precaution, because the whole inquiry is thereby rendered much simpler. That courts do add different risks is neatly demonstrated by Judge Hand's decision in *Carroll Towing,* the centerpiece of the conventional economic theory of negligence. . . .

Cause in Fact

The most common test of cause in fact is whether the harm would have occurred *but for* the defendant's failure to have taken the untaken precaution that constituted the breach of duty. In other words, viewed ex post, would the untaken precaution have *prevented* the accident? Indeed, in all situations except concurrent sufficient causation (for example, two converging negligently started fires), whether the untaken precaution would have prevented this harm is *the* test of cause in fact. Where the precaution identified would not have prevented the harm, the plaintiff can never recover.

Some writers have treated actual causation as a fact of nature, but this formulation ignores the plaintiff's creative role in specifying the breach of duty and the way in which the cause-in-fact issue depends on what breach-of-duty choice the plaintiff has made. While analysis of the breach-of-duty issue has an ex ante perspective, the test of cause in fact is whether the same untaken precaution, viewed after the accident, would have prevented it.

Carroll Towing itself illustrates how these two perspectives on the same untaken precaution can lead to one element being satisfied when the other is not. Under Hand's analysis, there were successive (that is, divisible) harms: the first, when the barge *Anna C* broke away and collided with other boats in the harbor, and the second, when it sank. Surprisingly, given the way the case is commonly understood, Hand concluded that the absent bargee's employer was *not* liable for the collision damage that immediately followed the breakaway. True, his famous language established that there was a breach of duty with respect to collision risk, because the burden of having a bargee aboard was less than the probability that the *Anna C* would break away multiplied by the gravity of resulting harm if she did. Nonetheless, as to the collision damage, there was no cause in fact because Hand concluded (in a little-heeded passage) that, if the bargee had been on board, the most he could have done would have been to protest to those who were moving the barges; there was no evidence that these people would have paid any attention to him. Judge Hand appears to have inferred that the harbormaster and his helper were very strong-willed individuals. Ex post the accident, having a bargee on board could not be seen as preventing the breakaway, and therefore this untaken precaution was not a cause in fact of the damages to the barge that came from the initial impact.

The question for Hand then became whether the absence of the bargee was a cause in fact of the sinking. Here, the answer was yes, because if the bargee had been aboard at the time of collision, he would have realized that the barge had developed a hole below the waterline. He both could and would have called for help, which was available and would have arrived in time to repair the barge before it sank.

It is instructive here to contrast the ex ante analysis of breach of duty with the ex post analysis of causation. The former necessarily focuses dominant attention on the usual and predictable types of accidents and is amenable to the identification of multiple risks, all of which may be reduced or eliminated by a single untaken precaution. In doing his breach of duty analysis in *Carroll Towing,* Judge Hand was probably thinking of the many different kinds of breakaway that could have been avoided if the bargee had been on board. In all likelihood, he thought that the dominant risk of breakaway was from collision. . . . At the breach-of-duty phase of the analysis, the harm from the barge sinking is a bit player—a risk reduced by having the bargee on board, but of lesser magnitude than collision.

Cause in fact, because of its reverse point of view, focuses on the oddity of the actual event, and in *Carroll Towing* the bit player becomes a costar as the drama progresses. Viewing the accident ex post, having a bargee would not have prevented the type of accident that mainly renders this precaution useful, because there were special circumstances (a willful harbormaster). Nonetheless, in the actual event, this precaution would have prevented a type of accident that it usually has no occasion to prevent, namely, sinking. Therefore, oddly, there was no liability for the collision, but there was for the sinking.

Judge Hand's real analysis in *Carroll Towing*—as opposed to its common idealization—illustrates not only how two parts of negligence analysis differ in the way the untaken precaution but also the key role played by multiple risks in most negligence cases. . . .

When courts make the same untaken precaution do dual service under rules that have opposite perspectives, they create an obvious tension. From a breach-of-duty standpoint, the plaintiff may wish to allege the defendant's most trivial failing because such precautions are often highly cost-effective if only they are taken. An example would be the failure to give a signal. Nonetheless, if the defendant chooses a precaution that is too trivial, although he may maximize his chance of prevailing on the breach-of-duty issue, he may seriously hurt his chances of prevailing on the cause-in-fact element. When the court shifts its inquiry to cause in fact, the plaintiff is bound by the untaken precaution selected in the duty phase of the trail. Now the question is whether this duty, if performed, would have prevented this particular harm that the plaintiff has suffered. Too often an obvious breach of duty will turn out to have no causal relation to the injuries suffered. . . .

Contributory and Comparative Negligence

The untaken precaution occupies a similar role when the attention shifts from the plaintiff's prima facie case to any defense based on the plaintiff's contributory negligence. As with the defendant's negligence, contributory negligence is not a unitary defense but consists of three separate elements. The plaintiff's breach of duty (owed to the defendant to take care of himself), cause in fact, and proximate causation—both as a matter of directness and of foresight— are all separate elements that must be proved. But again, through careful selection of the untaken precaution, the defendant can increase the chances of success. In *Berry v. Borough of Sugar Notch,* the plaintiff's untaken precaution was driving his streetcar too fast, and the defendant's was failing to cut down a rotten tree, which, in the actual event, scored a direct hit on the plaintiff. As the court reasoned, because the plaintiff's speeding did not increase the probability of a direct hit, his contributory negligence was not the proximate cause of his injury. Perhaps the defendant should have argued that if the plaintiff had been keeping a better lookout, he would have seen the tree toppling in enough time to get out of the way. Under this theory, the defendant's contributory negligence defense would not have been defeated by proximate cause.

The direct-consequences doctrine of proximate cause also applies to the plaintiff's negligence, but it generally goes by the name of "last clear chance" in this setting. Usually, it is the defendant's negligence, not some third party's, that intervenes between the plaintiff's negligence and his harm. In order to avoid liability under last clear chance, the defendant's best strategy is similar to that followed by the plaintiff in *Brauer*. The defendant picks an untaken precaution that was available to the plaintiff as late as possible—later than the defendant's own untaken precaution, or at least simultaneous to it. In this way, the plaintiff cannot argue that the defendant's negligence (the defendant's last clear chance) intervened between the plaintiff's negligence and the injury.

The plaintiff's strategy is, of course, just the reverse. He wants to find an untaken defendant's precaution that is later than his own untaken precaution. If the plaintiff succeeds, the defendant will be seen as having the last clear chance. Many negligence lawsuits turn into disputes over whose untaken precaution was last. A different strategy for the plaintiff is to find an untaken precaution that the defendant deliberately—as opposed to inadvertently— failed to take. . . .

Notes and Questions

1. The reading emphasizes the role of "specific" precautions as opposed to "global" actions. Can you think of any cases where the plaintiff's best argument is about the defendant's global behavior rather than specific precautions not taken?

2. Imagine that a plaintiff's attorney argues that the defendant should be liable for failure to provide a steel guard-rail because such a precaution, costing $5,000, would not only have prevented the plaintiff's car from leaving the road but also would have prevented other serious accidents for a total expected benefit of $6,000. The defendant's counsel responds that perhaps a wooden guard-rail should have been in place because, at a cost of $2,000, it would have provided an expected benefit of $4,000, whereas the extra $3,000 cost of a steel rail was not cost-justified because it would only have provided an additional $2,000 in expected benefits. In fact, a wooden guard-rail would not have prevented this particular plaintiff's damage. What result? The underlying problem of marginal precaution-taking is discussed in Richard Epstein, *Cases and Materials on Torts* 164 (5th ed. 1990).

Opting for the British Rule,
or If Posner and Shavell Can't Remember
the Coase Theorem, Who Will?

JOHN J. DONOHUE III

Consider a hypothetical involving the decision to settle or try a lawsuit. The plaintiff has brought a tort action against the defendant, and the parties have agreed to attempt to settle. Will the legal rules governing the assignment of trial costs affect the likelihood of settlement? In other words, will the parties be any more likely to settle the case if their suit is pending in a jurisdiction operating under the American rule, under which parties must bear their own legal expenses, rather than in one operating under the British rule, under which the losing party must pay the legal expenses of both parties?

There is now an immense literature analyzing this question. In his first foray into this field, Judge Richard Posner concluded that the greater risk associated with the British rule—parties can win more or lose more under this rule than under the American rule—would lead risk-averse litigants to settle more cases. Subsequently, Judge Posner and Professor Steven Shavell developed a model of the litigation decision to settle or try a case that produced the opposite result: the British rule would *reduce* the likelihood of settlement. But Posner and Shavell reached this conclusion because they neglected the analysis outlined by Ronald Coase thirty years ago in his path-breaking article, *The Problem of Social Cost.* Indeed, despite the number of economists and law and economics scholars who have exhaustively examined this question, not one has realized that the Coase theorem would govern under the assumptions that motivate the Posner and Shavell analyses.

. . .

*[It is plain that if the plaintiff's legal costs are greater than the defendant's, then there almost surely will be suits than the plaintiff will bring only under the British rule. On the other hand, a risk-averse plaintiff is more likely to be dissuaded by the British rule. If the defendant's costs are likely to be greater than the plaintiff's, then the British rule is likely to discourage more suits. It is thus hard to say which rule encourages more lawsuits.

There is also the question of how settlements will be affected by the rules allocating legal fees. We need to think of legal costs that are borne pretrial (so that if plaintiff P drops the suit, defendant D may be able to collect costs) and

---

*[Bracketed material is a condensation of the original article's arguments—Ed.]

Abridged and reprinted without footnotes by permission from 104 *Harvard Law Review* 1093 (1991).

then those that are run up by going to trial. It is these latter costs that can be saved through settlement.

Imagine a dispute over $1,000, with costs of going to trial amounting to $250 for each side. And imagine that P thinks she has a .8 chance of success at trial, but D thinks that P has a .35 chance of success. P is relatively more optimistic. With the American rule, the expected benefit to P of going to trial is .8($1,000) − $250 = $550; D's expected costs are .35($1,000) + $250 = $600. A settlement of between $550 and $600 will make both better off. (P expects less than D expects to pay.)

Under the British rule, expected benefit to P is .8($1,000) − .2($500) = $700. The expected cost to D of going to trial is .35 ($1,000 + $500) = $525. There is no settlement range. P wants at least $700, and D would rather go to trial than pay that amount. One might therefore argue that there is an important range of expectations where settlement is more likely under the American rule. If we switched to the British rule (with our cultural taste for litigation, and so forth) we might expect fewer settlements.

But this analysis, advanced by Professors Posner and Shavell, has ignored an important lesson from Coase's 1960 article. [See part I]

Introducing the Coase Theorem

. . .

One cannot simply examine the effects of legal regimes without considering the ability of the affected parties to reorder the apparently fixed environment. Regardless of which legal rule nominally applies, the parties can assess the expected benefits and costs from litigation under the two rules and agree to be governed by the one that is more favorable to them. If American-rule litigants find that they would be better off with a British rule, they can simply agree to shift fees to the losing party should the case go to trial. [Thus, in the previous example, P prefers $700 (the expected result in a trial with the British rule) to $600 (the maximum settlement D will pay) and to $550 (the expected result in a trial with the American rule). And D prefers to pay $25 (what D expects after a trial with the British rule) rather than $550 (best settlement) or $600 (after a trial with the American rule). Thus even in the United States, the parties can and should bargain to litigate under a fee-shifting (British-rule) arrangement.

It therefore seems that] . . . the nominal legal rule will not affect the rate of settlement. Any time a case would settle under the American rule but not under the British rule according to the Posner/Shavell analysis, the parties would be better off *litigating* the case under a self-imposed British rule. This result may seem counterintuitive, but it simply follows from the Coase Theorem. . . . Of course, one can immediately raise a number of objections to the preceding analysis

. . .

Risk Aversion. It is well settled that the British rule is more risky for litigants because the range of outcomes—both good and bad—is greater for both parties. If things go well for the plaintiff under the British rule, he can costlessly achieve his rightful award, whereas if things go badly, he receives nothing but bills for both his own and the defendant's legal expenses. Under the American rule the need to pay (only) one's own legal fees dampens the joy of victory but moderates the agony of defeat. Similarly, a defendant can walk away from British-rule litigation unscathed if she is successful at trial, but bears a larger burden in the event of defeat at trial than she would bear under the American rule. Risk-averse litigants will see the greater riskiness of the British rule as a disadvantage and one that encourages settlement.

My guess is that the effect of the greater riskiness of British-rule litigation is more powerful than some have assumed and that therefore bald claims such as "making the losing party pay the winning party's attorney's fees would reduce, not increase, the settlement rate" are quite likely to be flawed because they fail adequately to acknowledge the force of risk aversion. But risk aversion merely implies that the expected utility from going to trial under the British system is lower than suggested by the preceding analysis, which was based on risk neutrality. Thus settlements would occur under the British rule more often than the Posner/Shavell model indicates. Two conclusions follow: first, risk aversion will tend to diminish the difference between settlement rates under the American rule and the British rule in the static Posner/Shavell model; and, second, the greater similarity in the settlement rates under the two regimes reduces the number of times that parties will be tempted to opt for the British rule. In other words, the Coasean outcome is unchanged by risk aversion: regardless of the legal rule, parties will assess their preferred fee-allocation standard and shift to it if it is not already in place. Risk aversion will increase the proportion of settlements when compared to the case of risk neutrality, but, for any given level of risk aversion, the settlement rate in the absence of transaction costs will be the same under the British and American rules.

Rising Litigation Costs. Another objection to the standard Posner/Shavell analysis is that the greater riskiness of the British rule impels parties to invest more resources in litigating claims under that rule. This phenomenon has the identical effect as the introduction of risk aversion. It will induce more settlements under the British rule and make the British rule less attractive to parties in an American-rule jurisdiction. But it will not change the fundamental Coasean invariance prediction. The parties will still assess which rule they prefer and move in that direction if they are not at their preferred rule. The higher litigation costs will change the outcome vis-à-vis a model in which litigation costs are identical under the two regimes, but it will not make the rate of settlement dependent upon whether the litigation is taking place in a British-rule or an American-rule jurisdiction.

The Existence of Judgment-Proof Parties. Some losing parties with meager resources will be unable to bear the costs of any fee award under the British rule. Once again, this will not undermine the Coasean invariance prediction. Instead, the expected benefit to the plaintiff and the expected cost to the defendant under the British rule will be different from the level expressed above. In general, the size of the bribe to settle will be increased—leading to more settlements—because the optimism of the litigants is less likely to be converted into a belief that "someone else will foot the bill." As we saw in the previous two subsections, however, the parties will still be able to assess whether they are better off with one rule or the other in light of all of the information known to them. Their assessment will not depend on whether they reside in a British-rule or American-rule jurisdiction, and therefore their ultimate choice of which rule to use will be unaffected by the initial legal rule.

Implications for the Coase Theorem and Legal Scholarship

By now the reader should be convinced that, contrary to the assertion that the adoption of the British rule would lead to a fall in the settlement rate, the Posner/Shavell model implies that the settlement rate should be unaffected by the legislatively determined fee-allocation rule as long as the parties are free to adopt, through private contracting, their preferred cost-allocation rule. Simply because the Posner/Shavell economic model, when properly interpreted, implies that the Coasean invariance prediction will apply does not necessarily mean that the world will conform to the theory. Of course, we would expect departures from the invariate prediction because of the presence of positive transaction costs. . . .

The added negotiation costs that are incurred when side payments are needed will prevent some of the beneficial rule-shifts that fall into the second category from being realized. But it should be noted that the process of agreeing upon the appropriate side payment is identical to the customary task of settling a lawsuit, and therefore the negotiation costs should be roughly of the same magnitude. If the costs are prohibitively large, one might expect never to see agreements that require side payments, but one would also be unlikely to see settlement agreements at all in the typical case in which no rule-shift was contemplated. Because settlement is common, it is unlikely that the costs involved in negotiating a side payment are great enough to inhibit all rule-shifts that require such payments. Transaction costs may keep the zero-transaction-cost invariance prediction from applying, but they are unlikely to explain why there are *no* examples of rule-shifts.

This discussion leads to a very interesting puzzle: in light of the benefits of rule-shifting, why has there *never* been a case in which parties in the course of litigation contracted for a rule different from the one bestowed by the prevailing practice of the jurisdiction? Clearly, transaction costs will often keep parties from always reaching the otherwise optimal fee allocation rule, but it strains credulity to think that transaction costs alone could explain why rule-shifts in the course of litigation are entirely unknown. The answer must lie elsewhere.

Legal or Ethical Constraints

Because there is no evidence that anyone has attempted to shift fee-allocation rules during the course of litigation, no cases address the ethical or legal propriety of such behavior. Although no ethical constraint seems to exist, perhaps such a rule-shift would be struck down as contrary to public policy. The American rule represents an implicit legislative judgment that parties in litigation should bear their own legal fees, and perhaps courts will not allow them to overturn this judgment through private contract. Moreover, because the British rule involves an element of a gamble in that fee awards depend on the outcome in the case, the existence of state anti-gambling statutes may buttress the public policy against opting for the British rule.

There are three reasons why this argument cannot account for the complete absence of cases in which parties in mid-litigation contract for a rule-shift. First, the legal force of the public policy argument is highly tenuous. As noted above, parties frequently contract prior to litigation, with judicial and legislative approval, that disputes arising out of commercial and residential leases, mortgage agreements, and commercial loans will be adjudicated under a British rule. It is unlikely that a court would find such pre-litigation rule-shifting permissible but deem a mid-litigation rule-shift impermissible; the degree of gambling involved in both cases is identical.

Second, to the extent that the public policy argument against rule-shifting rests on an aversion to the gambling component of the British rule, this argument does not explain why there have never been any mid-litigation shifts from the British rule to the American rule. Such shifts would eliminate the gambling dimension, because both parties would be responsible for their own legal fees regardless of the case's outcome. As a general matter, explanations of why shifts from the American rule to the British rule have not been attempted are inadequate if they do not simultaneously explain why the reverse shifts have not occurred.

Third, it is unlikely that doubts about the enforceability of a contract on this basis would be sufficiently compelling to stop all experimentation in rule-shifting. When parties perceive possible profit opportunities, they have an incentive to test the limits of any potential legal impediment. Even when these apparent legal barriers are quite strong, the profitable activity will often be undertaken, thereby generating litigation. Indeed, the objection on public policy grounds seems so subtle that it might even be overlooked by the parties. Accordingly, the public policy argument seems too fragile an obstacle to have prevented every rule-shift. . . .

The Failure to Maximize Wealth or Welfare

This selection has argued that, if the Posner/Shavell model of the settlement decision is correct, litigants will expect to profit in certain cases by opting for their preferred rule of fee allocation. But the parties cannot act to circumvent the legislative rule unless they realize that this is a profitable possibility.

Because no other theory explains the complete absence of mid-litigation rule-shifting, the best explanation may be that no litigant has ever recognized this contractual opportunity.

But this conclusion rips at the heart of one of the core beliefs of the Chicago school—that profitable opportunities are never left unexploited for more than the shortest time frames. If the Posner/Shavell framework is a sound model for analyzing the decision to settle or try a case, the belief that "what is, is optimal" simply collapses. "What is" may be the best that can be done given the current state of knowledge, but only because profitable opportunities are being widely overlooked. But this means that Cosean outcomes might not be achieved even when transaction costs do not stand as a barrier and that unprofitable practices might persist for years, decades, perhaps centuries, without being driven from the marketplace.

How could this be? One might think that one litigant somewhere must have thought of the possibility of rule-shifting if it is indeed a potentially profitable strategy. Furthermore, if individuals and firms are truly utility and profit maximizers, they must take steps in pursuit of their best interests. Perhaps, however, the answer to these assertions is that they disregard the distinction between maximizing one's welfare (or wealth) while accepting one's environment. The human mind finds it far easier to make the best out of the current state of the world than it does trying to conceive all of the ways in which the state of the world itself can be altered. Indeed, this distinction is analogous to the point that Coase deemed most significant in his 1960 article: one cannot consider only marginal changes and hope to maximize welfare; one must compare the total welfare under differing social or legal arrangements. Both Posner and Shavell, along with their supporters and critics, have focused on the marginal analysis and overlooked the crucial comparison between welfare under the two different—and always available—regimes.

The very fact that the rule-shifting option eluded Posner, Shavell, and all of the other law and economics scholars working in this field buttresses the claim that a profitable opportunity has indeed been overlooked by litigants. Of course, the supporters of Dr. Pangloss might rush to explain that Posner and Shavell would have nothing to gain from spotting the applicability of the Coasean invariance prediction, whereas litigants and other market participants have real money at stake, making it impossible for them to overlook profitable opportunities for long. Therefore, they would argue, the fact that academics have missed the point cannot lend credence to the view that litigants have overlooked this contracting strategy (instead of merely rejecting it as involving overly high transaction costs or because of its incompatibility with public policy). Although there may be some force to this argument, we should explore it more carefully because, in essence, it implies that there is little incentive for legal academics to analyze issues correctly. This also may be true, but it is a serious charge.

Judge Posner has commented on the Popperian conception of the scientific method: "The big thing is to come up with hypotheses that have a sufficiently low antecedent probability of being true to be interesting, but that are not so ridiculous that the results of testing them empirically are a foregone conclusion, and to get on with the testing of them. Because, on the whole, legal scholars do very little testing, the central task in the legal academy may be merely to come up with interesting hypotheses that have a sufficiently low probability of being true. The Posner/Shavell assessment on the inhibiting effect of the British rule on the rate of settlement is counterintuitive—indeed contrary to Posner's initial view based on risk aversion—but it would seem no more counterintuitive than the invariance prediction discussed here. On the other hand, the Posner/Shavell model is counterintuitive in a new way, whereas the invariance point is counterintuitive in a way that has already brought celebrity to Ronald Coase. Therefore, the extreme Panglossean view might be that it was in Posner's and Shavell's self-interest to overlook the Coase Theorem. Nonetheless, the parade of scholars who continued the analysis of the British rule might have profited by heralding the Coasean insight. But none did.

It is possible, then, that we have never heard of a single case of contracting around the fee-allocation rule because the parties never thought of it, and that we have never read about the idea in the legal or economic literature because no one—regardless of how familiar with the Coase Theorem—has ever conceived of the idea? At one level, both lapses may be somewhat surprising. After all, we needed just one litigant to realize that opting for the British rule would make sense, and if it were a good idea, it would quickly catch on. It is also surprising in the academic setting, because the individuals who have neglected the point are clearly those who are most knowledgeable about the Coase Theorem and therefore least likely to overlook its dictates. If renowned law and economics scholars can overlook the Coase Theorem in publishing repeated articles on a theme that cries out for Coasean treatment, what hope is there that some harried litigant will perceive the opportunity?

The lapse among litigants should give Coaseans pause in suspecting that parties will always perceive Coasean bargaining opportunities, particularly those that require reshaping the legal structure of the world that confronts us. That Posner and Shavell could overlook such an opportunity dramatically confirms that exceptional intelligence and thorough familiarity with the Coase Theorem cannot guarantee that Coasean bargains will be perceived and struck. Consequently, we should be cautious in asserting that Coasean invariance predictions will universally be achieved—even without giving any consideration to transaction costs.

Indeed, this caution is particularly appropriate in light of the fact that American litigants might be expected under Coasean logic to opt for a fee-shifting rule even if no model of fee-shifting were available to inform their enterprise of wealth maximization. But there *is* such a model. Almost every country in the world except the United States has adopted some form of the British rule. The comparative evidence cries out for litigants in America to see what much of the rest of the world seems to know—that the British rule has some considerable advantages. But even with this tip, the message has not been heard. A very important lesson may emerge from this discussion: if *all* profitable steps of equal opacity could suddenly be recognized, the gross national product could be vastly higher than it is today.

Notes and Questions

1. Which innovation do you think more likely: a litigant in the United States offering to abide by the British rule or a litigant in Britain bargaining to employ the American rule for fee-shifting? What message might be read into an offer by one party to switch fee-shifting rules? Can you think of a way to make such an offer without creating a perception of weakness?

2. One of Donohue's messages is that it is easy to miss bargains that effectively alter or escape legal rules. There is, for example, the rule in contract law that penalty damage provisions will not be enforced. If firm A agrees to pay $1,000 a week for summer legal work, and student X agrees to work for A and agrees to a provision that X will pay A $15,000 if X fails to report for work on June 1, a court is unlikely to enforce this provision against X if X first reports for work on June 4. Knowing that a court is likely to regard the damage clause as imposing a penalty, could X and A have fashioned the agreement in a way that is likely to be enforced? Have you ever seen such bargaining around this legal rule? Returning to tort law, can parties subject to a negligence rule ever contract in advance to be governed by a rule of strict liability? Can parties governed by strict liability ever convert to a negligence regime?

3. Contingency fees, such as arrangements in which plaintiff's lawyers receive one-third of any award or settlement (but nothing if the defendant prevails), are quite common in American tort law. Very high commissions or percentages, however, are not permitted, as our next selection emphasizes. Thus a plaintiff's lawyer is probably unable to ask for and receive 30 percent of whatever amount the defendant is made to pay, *plus* another 65 percent of everything awarded above $1 million. It is interesting that when some version of the British rule is in effect, requiring the losing party to pay the winner's costs, there is generally a ban on contingency fees (of even 30 percent). What might explain the correlation between rules on fee-shifting and the legitimacy of contingency fees? If a jurisdiction operating under the British rule on fee-shifting decided to permit contingency-fee arrangements, would these fees need to be higher or lower than under the American rule? Even in the United States, contingency fees are not permitted in criminal defense work. If such arrangements were permitted, how would they be formulated? What might explain the rule forbidding those arrangements?

A Market in Personal Injury Tort Claims

MARC J. SHUKAITIS

Introduction

Millions of Americans suffer each year from personal injuries caused by tortious acts. Under today's law, a tort victim may not sell his personal injury claim to a third party who would then pursue the claim against the tortfeasor. To receive compensation, a personal injury tort victim must choose between settling with the tortfeasor or litigating to a court judgment.

Both litigation and settlement, however, have serious disadvantages for a tort victim. Delay and uncertainty reduce the attractiveness of litigation for risk averse victims who need immediate compensation. While settlement may provide quick and certain compensation, its price is usually a smaller recovery because settlement places risk averse tort victims who need immediate compensation in a poor bargaining position.

Allowing victims to sell their claims to third parties, that is, allowing a market in personal injury tort claims, would have significant advantages for tort victims. Compared with litigation, tort victims would be able to receive immediate and certain compensation by selling their claims to purchasers in the market. Compared with settlement, tort victims would receive compensation at a market price closer to what they would expect from a court judgment. Thus compensation by the tort system would be made less dependent on the tort victim's ability to withstand delay and uncertainty. Victims who now do not pursue their claims because of ignorance of their rights or a lack of resources might receive compensation. Some of the incentive problems of hiring an attorney on either a contingent fee or an hourly arrangement would be lessened. And, by increasing the costs of tortfeasors, a market would increase deterrence against harm-causing activities.

Yet there are also possible difficulties with a market. An active market might increase the total volume of litigation. Litigation of spurious claims might increase. Unscrupulous claim purchasers might take advantage of unsophisticated tort victims. Finally, some people might find the very act of selling personal injury tort claims offensive. . . .

The Law Today

Following a long common-law tradition, state laws today bar the operation of a market in personal injury tort claims. Most states prohibit the assignment of

Abridged and reprinted without all original footnotes by permission from 16 *Journal of Legal Studies* 329 (1987). Copyright © 1987 by University of Chicago Press.

personal injury tort claims either out of a professed fear of maintenance, champerty, and barratry* or on the theory that personal injury claims do not "survive" and only actions that survive can be assigned. In most states, the prohibition against assignment of personal injury tort claims flows from judicial interpretation of common law and public policy, but a few states explicitly ban the assignment of personal injury actions by statute. Even where transfer of personal injury tort claims is permitted, the limits on such transfer prevent the formation of a market.

Common-Law Fear of Maintenance, Champerty, and Barratry

Prohibitions against maintenance, inherited from English common-law doctrines developed in the fifteenth and sixteenth centuries, arise out of the fear that otherwise disinterested parties would stir up litigation for harassment and profit. Conditions in England at that time gave courts considerable reason to bar third parties from participating in the litigation process. Resourceful lords commonly profited from supporting litigation in a system that gave the rich tremendous advantages in litigation. Witnesses could be bought because perjury was not a crime, the judicial system was expensive and highly technical, and direct corruption of juries and judges was widespread.

To discourage these injustices, the common law broadly prohibited the assignment of choses in action, a category including personal tort actions as well as most contract and property tort actions. The common-law condemnation of maintenance was also an effort to discourage both the remnants of feudalism and the spread of capitalism in England. In recognition of business needs, the English common law slowly loosened the prohibition against maintenance to allow the assignment of debt actions. Judicial integrity and social conditions, however, have improved markedly since the development of the common-law prohibitions. Today one must question whether any reason remains for prohibiting assignments to avoid maintenance.

Common-Law Assignability/Survivability Equivalence

At common law, a person's death terminated any cause of action for a personal tort that the victim could have asserted before his death—a cause of action for personal tort was thus said not to "survive" the injured party's death. The common law's reasons for refusing to allow the survival of personal tort actions are obscure but involve notions of a tort remedy as meting out punishment and revenge. Revenge was thought possible only by the injured person; after the tort victim's death, the crown was expected to punish the tortfeasor.

While the survival doctrine originated in English common law, the equivalence of survival and assignability of causes of action is peculiarly an Ameri-

---

*"Maintenance" exists when a person "without interest" in a suit assists a party in litigation. "Champerty" is maintenance plus an agreement to share in the proceeds of the suit. "Barratry" is repeated maintenance or champerty.

can doctrine. . . . Many American courts accepted the equivalence of survivability and assignability, although "[t]he reason for this equation between
survivorship and assignability is seldom explained."

Survivability restrictions have been legislatively eased in many states to
allow personal injury tort claims to survive. Courts in some of these states
have blindly followed the equivalency doctrine, holding that a statute that
permits survival authorizes assignment. Other courts have rejected the surv-
ivability/assignability equivalence where state laws permit survival, analyzing
assignability as a separate policy choice.

Subrogation and Other Lawful Claim Transfers

Some courts allow the functional equivalent of assignment of personal injury
claim rights in the limited context of subrogation by an insurer. A tort victim
compensated for his injuries by his insurer retains a cause of action against the
tortfeasor for the harm done. Either contractually or equitably, the cause of
action may be subrogated to the insurer to the extent of the insurer's payment. While some courts allow subrogation of personal injury claims, other
courts do not, arguing that subrogation of a personal injury claim is equivalent
to assignment of the claim and, similarly, should be prohibited.

In a few states, courts have allowed the assignment of personal injury tort
actions, typically through the interpretation of state survival statutes. . . .
Even in the states that permit assignment of personal injury tort claims, courts
and legislatures have erected significant barriers to the development of a
market in tort claims. . . . Other states that permit assignment of tort claims
bar the purchase of tort claims by market participants through maintenance,
champerty, and barratry statutes. Courts in two states appear to prohibit an
assignee of a personal injury claim from recovering more than the consideration the assignee paid to the tort victim, thus effectively preventing a market
by removing the profit incentive of a potential claim purchaser.

The Potential Benefits of a Market

Compensation to Tort Victims

The creation of a private market in personal injury claims would provide tort
victims with the advantages of a competitive market. No longer would the sole
source of compensation come from the tortfeasor, who has every incentive to
delay settlement of the claim. Instead, the ability to sell claims would improve
the level of compensation received by tort victims, increase the number of
meritorious claims that are brought within the system, and obviate some of
the contrasting difficulties associated with hiring attorneys on either contingent fee or hourly arrangements.

Increased Value of Compensation. A tort victim seeking compensation
from the tortfeasor is frequently in a very poor bargaining position. He may

need money immediately to pay medical bills, to replace lost wages, and to cover living expenses. If he seeks compensation through the court system, however, he may wait as much as several years before trial. The tort victim who is unwilling or unable to wait for a court judgment must attempt to settle with a resistant tortfeasor who may insist on litigating so as to discourage future claims. If the tortfeasor insists on litigating, the tort victim will receive no compensation except through the legal process no matter how pressing his immediate need.

Unlike the tort victim, the tortfeasor typically has little incentive to settle quickly. Indeed, the tortfeasor prefers delay because a payment in the future costs him less than an immediate payment. Delay also helps the tortfeasor, who is better able to bear the ongoing expenses of litigation. The tort victim's need for immediate compensation and the tortfeasor's incentive to delay thus reduce the level of settlement.

The parties' differing attitudes toward risk also reduce the settlement. If both the victim and the tortfeasor had the same attitude toward risk, that is, if both were risk neutral or equally risk averse, then neither party would have a bargaining advantage based on his ability to subject the other party to risk. The tort victim, however, is probably risk averse rather than risk neutral. Typically, the tort victim has a very high discount rate and is unable to borrow money inexpensively from a bank or other institution, especially if his only substantial asset is the tort claim itself. Banks are not well equipped to evaluate tort claims as collateral for loans; in any case, the maximum amount a lending institution would be willing to loan on the security of a tort claim would be the discounted expected value of the claim (that is, the price a tort victim could sell his claim for in the market).

The tortfeasor, on the other hand, is probably much less risk averse. Since most personal injury defendants are insured, one would expect a tortfeasor to be risk neutral. A bargaining situation between the risk averse tort victim and the risk neutral tortfeasor favors the tortfeasor who can threaten the victim with the uncertainty of the litigation. By improving the bargaining position of risk averse tort victims, a market in tort claims would raise the amount of compensation to a tort victim closer to the expected court judgment.

A market would also provide many potential buyers for tort claims and thereby eliminate the monopsony advantage that a tortfeasor now enjoys. Thus a tortfeasor would no longer be able to threaten a tort victim with the choice between a low settlement and an interminable court battle. Rather than bargaining with one intransigent tortfeasor, a tort victim could simply sell his claim to one of many potential buyers in the market.

A tort claim market would similarly reduce the amount of compensation that the victim must now forgo to obtain an immediate cash payment. Once claims are freely alienable, the tort victim could sell his claim to a third party with a lower discount rate, who should therefore attach a higher value to the claim.

A tort claim would also be worth more to a market purchaser than to the victim because a purchaser would hold a diversified portfolio of claims. A diversified portfolio would include many (perhaps a dozen or more) claims from different victims that would involve different accidents and types of injuries and that would be presented in different courts using different legal theories. Such a diversified portfolio of tort claims would be worth more than the sum of the expected monetary values of the individual claims because diversification would reduce some of the risk associated with the claims. A purchaser of a diversified portfolio would eliminate the "unsystematic" risk associated with claims and would thus discount his purchase price of a claim only for "systematic" risk.* The diversified portfolio holder could thus pay the tort victim more than the victim's valuation of the claim, which would be reduced for both systematic and unsystematic risk.

Collecting a diversified portfolio would be to some extent inconsistent with increasing value through expertise. The purchaser of the claims would therefore be required to trade off expertise against diversification. This trade-off exists in all separate markets, and there is no reason to believe that professional purchasers of tort claims would be less able to make this trade than are, say, professional managers of mutual funds. A tort victim should be able to receive greater compensation from a market purchaser than from the tortfeasor if the market purchaser is an expert in the sort of claim that the victim has. For example, a purchaser might specialize in automobile accident claims or in airplane crash claims in much the same way some law firms specialize in particular types of claims. The claim will be worth more to the expert (who will thus be willing to pay more for the claim) because his knowledge or experience allows him to value the claim more precisely and to recover a larger judgment in court. A purchaser might also have his own legal staff, which could reduce the costs to him of litigating a claim. The net effect of a well-functioning market would be to raise the compensation received by a tort victim toward the expected value of the claim discounted at a market interest rate appropriate for the riskiness of a diversified portfolio of personal injury tort claims.

Increased Access to Compensation. Many tort victims now receive no compensation from tortfeasors because they do not pursue their claims. Even tort victims who both are aware of their legal rights and have the resources to pursue those rights may choose not to do so; some may decide that the discounted expected value of the claim is less than the expected costs of pursuing the claim. Presumably, however, many tort victims fail to pursue valid claims simply because they are unaware that they have legal rights to compensation. Tort claim purchasers in the market could inform these potential claimants about their legal rights when soliciting to buy their claims.

Other tort victims presumably remain uncompensated because they lack the money needed to pursue their rights. Given their expected risk averseness, poorer tort victims may be especially dissuaded from pursuing valid

*"Unsystematic" or "unique" risk is risk peculiar to the individual claim (for example, the risk of a key witness failing to testify). "Systematic" or "market" risk is economy-wide risk that affects all claims (for example, the risk of inflation or a shift in public sentiment toward the appropriate level of compensation for personal injury victims).

claims because of the costs involved. In the United States, personal injury tort victims have access to legal counsel through contingent-fee representation. While a contingent fee arrangement shields the tort victim from paying attorneys' fees if he does not prevail in the suit, the tort victim at least in theory remains liable, win or lose, for other expenses of litigation such as court costs, witness fees, and so on. The contingent fee arrangement, unlike a straight sale, leaves the tort victim with some risk. As with contingent fee lawyers, there is no reason to expect that all purchasers would specialize in the same type of claims. There may be some who would take claims with little risk and others who would specialize in high-risk cases. As long as there is a broad band of purchasers, the same range of talents found today in contingent fee lawyers should be found in the purchasers of tort claims.

· · ·

Arguments Against Allowing a Market

A number of arguments have traditionally been offered to explain why permitting the free alienation of personal injury claims is undesirable. These arguments can be grouped into four categories: first, that purchasers would harass defendants with spurious claims; second, that the volume and duration of litigation would increase; third, that "personal" injury claims are inherently inalienable; and fourth, that unsophisticated tort victims might be taken advantage of by unscrupulous professional claim buyers. These considerations, however, do not overcome the case for free alienation.

Nuisance Suits

One potential danger of a market in tort claims would be that unscrupulous people could purchase groundless claims and pursue them for their nuisance value. Several courts have cited this as a reason for prohibiting the assignment of personal tort claims, with the Supreme Court of Rhode Island commenting that "to hold otherwise would permit the pernicious and somewhat profitable practice of allowing a person to purchase these claims with the consequent harassment and annoyance of others." In an ideal, costless judicial system, a defendant would never be willing to settle when threatened with a groundless suit. The considerable time, expense, and possible publicity engendered by real world litigation, though, offer powerful incentives for defendants to settle even groundless suits for hard cash.

The risk of abuse in an uncontrolled market is evidenced by the Texas saga of Frank McCloskey. After Texas passed a survival statute in 1895, Texas courts allowed causes of action to be "bartered, sold, and contracted for like personal property." State barratry statutes prohibited only *attorneys* from soliciting claims. McCloskey, who was not an attorney, set up a business in San Antonio, where he and his employees solicited accident victims for their tort claims. McCloskey's business prospered: a San Antonio railroad testified that

60 percent of the claims against it were presented by McCloskey; the San Antonio Public Service Company testified that 38 of the 67 suits filed against it in 1931 were presented by McCloskey.

McCloskey's representation of tort victims went well beyond acceptable bounds. Courts found that McCloskey "counseled malingering to enhance damages" and insisted on the injured parties "using crutches when not necessary." Efforts to curtail his practice were frustrated. The Texas legislature tried unsuccessfully to shut down McCloskey's business in 1917 by amending the state barratry statute to prohibit solicitation of claims by nonlawyers as well as lawyers. McCloskey's business must have been a profitable one; despite a U.S. Supreme Court decision against him in 1920, he continued soliciting claims in Texas at least until 1935.

Increased Volume and Duration of Litigation

Another argument against the free alienation of personal injury actions is that it would increase the volume of civil litigation. In the short run that conclusion seems correct because the total level of tortious behavior should remain constant, while the fraction of claims prosecuted would increase by virtue of sale. Nonetheless, in the long run the situation is not as clear. While a larger fraction of claims would continue to be pursued, the increased efficiency of the system should improve deterrence and therefore decrease the level of tortious activity. The two effects cut in opposite directions; their relative size is largely an empirical question.

Similar ambiguities pervade the settlement process. Here at first glance it appears that the fraction of claims settled would be reduced once assignment is allowed: risk neutral professionals with a diversified portfolio of claims are more resistant to delay. Professional claim purchasers may well be more astute in the valuation of claims than their sellers, however, as well as more efficient in handling them since they do not have to be educated in the claims process generally. Their willingness to take delay may well spur a quick settlement where both sides benefit by saving additional administrative costs. In light of these factors, professionals may respond more quickly to settlement offers.

The problem of the increased volume and duration of litigation thus appears to be overrated. In addition, any effort to solve it faces serious difficulties of its own. For instance, the law could give the tortfeasor an absolute option to settle the claim purchased within some short period, say 30 days, for some fixed premium, say 10 percent, above what the claim purchaser paid the injured party. The idea behind the system is twofold. The claim purchaser would have less incentive to underpay the injured party, and settlement would be encouraged because the tortfeasor would always be offered the opportunity of settling for the "market price" plus the small premium.

Yet adopting such a procedure would spawn new difficulties. First, the calculation of the settlement price will not be straightforward where the purchase is not in the form of a straight cash sale. Second, and more important, the option would confer powerful strategic advantages on the tortfeasor, who frequently has superior information about the potential value of the plaintiff's claim. When the tortfeasor knows that the claim is a valuable one, he will exercise the option to settle for the stated premium. When he knows that the claim is weak, he will resist settlement. The automatic settlement option of the tortfeasor therefore eliminates all potential upside gain of the purchaser while exposing him to all the downside risk. The rule thus reduces both the value of each claim in the portfolio and the gains from diversification that the purchaser hopes to obtain. As it seems likely that defendants have good information on claim aspects, it would be inadvisable to institute any reform that confers such an option on the tort defendant.

In conclusion, there is no obvious reason to think that the assignment of claims would in the long run increase appreciably the burdens imposed on the tort system. And any net cost that assignment imposes should be more than overcome by the increased benefits of deterrence and compensation that the free alienation rule offers.

Inalienability and Trafficking in Claims

A further objection to a market in tort claims is the notion that personal injury tort claims are so inherently "personal" that society should not allow them to be bought or sold. This was one of the rationales for the early common-law prohibition against assignment of choses in action. . . .

From the perspective of economic welfare, restrictions on alienability of tort claims initially seem allocatively inefficient. A person selling a cause of action believes the sale improves his welfare, or he would not sell; a person buying a cause of action similarly believes the sale improves his welfare, or he would not buy. This simple analysis ignores two reasons why an inalienability rule might still be desirable: people other than the buyer and the seller may be affected by the transaction, and victims may be incapable of making wise decisions whether to sell.

Courts have observed that a market in tort claims offends public "sensibilities" so as to outweigh any advantages it might otherwise create. . . . In addition, allowing a market in tort claims could change the way tort victims think about their claims and their decision whether to pursue them. One might not want tort victims to act as though they were the owners of a commodity, presented with an economic decision to make.

What are being sold, however, are not human injuries and pain but rights to compensation for injuries that have already been suffered. There is, after all, little difference between a tort victim selling his claim to a claim purchaser and a tort victim settling his claim with a tortfeasor. A market in tort claims may seem unnatural to many people simply because a market does not exist now—tort liability insurance is widely accepted today but was strongly attacked at its inception as an immoral sale of the right to injure. Certainly, the desire to compensate tort victims and to set optimal deterrence levels should outweigh any value placed on preserving the "sensibilities" of people who would be offended by the thought of people profiting by the purchase of personal injury tort claims.

Protection of Unsophisticated Victims

Refusal to sanction a market might also be justified with the paternalistic argument that tort victims are incapable of making wise decisions whether to sell their claims. To the extent that a tort victim may be ignorant about his legal rights, one might worry that unscrupulous tort claim buyers would coerce victims into selling their claims too cheaply. A tort victim unfamiliar with the legal system might have little basis on which to value his claim, and a prospective purchaser might have little incentive to inform the injured person accurately. Similarly, to the extent that many tort victims are unsophisticated and relatively ignorant about their legal rights, one might worry that quick-talking purchasers would be able to purchase many claims for a fraction of what they are "worth." One can imagine purchasers applying high-pressure tactics to close a sale before the tort victim has the opportunity to negotiate with other potential purchasers: for a competitive market price to develop, competing potential buyers must be informed and have an opportunity to purchase.

Certainly, abuses would occur—there would be instances of unscrupulous buyers making unconscionable purchases from unsophisticated victims. Even in the absence of regulation, however, buyer behavior would be constrained by purchase price competition of other potential buyers. Even where only one market purchaser is interested in the claim, the purchaser would always have to offer the tort victim more than the tortfeasor (or the tortfeasor's insurer) is offering to settle. Careful count scrutiny of purchase arrangements would also restrain buyer behavior. A court would always have the power to void an unconscionable assignment and award part of the judgment to the tort victim.

Regulation of the market would also reduce the danger from unscrupulous buyers. Ethical standards similar to those for the legal profession could be enforced. The fear of quick-talking purchasers buying a tort victim's claim before the victim has had a chance to receive advice or competing offers could be lessened by giving tort victims an absolute option to void their sale contracts within a stated period—perhaps one week—after sale. One might also require central registration of purchased tort claims and publication of purchase.

Another objection to a market is the argument that tort victims would receive less than their claims are "worth" because a market purchaser could not capture the "sympathy" of a court or jury, that is, "the considerations urged to a jury in a personal injury case are of such a personal nature that an assignee cannot urge them with equal force." But this is not a strong argument against allowing a market. Tort claim purchasers, presumably, would require the seller to covenant cooperation in pursuing the claim; purchasers might also purchase only a percentage of the claim so as to leave the tort victim an incentive to cooperate. In any event, a purchase would take place in the market *only* if the value of the claim to the purchaser, after any reduction in value for lost sympathy, were greater than the value of the claim to the seller. Thus a market would function only if claims in fact were not so reduced in value as to offset the increase in value of a tort claim from purchase by a market participant. . . . The chief advantage of assignment is that it moves the claim from the victim, who assigns it a low value, to the professional claim purchaser, who assigns it a higher value. By itself, that gain in welfare should be decisive unless it can be shown that the purchase has negative effects on third parties, in this case tortfeasor. . . . [T]hese external risks are typically overstated and are better controlled by rules directed toward the actual conduct of litigation itself.

State legislative action is needed to authorize and regulate the development of a market in personal injury tort claims. There are powerful interest groups that might be opposed to such legislation, however. Insurance companies might oppose the change if it increases their overall defense and indemnity costs. Beleaguered defendants in product liability and medical malpractice areas are likely to resist the proposal as well, especially if there are no concomitant changes in the substantive and procedural rules governing tort litigation. Even plaintiff's lawyers might join the opposition if they thought that nonlawyer purchasers might diminish their now dominant position in the contingent fee market. Nonetheless, if the arguments set out above are correct, a market in tort claims promises substantial benefits that more than outweigh the costs imposed.

Notes and Questions

1. What other rights are inalienable in our legal system? Does the right to bring a tort suit for personal injury have anything in common with some of these other inalienable rights?

2. The reading observes that insurers generally obtain the right to step into the shoes of the insured-victim-plaintiff in order to bring suit against a tortfeasor. Many victims recover their medical and other expenses from insurers who have promised to pay these costs in return for premium payments and the possibility of recovering costs from a tortfeasor should the court determine that the insured's injury resulted from a tort. It might therefore be said that while personal injury claims cannot be sold after they arise, they can be sold earlier in time, behind the veil of ignorance, through the institution of insurance. Can you think of reasons why this kind of insurance should be forbidden? Does this restatement of the prevailing rule, forbidding the selling of claims after (but not before) they arise, offer any insight as to the role of the rule?

3. Contractual claims are often sold. Indeed it is often the case, when a business is sold, that the bulk of the value transferred is the right to collect for contractual services already performed. And it is common for businesses to sell claims, uncollected bills, mortgage contracts, and so forth to other parties, many of whom specialize in creating markets in these claims. How are contractual claims different from personal injury claims?

V

Liability Arising Out of the Workplace

The relationship between employer and employee, and that among employees, has generated a unique set of tort doctrines. The most important of these is less a doctrine than a departure from tort law. Worker's compensation acts seek to impose liability on employers even when there is neither fault on the part of the employer nor a causal connection between the employer's conduct and the employee's injury. Such workplace insurance systems guarantee the employee relative certainty of recovery, but only for specified, limited damages. An excellent, brief introduction to this area of tort law, with citations to the scholarly literature, can be found in Richard Epstein, *Cases and Materials on Torts,* Chap. 13 (1990).

Other developments in tort law relating to the workplace provide particularly rich opportunities for comparing, or simply enjoying and learning from, a variety of interdisciplinary approaches to law. The first selection takes an economic approach to the question of the employer's liability for an employee's tortious actions. It may be useful to think of two kinds of vicarious liability for an employee's acts. First, it is possible that the employer has been negligent in hiring the employee. If a crime is committed when the opportunity arises out of the employment arrangement (as when a repair person working for a utility company gains access to a home and then steals from or murders the occupant), and it can be shown that the employer did not inquire into the employee's startling criminal record, the question is whether the employer shares vicarious liability for a negligent act despite the intervening and more serious act of the criminal. Plaintiffs are interested in such liability because the employer is normally wealthier and more available for suit than the employee. A second, more common and perhaps more interesting type of vicarious liability involves a claim when there is no negligence on the employer's part. The doctrine can be seen either as strict liability for the negligence of the employee or as the result of a conceptual leap in which the employer and employee are treated as a single unit.

The second reading offers a sophisticated, historical approach to an exception to an employer's vicarious liability. The "fellow-servant rule" relieved employers of liability for negligence inflicted by an employee on a fellow employee. The evolution of this doctrinal exception can be viewed as a function of the law's role in serving the interests of the emerging industrial sector (a topic explored in the article by Professor Schwartz in part II), but it can also be seen through the lens of intellectual history and from the economic perspective set out in the first reading.

The third selection considers the rise and fall of the fellow-servant rule and the emergence of workers' compensation systems from a sociological perspective, which views social change as a complex chain of group bargains. At the same time, it allows a role for cross-cultural borrowing and recognizes the importance of individuals in effecting social change.

The Boundaries of Vicarious Liability:
An Economic Analysis of the Scope of Employment Rule and Related Legal Doctrines

ALAN O. SYKES

"Vicarious liability" may be defined as the imposition of liability upon one party for a wrong committed by another party. One of its most common forms is the imposition of liability on an employer for the wrong of an employee or agent.

The imposition of vicarious liability usually depends in part upon the nature of the activity in which the wrong arises. For example, if an employee (or "servant") commits a tort within the ordinary course of business, the employer (or "master") normally incurs vicarious liability under principles of respondeat superior. If the tort arises outside the "scope of employment," however, the employer does not incur liability, absent special circumstances. Roughly speaking, this "scope of employment rule" restricts vicarious liability to tortious conduct that "should be considered as one of the normal risks to be borne by the business," thereby excluding an employee's "personal" torts.

. . .

The Significance of the Choice Between Personal Liability and Vicarious Liability. Economic theory suggests that between any employer and employee there exists an optimal allocation of the risk of financial loss attendant upon any judgment against the employee. This allocation must take into account each party's attitude toward risk bearing, the employee's incentives to avoid whatever conduct might lead to a judgment against him, and the incentives available to the employer to monitor the employee or otherwise to guard against the occurrence of a wrong.

If the assumption of liability by the employer is efficient from the perspective of the employer and the employee, and the employer agrees to bear all or part of the risk of a particular judgment, that agreement represents part of the employee's total compensation package, and other components of total compensation are reduced commensurately. In general, the employer will agree to bear all or part of the risk of a judgment if and only if the expected cost to him of doing so (including perhaps a risk premium if the employer is risk averse) is less than the reduction in the other components of the employee's compensation package that the employee will accept in exchange for the employer's promise.

Absent transaction costs for private risk-allocation agreements, the optimal allocation of risk does not depend upon where the law initially places liability.

Abridged and reprinted without footnotes by permission from 101 *Harvard Law Review* 563 (1988).

. . .

Regardless of who bears liability initially, private contracting between the employer and the employee can ensure that the ultimate liability of each party accords with the optimal allocation of risk. Such analysis suggests that the choice between a rule of personal or vicarious liability may be unimportant.

This conclusion rests, however, upon two critical assumptions. First, it assumes that the sum of the payments to the plaintiff by the employer and the employee is the same under both rules of liability. If the plaintiff's judgment exceeds the employee's assets, however, as may often be the case, then the plaintiff will collect more under vicarious liability.

. . .

Second, as suggested at the outset, the proposition that the choice of liability rule does not affect the ultimate incidence of liability depends upon an assumption that the transaction costs of negotiating and enforcing a private risk-allocation contract are small enough to permit the employer and the employee profitably to enter into such a contract. In fact, the costs of risk-allocation agreements may be considerable, and a reallocation of liability by contract thus may be infeasible or at least quite costly.

Hence, the choice between vicarious liability and personal liability is a significant one whenever the employee is unable to pay judgments in full under a rule of personal liability or the costs of negotiating and enforcing a private risk-allocation contract between the employer and the employee are significant.

. . .

The possibility that employees may be unable to pay judgments against them in full creates three possible inefficiencies under a rule of personal liability. First, because the employee does not bear the full (expected) cost of his wrongs, his incentive to avoid committing a wrong, either accidentally or intentionally, often (though not always) will be inefficiently small. As a result, the incidence of wrongdoing increases to an inefficiently high level.

Second, because the employer's business does not bear the full cost of the compensable wrongs attendant upon its operation (either directly through liability payments or indirectly through wages paid to employees who make liability payments), its profitability is inflated relative to what it would be if the employee could pay judgments in full.

Third, when employees are potentially insolvent, they may be dissuaded from entering optimal risk-allocation agreements with their employers. An effective risk-sharing agreement requires the employer and the employee to pay the full value of the judgment against the employee, which clearly reduces their expected wealth relative to a regime in which part of the judgment goes unsatisfied. The existence of this added "cost" to risk-sharing may dissuade risk-sharing—indeed, it may motivate employers to seek out insolvent employees to bear the risk of dangerous activities—despite the fact that the employer is often the better risk bearer.

In addition to these three inefficiencies, inefficiencies may arise under a rule of personal liability even when employees *are* able to pay all judgments in full if there are significant costs to employers and employees in negotiating and enforcing risk-allocation agreements. Specifically, if the employer is the better risk bearer, such costs may prevent the employer from assuming the risk of judgments against the employee that exist under personal liability. Alternatively, the employer and the employee will have to incur the costs of an agreement to shift the risk of liability to the employer.

The Effects of Vicarious Liability. Vicarious liability reduces or eliminates *some* of the inefficiencies that can arise under personal liability. First, because vicarious liability ensures that any judgment against the employee will be paid to the limit of the combined assets of the employer and the employee, any inefficient expansion of the scale of business activity that results when the employee cannot pay judgments is avoided, or at least lessened. Second, vicarious liability often will improve the efficiency of risk-sharing by eliminating the incentive that may exist under personal liability to leave liability on the employee in order to take advantage of his inability to pay judgments. Moreover, if the employer is the better risk bearer, vicarious liability can eliminate transaction costs that the employer and the employee might otherwise incur in negotiating and enforcing a private agreement under which the employer assumes the risk of liability.

The ultimate efficiency or inefficiency of vicarious liability also depends, however, on its effect upon employees' incentives to avoid wrongful conduct. The effect of vicarious liability on such incentives depends in turn upon the devices available to the employer to induce careful behavior and the costs of those devices.

One such device is direct observation of the employee's activities. If the employer can cheaply observe precautionary behavior (or deliberately tortious behavior) by each employee, the employer can simply announce a desired standard of conduct, observe subsequent employee behavior, and penalize employees who fail to meet the desired standard.

· · ·

The employer's influence over advancement and compensation decisions often provides another important incentive device. If the employee desires a long-term relationship with the employer, the employer can exploit this desire through policies under which the employee's prospects for advancement and pay raises, and even the employee's continued employment, depend in part upon the employee's avoidance of any misconduct that might result in harm.

· · ·

The ability of the employer to affect the employee's incentives is much more limited, however, when the employee's behavior is not easily observed or when the employee has no interest in maintaining a long-term employment relationship. Under these conditions, perhaps the only device available to dissuade misconduct is the threat of an indemnity action against the employee in the event that the employer incurs vicarious liability. Such actions can be quite costly to pursue, however, and may not be cost effective if the employee's assets are modest. When the threat of an indemnity action is the only device for the maintenance of incentives, therefore, a rule of vicarious liability actually may reduce employees' incentives to avoid wrongful conduct and reduce economic welfare.

. . .

The analysis to this point has assumed that the wrong committed by the employee is properly viewed as a cost of the employer's business. For example, the proposition that the evasion of liability judgments by an insolvent agent may lead to an inefficient expansion in the scale of business activity clearly rests on this assumption or, to put it slightly differently, on an assumption that the business is the "cause" of the wrongs that lead to the judgments. These notions warrant further elaboration.

A Definition of "Enterprise Causation." The legal meaning of the term "cause" depends upon context. . . .

This section introduces a definition of causation that captures the relationship between the existence of an employer's business and the occurrence of a wrong by an employee. To distinguish this concept of causation from similar concepts in the law and in the literature, it will be termed "enterprise causation."

. . .

> Definition: An enterprise "fully causes" the wrong of an employee if the dissolution of the enterprise and subsequent unemployment of the employee would reduce the probability of the wrong to zero.

For example, consider the case of a service station attendant who negligently performs improper repairs on a customer's vehicle, resulting in an accident and a judgment against the attendant. The accident is fully caused by the service station enterprise under the above definition if the dissolution of the service station enterprise and the subsequent unemployment of the attendant would reduce the probability of an accident attributable to negligent repairs by the attendant to zero.

By contrast, suppose that the same attendant occasionally commits acts of wife-beating and that the incidence of such acts bears no relation to the attendant's employment or unemployment. Then, the tort of wife-beating is not caused by the service station enterprise because its dissolution and the employee's subsequent unemployment would not reduce the probability of the tort.

To be sure, intermediate cases exist. The probability of wife-beating, for example, might be increased by certain stressful occupations. Such cases motivate a further refinement of the definition of enterprise causation.

> Definition: An enterprise "partially causes" the wrong of an employee if the dissolution of the enterprise and subsequent unemployment of the employee would reduce the probability of the wrong but not eliminate it.

. . .

The importance of enterprise causation may be illustrated with variations on the following hypothetical. An enterprise produces a good that many consumers value well in excess of its marginal capital, labor, and materials costs. . . . Suppose, however, that some of the employees who produce the good tend to commit assaults that result in civil liability. By hypothesis, neither the probability that an employee will commit an assault nor the magnitude of the resulting liability depends upon the employee's wealth, upon whether he is employed or unemployed, or upon any characteristic of his employment. In addition, by hypothesis, no incentive mechanism exists that would enable the employer to reduce the incidence of employee assaults.

To hold the employer vicariously liable for such employee assaults, regardless of when or where they occur or the identity of the assault victim, is almost certainly inefficient. The employer cannot affect the number of assaults or the total amount of damages. Thus, to the extent that vicarious liability imposes additional costs on the enterprise, it may drive the enterprise out of business or at the very least cause it to shrink as the marginal costs of production increase. Society then will lose at least some of the economic surplus from the production of the enterprise.

Employers might recoup some of their losses through indemnity actions, and when they did not, the costs of vicarious liability to the employer would be mitigated by a decline in the wages paid to employees—if an employer were forced to "insure" his employees against liability for assaults that are unrelated to the enterprise, employees would reduce their wage demands by the expected value to them of the "insurance" provided by their employers. In a perfectly informed labor market in which employees could pay all judgments against them, and neglecting the effects of risk aversion, this reduction in wage demands would result in a wage that was lower by precisely the expected value of the liability borne by the employer. The marginal costs of production would be unaffected, and no loss of surplus would arise.

This conclusion, however, assumes away the familiar problems of employee insolvency and transaction costs. Employees who cannot pay judgments in full may not value "insurance" provided by employers at its full expected cost to the employer. Their wage demands thus will not fall by the full expected value of the "insurance," and a decline in wages will not fully compensate the employer for increased liability. In addition . . . various transaction costs tend to increase when employers are held vicariously liable, and these costs will add to the marginal costs of production. Hence, the imposition of vicarious liability under the circumstances hypothesized here appears inefficient because it leads to higher costs of production, higher prices, and a loss of consumer and producer surplus, but in no way reduces the expected costs of employee misconduct.

This analysis suggests the general proposition that vicarious liability for a given wrong is probably inefficient if (a) the enterprise does not cause the wrong, even partially; and (b) the enterprise cannot reduce the probability of the wrong through incentive contracts with its employees.

A second proposition can be illustrated with a slight modification of the [previous] hypothetical. Continuing with the assumption that the employer cannot reduce the incidence of assaults through incentive devices, suppose that the employee is more likely to commit assaults in the workplace than elsewhere, so that his employment increases the probability that an assault will occur. The probability of a wrong thus depends upon whether the employee is employed or not, but (by hypothesis) cannot be affected by the employer once the employee has been hired. An employee with an intrinsically stressful job may be more likely to commit an assault, for example, but the employer's only effective means of reducing the probability of an assault may be to terminate the employment relationship.

In this situation, it is inefficient to impose vicarious liability on the enterprise for the full value of the wrong, because the scale of the enterprise's activity will tend to contract excessively. Yet it is also inefficient to allow the enterprise to escape liability altogether—if the employee is insolvent, the resulting scale of enterprise activity will be excessive. Efficient resource allocation thus requires the enterprise to bear liability for part, but not all, of the cost of the wrong.

. . .

Other things being equal, the greater the extent to which the employment relation "causes" the wrong, the more likely it is that vicarious liability for the entire judgment will be second-best efficient.

A third proposition may be illustrated with a further variation of the [previous] assumptions. . . . Again suppose that the employer cannot reduce the incidence of assaults with incentive devices and that the probability of an assault depends solely upon the employment or unemployment of the employee. Assume further that the probability of an assault would be *zero* if the employment relationship were dissolved and the employee remained unemployed thereafter. Thus, the enterprise fully causes the assault. It follows that the injury attributable to the assault is a cost that is fully attributable to the operation of the enterprise—the enterprise will operate at an efficient level of output only if it bears, directly or indirectly, all liability for the employee's assault. The imposition of vicarious liability is plainly efficient under these circumstances because, by hypothesis, it can have no adverse effect on the incentives of the employee to engage in tortious conduct.

The analysis becomes more complex if incentive devices can reduce the probability of the wrong or if the assumption of liability by the employer may affect the propensity of the employee to commit the assault. To continue with the previous hypothetical, suppose that the probability of an assault is not directly affected by the employment or unemployment of the employee—and thus the enterprise does not even "partially cause" the assault—but that the employer can reduce the probability of an assault through some appropriate incentive device, for example, by threatening to discharge the employee if he commits an assault or by observing the employee's behavior in advance of an assault and intervening to prevent the assault before it occurs. Obviously, the employer has no reason to use such an incentive device under a rule of personal liability. The question thus arises whether vicarious liability may enhance economic welfare even though the enterprise does not cause the wrong.

To examine this issue, suppose first that the costs to the enterprise of using the incentive device are zero and that the incentive device always prevents the assault—thus, the enterprise never actually pays damages under a rule of vicarious liability. Vicarious liability is clearly efficient under these circumstances, because it eliminates any possibility of an assault at no cost to the enterprise.

Now suppose that the incentive device reduces but does not eliminate the probability of the assault and that the reduction of the expected damages from assaults when the incentive device is utilized exceeds the costs of the incentive device. If the prospective victims knew of their impending assault, they would offer the enterprise, and the enterprise would accept, a side payment to use the incentive device. The marginal costs of production for the enterprise would not increase, and the incidence of assaults would decline with no adverse effect on the level of enterprise activity.

Of course, such side payment schemes are unlikely to arise in practice. In their absence, perhaps the only way to motivate the employer to utilize the incentive device is to hold him liable for any judgment against the employee. Under a strict liability theory, the employer would always incur joint and several liability with the employee. Under a negligence theory, the employer could avoid liability with a showing that the assault occurred despite the fact that he utilized the appropriate incentive device.

Usually, vicarious liability in employment relationships (conventional respondeat superior liability) is "strict"—the employer cannot successfully defend the action by proving that he exercised all reasonable means to dissuade the employee from committing the wrong. In the present hypothetical, strict liability will indeed induce the employer to utilize any and all incentive devices that are cost effective in reducing his liability. Because the employer is not compensated for the costs of the incentive devices as he would be under an idealized side-payment scheme, however, and because he bears liability for judgments against the potentially impecunious employee, the marginal costs of production for enterprise output increase and the scale of enterprise activity contracts to a point at which the prevailing price exceeds the social marginal costs of production. To determine whether vicarious liability has reduced or enhanced economic welfare, the resulting loss of consumer and producer surplus must be weighed against the reduction in the expected value of losses from assaults. The welfare effects of vicarious liability are more favorable the greater the degree to which the incentive device reduces the probability of assaults and the lower the costs to the enterprise of implementing the incentive device.

An alternative approach to employer liability, as noted, is to impose liability on the employer only if he is negligent and fails to use the desired incentive device. Hereafter, this policy will be termed "vicarious liability based on negligence"—a term that may suggest that liability is not really vicarious at all, but that I employ here for expositional convenience. In the present hypothetical, this approach is unambiguously superior to strict vicarious liability because the employment of the tortfeasor does not increase the probability of the wrong, and hence the associated liability is not properly included in the social marginal costs of enterprise production. If the costs to the enterprise of using the incentive device are significant, however, the welfare consequences of vicarious liability based on negligence are still ambiguous. Although the enterprise escapes liability if it uses the incentive device, the costs of using it will become part of the marginal costs of production for the enterprise. Price will rise and consumer and producer surplus will fall. Again, to determine the net effect on social welfare, this reduction of surplus must be weighed against the benefits of the incentive device to potential assault victims. Other things being equal, the lower the cost of using the incentive device, and the greater its effectiveness, the more favorable are the welfare effects of vicarious liability.

. . .

The law places agency relationships into several categories, and the rules governing the vicarious tort liability of an employer or "principal" depend critically upon the category into which the agency falls. Most employer-employee relationships are "master-servant" relationships, in which the master controls or has the right to control the physical conduct of the servant in the performance of the servant's duties. Absent the requisite degree of control, however, the employee may be classified as an independent contractor or other nonservant agent. Under the doctrine of respondeat superior, masters are liable for torts that their servants commit within the scope of employment. If the tortfeasor is an independent contractor or nonservant agent, however, with some exceptions, the principal or employer is not liable for the tort.

The discussion to follow focuses upon respondeat superior and specifically upon the scope of employment limitation to the master's liability. The *Restatement (Second) of Agency* defines the "scope of employment" as follows:

(I) Conduct of a servant is within the scope of employment if, but only if:
    (a) it is of the kind he is employed to perform;
    (b) it occurs substantially within the authorized time and space limits;
    (c) it is actuated, at least in part, by a purpose to serve the master; and
    (d) if force is intentionally used by the servant against another, the use of force is not unexpectable by the master.
(2) Conduct of a servant is not within the scope of employment if it is different in kind from that authorized, far beyond the authorized time or space limits, or too little actuated by a purpose to serve the master.

According to the *Restatement,* masters are not liable for torts committed by their servants outside the scope of employment unless:

> (a) the master intended the conduct or the consequences, or (b) the master was negligent or reckless, or (c) the conduct violated a nondelegable duty of the master, or (d) the servant purported to act or to speak on behalf of the principal and there was reliance upon apparent authority, or he was aided in accomplishing the tort by the existence of the agency relation.

. . .

To facilitate concrete discussion of the rule and its economic consequences, this section considers two groups of cases: cases involving torts by servants during a purported "frolic and detour" from their employment and cases involving intentional torts by servants. . . . A "frolic" or "detour" from employment occurs when an employee, while on travel for the employer, deviates from his assigned tasks for personal errands or other personal business. Not surprisingly, most frolic and detour cases involve motor vehicle torts. . . . Suppose, for example, that a truck driver negligently causes an accident while making a delivery for his employer. If the driver precisely followed the route suggested by the employer, and undertook no personal business during the trip, then the accident is fully caused by the employer's enterprise. . . . In many instances, the driver will not have insurance, or the insurance that he does have will be inadequate to pay the judgment in full, particularly if the judgment is large. Vicarious liability then has two obvious benefits. First, it will force the enterprise to bear a greater proportion of the cost of the accidents that it "causes." Second, it will shift risk from the driver and accident victim, who are probably relatively poor risk bearers, to the employer or his insurance company, who are likely to be superior risk bearers.

. . .

How does the analysis change if the tort is not "caused" by the enterprise? Suppose, for example, that an employee has an accident while shopping for groceries on his own time over the weekend—an activity clearly unrelated to the enterprise. Under these circumstances, the imposition of vicarious liability would place a cost on the enterprise that was not related to any wrong "caused" by its activities. The price of the enterprise's goods or services would rise inefficiently, and consumer and producer surplus would decline. Vicarious liability would then be inefficient *unless* (a) it caused the employer to adopt incentive devices that produced a reduction in the incidence of motor vehicle torts by employees that was substantial enough to offset the inefficiencies of imposing added costs on the enterprise, or (b) it resulted in an increase in the efficiency of risk-sharing that offset those inefficiencies.

As to possibility (a), it is doubtful that an employer could reduce significantly the number of motor vehicle torts committed by employees on personal business. . . . It thus seems improbable, or at least highly conjectural, that the benefits of a reduced-accident rate under vicarious liability would be substantial enough to justify the imposition of added costs on the business enterprise.

With regard to possibility (b), any risk-sharing benefits would also be offset at least in part by the inefficiency of imposing costs on the enterprise that are unconnected to its operation and by added costs of litigation. Moreover, many (though not all) of the costs of personal injuries are covered by first-party insurance policies and, to that extent, risk-sharing is reasonably efficient even without transferring losses to the employer. Thus, the imposition of vicarious liability probably could not be justified simply as a means of increasing the efficiency of risk-sharing.

The law is basically consistent with this analysis. A motor vehicle tort committed by an employee while on a personal shopping trip on his day off, for example, is unquestionably outside the scope of employment, and his employer will escape liability under the scope of employment rule. In most reported frolic and detour cases, however, the tort occurs while the employee is on "company time" but is at least arguably engaged in personal business. The analysis of such cases is only slightly more difficult.

For example, assume that an employee runs a personal errand during a business trip and commits a motor vehicle tort during the personal errand. Assume further, however, that the employee would run an identical personal errand on his own time even if he had no affiliation with the employer. Under this assumption, the existence of the employment relationship has no effect on the ex ante probability of the tort. For the reasons given above, therefore, it is probably inefficient to impose liability on the employer. Indeed, in reaction to the imposition of vicarious liability, an employer might attempt to discourage employees from performing personal errands while on company business. Yet society may actually benefit if employers permit employees to perform such errands, because the employee's total company and personal travel will decline, thereby reducing the overall likelihood of a tort, as well as the cost of travel.

On the other hand, some torts committed during personal errands are "caused" by the enterprise. If an agent travels on business to a distant city, for example, and his lack of familiarity with the city streets considerably increases the probability of an accident, any resulting tort is to a considerable extent "caused" by the enterprise.

Hence, the following principle would, in theory, provide a reasonably efficient disposition of the frolic and detour cases involving motor vehicle torts. If a motor vehicle tort occurs during a personal errand of the employee, the employer incurs vicarious liability for the tort only if (a) the tort arose during an errand of the sort that probably would not have occurred absent the existence of the employment relationship, or (b) for some other reason, the probability of the tort was substantially increased by the existence of the employment relationship—for example, the tort occurred because the employee was unfamiliar with the area in which he was required to travel.

The courts' approach to frolic and detour cases, however, is often quite different. In many cases, the courts focus on the length of the detour from the employee's business travel route as the test for vicarious liability—the employer is only liable for torts committed when the employee makes "short" detours. Such a focus seems inefficient because the relationship between enterprise causation and the magnitude of the departure from an assigned route appears quite weak. An employee may stop for personal business that he would transact anyway without any significant departure from his route, and an employee who travels across country on business may make an extensive detour for sightseeing that he would not undertake absent his business travel.

. . .

The basic principles that govern the efficiency of vicarious liability for intentional torts are developed through the various hypotheticals involving employee assaults. In analyzing the case law with respect to these principles, it is instructive to divide intentional torts by employees into two categories: torts that arise from the employee's purely personal motivations, and torts that occur when the employee acts to serve the principal. This division highlights important distinctions in existing law as well as important distinctions in the effects of vicarious liability on resource allocation.

As to the first category, vicarious liability is likely to be efficient even if an intentional tort has a seemingly personal motivation when two conditions hold. First, the tort must be "caused," at least in large part, by the employment relationship. Second, the imposition of liability on the employer must not excessively reduce the employee's incentives to avoid wrongful conduct.

The second of these conditions for efficiency is familiar. The first condition, however, requires further discussion. Suppose, for example, that an intentional tort occurs because the existence of the enterprise causes the employee to encounter unusual circumstances that he would not otherwise encounter. Intrinsically stressful occupations, for example, may precipitate intentional torts. In such cases, the business enterprise "causes" the tort even though the employee's tortious behavior may evince a purely personal motivation.

Most hornbook statements of the law suggest that vicarious liability will not apply if an agent acts out of personal ill will. Rather, the employee must act out of a purpose to serve the employer. Recent case law, however, reflects a more flexible approach that appears consistent with the efficiency analysis. For example, one California court upheld the imposition of vicarious liability against a building subcontractor for his drunken employee's assault of two employees of the general contractor. The court reasoned that the tort resulted from the tortfeasor's perception of his rights as an employee.

Similarly, the D.C. Circuit has held that a jury could impose vicarious liability on a trucking company whose deliveryman raped a customer. Crucial to the holding was the court's observation that the deliveryman's "badge of employment" enabled him to secure entry to the victim's premises. This result appears efficient if one assumes that the probability of a rape was considerably enhanced by the deliveryman's employment status.

Finally, in *Ira S. Bushey & Sons, Inc. v. United States,* the Second Circuit affirmed the imposition of vicarious liability upon the Coast Guard for vandalism by a drunken sailor. Observing that "the proclivity of seamen to find solace for solitude by copious resort to the bottle while ashore has been noted in opinions too numerous to warrant citation," Judge Friendly suggested that vicarious liability was appropriate because the tort was caused by the nature of the tortfeasor's employment.

The other major class of intentional tort cases—in which the tort occurs at least in part because the agent seeks to further the interests of his principal—raises entirely familiar issues. In these cases, there is rarely any question that the existence of the agency "causes" the tort. Hence, subject to the previously discussed danger that vicarious liability may reduce the incentives of employees to avoid wrongful conduct or necessitate a spate of costly indemnity actions, the greater willingness of courts to impose vicarious liability in these cases is economically sound.

Notes and Questions

1. A hires B to work one evening as a babysitter for A's child. A asks B to take the child for a drive in A's car, carefully and slowly, in order to purchase hamburgers for dinner. B drives recklessly, injuring C, and then in a fit of frustration B strikes a police officer who comes to the scene of the accident. Who should be able to collect from A? Is your answer informed by or contradicted by the author's analysis?

If A is liable in the preceding example, is the same true of a study group consisting of four law students that delegates one of its number to pick up a pizza when that student drives recklessly, injuring C and striking a police officer? Does it matter if the driver is compensated by the other members of the group with free pizza? If A is not liable at all in the first example, then why is A different from a more permanent employer?

2. As suggested in the introduction to this reading, one way to think of respondeat superior, or a kind of vicarious liability under which A is liable, is as an example of strict liability in a regime that normally looks for fault as a precursor to liability. This perspective raises several questions. Is there some similarity between this exception to the negligence norm and others? Where else have you seen strict liability, and is there anything these other strict liability areas have in common with vicarious liability?

3. If vicarious liability is strict liability, then should A be liable for every injury caused by B, the employee, or only those injuries caused by B's negligence? Would it ever make sense to hold an employer strictly liable for an employee's entire behavior when the employee would only be directly liable for negligent behavior? Would it ever make sense to hold an employer vicariously liable for an employee's torts, and excuse the employee from liability, even if that employee has sufficient funds to cover the tort liability in question?

4. If vicarious liability is strict liability, then should the limit of the employer's liability be the same as that imposed by the requirement of proximate cause?

5. Returning to the first example offered, it happens that in many jurisdictions the police officer will have trouble collecting, if only because of the "fireman's rule." Under this doctrine, public safety employees are considered to be servants of the taxpayers, all of whom expect there to be risks on the job and no prospect of reimbursement from individual risk-creating citizens. Thus, police and firefighters need no extra compensation (that is, no prospect of tort liability) for the risks encountered in the line of duty. Note that there will be some insurance coverage, provided through the employer, for losses suffered in the line of duty. How might you argue on behalf of the police officer, and against A, that the fireman's rule should be a weak rule, not applicable in this case?

6. How do you expect the extent of vicarious liability to affect employment contracts between employers and their agents?

Comment—The Creation of a Common-Law Rule: The Fellow-Servant Rule, 1837–1860

Introduction

Searching for a comprehensive explanation of the nature of legal change is like searching for an honest man: the quest is often more illuminating than the conclusion. The quest has recently been undertaken by a new generation of legal historians who have focused upon the development of modern private-law doctrines in nineteenth-century America. One such doctrine, the fellow-servant rule, presents a stark and self-contained example of common-law development. This selection examines the development of the rule to see if such an examination can clarify the interplay between competing explanations of nineteenth-century legal change.

The fellow-servant rule was a rule of tort law created in the mid-nineteenth century. It carved out an exception to the well-established rule of respondeat superior, and relieved employers of liability for injuries negligently inflicted by an employee upon a "fellow servant."

. . .

Commentators have used the rule to illustrate conflicting theories concerning the nature of legal development. These writers fall into two broad categories: those who view legal change largely as a function of the legal community's response to social and economic changes in society, and those who see it primarily as a response to forces within the legal community.

The clearest, and perhaps simplest, explanation is that of the economic and social determinists. They view legal development, in general, as a product of the legal system's close adaption to changes in society. To such theorists the fellow-servant rule is an example of the general development of tort law in the direction of protecting nascent industry in an industrializing society.

The major new version of this interpretation is that of Morton J. Horwitz. While previous economic and social determinists depicted legal change largely as a passive response to changes in society at large, Horwitz's exhaustive study of the changes in American private law in the nineteenth century describes a conscious effort by the legal establishment "to promote economic growth primarily through the legal . . . system." To Horwitz, nineteenth-century common-law judges were deliberate and conscious sources of legal change, purposefully shaping private law to accommodate the needs of their social and economic partners in the entrepreneurial classes. To these "instrumentalist" judges, the private law was malleable. Their role was to mold it by discarding or reshaping ancient common-law doctrines and English precedents and

Abridged and reprinted without footnotes by permission from 132 *University of Pennsylvania Law Review* 579 (1984). Copyright © 1984 University of Pennsylvania Law Review.

thereby to render the nineteenth-century legal system consonant with the social and economic needs of an industrializing America.

Central to Horwitz's thesis is a stringent periodization; Horwitz separates the early nineteenth-century period of dynamic legal change from the static eighteenth-century world which preceded it and from the late nineteenth-century era of legal formalism which followed it when the triumphant entrepreneurial classes sought to consolidate the gains they had won earlier.

· · ·

A far different perspective on nineteenth-century legal change, and on the fellow-servant rule, is presented by those scholars who view legal development not as a reaction to social and economic change, but rather primarily as a response to intellectual forces within the legal community itself. While social change may have provided the occasion for legal change, they believe, it cannot provide a sufficient explanation for the direction such legal development took. Thus, for example, one modern scholar, G. Edward White, has attributed the development of nineteenth-century tort law largely to "changes in jurisprudential thought." In such a scheme, the intellectual force of the arguments of premier judges like Lemuel Shaw assumes major significance.

When considering the role of intellectual factors in impelling legal change, it is important to remember that the intellectual process need not result in a set of legal rules that are either consistent or coherent.

· · ·

This selection examines the sufficiency of these views to account for the development of the fellow-servant rule. It traces the migration of the rule through various jurisdictions through a study of cases of first impression. Such cases are excellent vehicles for examining the migration of any doctrine, for in adopting or rejecting any particular rule, judges feel a particular burden of explanation. . . .

The Development of the Rule

Priestly v. Fowler *and Its Influence Upon American Courts*

It is not strange that economic determinists detected a conspiracy to aid the industrial classes in the fellow-servant rule because the law out of which the rule developed had definitively imposed liability on a master for the consequences of a servant's negligence. In the eighteenth century, the rule of respondeat superior was extremely broad and firmly entrenched. As one of the earliest compilations of American law restated the rule: "the master is responsible for the acts of his servant, if done by his command, expressly or impliedly given." There were few exceptions to the rule, noted one treatise writer, for the law imposed upon the master the absolute duty to "employ servants who [were] skilful and careful."

This broad principle could have covered the situation in which an employee negligently inflicted an injury, not upon a stranger, but upon a fellow employee. . . . By the end of the eighteenth century, English and American courts had begun to set some limits to a master's liability for his servant's negligence, especially in those situations where the master actually exercised little control over the servant's behavior. He would not be held liable, for example, for a servant's willful, as opposed to negligent, wrong. Similarly, there were limits placed upon the extent of a shipowner's liability for the captain's negligence. Both limitations had as their rationale the protection of the master-servant relationship. The law could not, courts held, "allow to servants a power to ruin their masters."

A similar spirit animated Lord Abinger, sitting in the Court of Exchequer, when in 1837, he announced the court's decision in *Priestley v. Fowler,* the case which first enunciated the fellow-servant rule. *Priestley* was an unlikely case to carry its heavy precedential burden. Fowler, a butcher, ordered his assistant Priestley to make deliveries in an overloaded van driven by another assistant. The van capsized, fracturing Priestley's leg. He sued his master for damages, won, and the master appealed.

Lord Abinger's opinion is scarcely a model of judicial clarity or authority. He began by making the questionable decision to treat the case as one of first impression, which he was "therefore to decide . . . upon general principles."

One general principle relied upon by Abinger was assumption of risk. During the argument at court, Abinger stated that "[t]he plaintiff was not bound to go by an overloaded van; he consents to take the risk." In his opinion for the court, he stated that "the plaintiff must have known as well as his master, and probably better" about the dangerous condition of the van.

. . .

Abinger also rested his decision upon general principles of public policy. He stated that a servant had a duty to exercise "diligence and caution" to protect himself from the negligence of a fellow servant, and that allowing the plaintiff to recover would discourage the exercise of that caution. He concluded that "diligence and caution . . . are a much better security against any injury the servant may sustain by the negligence of others engaged under the same master, than any recourse against his master for damages could possibly afford."

Perhaps the true basis of Lord Abinger's opinion lies in his "parade of horrors"—the consequences he feared would ensue from holding the master liable. Should Priestley recover, he warned, "the principle of that liability will be found to carry us to an alarming extent." He predicted,

> The master, for example, would be liable to the servant for the negligence of the chambermaid, for putting him into a damp bed; for that of the upholsterer, for sending in a crazy bedstead, whereby he was made to fall down while asleep and injure himself; for the negligence of the cook, in not properly cleaning the copper vessels used in the kitchen; of the butcher, in supplying the family with meat of a quality injurious to the health; of the builder, for a defect in the foundation of the house, whereby it fell, and injured both the master and the servant by the ruins.

. . .

Although the industrial contexts of most American cases did not provide good factual analogies to *Priestley's* master-servant situation, *Priestley* was continually cited in America for the broad proposition that an employer should not be liable for negligent injuries inflicted by one of his employees upon another. Even those judges who struggled to narrow the rule's application felt compelled to explain at length why *Priestley's* rationale did not cover the case before the court, rather than simply distinguishing the case on its facts, or dismissing it as unauthoritative "foreign" authority. What, then, explains the continuing disposition of American courts to acknowledge the authority of this recent English decision?

One explanation . . . recognizes the continuing authority of English precedent in mid-nineteenth century American courts. This explanation, however, conflicts with one central proposition of the Horwitz thesis. For Horwitz, the transformation of American law depended upon a rejection by instrumentalist judges of the constraints of English form and authority and an acceptance instead of a style of judicial reasoning which was "derived from a new conception of common law as a self-conscious instrument of social policy." For evidence, Horwitz relied on state statutes forbidding the continued citation of recent English cases and on statements by American scholars that much of English common law was "as inapplicable to [American] concerns as the laws of Germany or Spain."

In fact, such evidence may prove only that the fears of those jurists who wished to free American law from common-law constraints were well grounded. American lawyers, despite the best efforts of these reformers, remained wedded to the English sources of authority that their training had taught them to revere. . . .

This continuing hold of English authority is not surprising. The apprenticeship training of most American lawyers practicing in the early nineteenth century depended heavily upon an almost ritual reading of English authorities. . . .

The continuing weight given precedents of authoritative English courts had an important influence on the development of the fellow servant rule in America. . . .

Farwell v. Boston & Worcester Rail Road: *Establishing the Rule in America*

The first American cases to adopt the fellow-servant rule were *Murray v. South Carolina Railroad* and *Farwell v. Boston & Worcester Rail Road*. Both cases involved railroad accidents, which would come to be the typical fellow-servant context. Both cases squarely held that masters would not be held liable for negligent injuries inflicted by one employee upon another.

. . .

Farwell came before Chief Justice Lemuel Shaw of the Supreme Judicial Court of Massachusetts in March 1842. The facts presented a sympathetic case for the plaintiff. Farwell was an engineer on the new Boston & Worcester.

While the engineer was driving his engine with all due care, the train jumped the track because a depot employee had neglected to throw the appropriate switch. Farwell was "thrown with great violence upon the ground" and his right hand was crushed under the passing railroad car.

Counsel for both plaintiff and defendants presented a narrow case to Shaw for decision. Plaintiff's counsel conceded that respondeat did not cover the case, and that *Priestley,* which had been "rightly decided," was the relevant precedent. He believed, however, that his client's case must be distinguished because an engineer and a switch tender did not share a "common purpose," and therefore could not be considered to be "fellow servants." Defendants' counsel, predictably, argued that *Murray* and *Priestley* precisely covered the case, but that, in any event, it would "not be necessary for the court to lay down a general rule, in order to decide this case for the defendants."

The arguments of counsel are particularly interesting because they demonstrate clearly the narrow way in which *Farwell* might have been decided by a judge other than Shaw. They epitomize the model of decision in fact followed in later fellow servant cases.

Shaw, however, viewed the case in a much broader light and self-consciously laid out and justified a general rule for regulating a master's liability for negligent injuries inflicted by employees upon their co-workers. He framed the question for decision in the broadest possible terms. "The question is," he noted, "whether, for damages sustained by one of [a railroad employees] by means of the carelessness and negligence of another, the party injured has a remedy against the common employer."

Shaw's *Farwell* opinion is well known and his reasoning has been extensively analyzed. He simply brushed away any suggestion that the case could be decided by extending respondeat, with a grateful nod to plaintiff's counsel for conceding the point, and proceeded to justify his rule on the intertwined theories of implied contract and assumption of risk. When a servant voluntarily accepts employment, Shaw concluded, he assumes all risks "incident" to his employment, among which must be included the risk of "carelessness and negligence of those who are in the same employment." Such a rule best ensures industrial safety, Shaw reasoned, for it gives employees, who are in the best position to observe and guard against their co-workers' negligence, the incentive to do so.

A striking aspect of *Farwell,* to a modern observer, is its self-consciousness. Shaw was acutely aware that he was creating law. He stated his conclusions broadly, and dismissed any attempts to narrow his rule as "supposed distinctions" on which "it would be extremely difficult to establish a practical rule." This self-consciousness is apparent in Shaw's cavalier reference to the only precedents which existed, *Priestley* and *Murray.* In contrast to his elaborate exposition of his assumption of risk and implied contract theories, Shaw simply observed that the "authorities, as far as they go," supported his position.

. . .

The entrepreneurs who owned the railroads, then, were interested in reducing safety hazards as well as in turning a profit. Shaw responded well to their concerns when in *Farwell* he relieved the railroads of potentially massive liability for fellow-servant injuries and thereby placed the burden of avoiding accidents on the employees who were already paid to prevent them.

Shaw, then, by attending so well to the needs of the entrepreneurial classes, is the very model of Horwitz's instrumentalist judge. Because many of his contemporaries on the American bench and bar were motivated by far different imperatives, however, they simply followed *Farwell* as precedent, rather than emulating Shaw's style and examining his premises.

Following Farwell: *Jurisprudence by Discovery*

With the spectacular growth of railroads in the 1840's came the predictable rash of railroad accidents. One American jurisdiction after another faced the question whether to adopt the fellow-servant rule. Many recognized that it was "a question of very great importance, from the frequency with which accidents of that kind are liable to happen upon the numerous railroads of the state." Most courts decided to follow *Priestley* and *Farwell* and broadly adopt the rule.

New York was typical. A lower court had declined an opportunity to adopt the rule in 1844 and the question did not reach an appellate court until 1849 in *Coon v. Syracuse & Utica Rail-Road.* The plaintiff, a track repairman operating a hand car, was "run over" by a repair train. While the case presented questions of contributory negligence, the court ignored them and rested its opinion squarely on *Priestley* and *Farwell,* in which "the subject . . . would seem to be exhausted." "[I]t would be presumption in me," Judge Pratt wrote, "to attempt to add anything to the force of those decisions, by way of argument or illustration." His opinion closely followed Shaw's implied contract theory, and adverted to Lord Abinger's examples in discussing the policy consequences of a contrary rule. *Coon* is striking, and typical, in the unqualified manner in which it adopts the fellow-servant rule. On the authority of *Farwell,* Pratt simply discussed any possible limitations on the rule. The New York fellow-servant rule, therefore, came close to exempting employers from all liability for industrial accidents caused by an employee's negligence.

Other jurisdictions emulated the New York approach; some opinions did no more than observe that the question was a new one in the state, that "the whole argument upon the question is embodied in the opinion of Chief Justice Shaw," and that therefore the rule must be adopted. As more jurisdictions followed this pattern, the string citations following the adoption of the rule became longer.

Some scholars have simply correlated the adoption of the broad rule in so many jurisdictions with the contemporaneous rapid expansion of railroads. The two phenomena undoubtedly coincided. It is also true that railroad accidents provided the factual context for most fellow-servant cases.

Any such explanation, however, standing alone, is insufficient. Some jurisdictions that followed the rule were highly developed railroad states, but others were not. Moreover, some jurisdictions that struggled against the rule had a heavy concentration of railroads. While it would be wrong to ignore the role played by railroads in providing both the occasion, and, in some instances, the impetus for judges to adopt the fellow-servant rule, such an explanation cannot explain the differences in the rule adopted by jurisdictions exhibiting a generally similar pattern of railroad development.

A plausible explanation for these results is that vastly divergent legal styles coexisted in the nineteenth century and created the mass of conflicting jurisprudence that was the fellow-servant rule. That is, some judges may have been disposed to follow *Priestley* and *Farwell* not because they were instrumentalists attuned to economic developments within their jurisdictions as Horwitz would suggest, but rather because their training and intellectual bias required them to adhere to certain common-law myths. This explanation requires recognition that instrumentalist judges like Shaw coexisted with judges like Pratt whose model of proper judicial behavior included adherence to these myths.

According to these myths, the common law grew, slowly or quickly, out of the customs and practices of any particular society. The role of judges, in any particular case, was to "discover" what that rule or custom might be. To any particular legal question, the common law posited but one correct answer, and those judges who rightly "discovered" the correct rule were entitled to respect and adherence. These myths put a premium on judicial harmony and consistency across jurisdictions. A corollary of the concern for harmony was a reverence for unanimous opinions and a suspicion that divided courts had probably "discovered" wrongly.

Horwitz has argued, however, that the old common-law myths of judicial discovery and resultant harmony were discarded by nineteenth-century judges as unwarranted constraints upon their ability to make new law. The fellow-servant cases, however, including those in which the *Farwell* rule was broadly adopted, demonstrate that these myths did not disappear completely in the nineteenth century.

Shaw's opinion in *Farwell* contained all the characteristics to indicate that its legal solutions were rightly discovered, and thus could be unquestioningly followed by subsequent courts. First, the opinions of certain judges carried more weight than others, and Shaw individually was preeminent. Moreover, as chief justice of the highest court of Massachusetts, he spoke from a forum which, like the Court of Exchequer, commanded particular respect.

Shaw's method of reasoning, furthermore, reinforced his authority. He had very carefully considered the nature of authoritative legal argument, and his *Farwell* opinion demonstrated his method. He clearly stated the theories of implied contract and assumption of risk upon which the opinion rested, considered the policy consequences of an alternative decision, drew proper analogies to existing law, and stated a broad and clear general rule for exempting a master from liability. This style of analysis carried great persuasive force, if not binding authority. Because the efficient new legal information system put this case quickly into the hands of railroad lawyers, subsequent judges would have been required to go to elaborate lengths to refute it. It was much easier to rule that *Farwell* "appear[ed] to have been thoroughly examined and considered," and therefore one need "entertain no doubt of its correctness."

Finally, as the years passed, and the number of decisions following *Farwell* began to mount, judges disposed to hold masters liable for negligent injuries committed by one servant upon another began to be overwhelmed by the sheer number of precedents against such liability. Railroad lawyers were quick to argue that the fellow-servant rule was "well settled law. . . ."

The "Exceptions" to the Fellow-Servant Rule

The broad fellow-servant doctrine was not universally accepted. While those jurisdictions that followed the rule recognized that the principle was a "comprehensive doctrine, and applies to all agents engaged in the business of the principal," several jurisdictions so severely qualified the rule that its effect in exempting an employer from liability was virtually nullified. The technique by which they did so was to create exceptions which came close to swallowing the rule. These "exceptions," which came in a variety of forms, gave the late nineteenth-century rule its characteristic incoherence which contrasted so sharply with the clarity of Shaw's rule. No account of the development of the rule, therefore, would be complete without an examination of the "exceptions."

Those scholars who explain the rule as a response to economic needs have a ready corollary to explain the growth of the exceptions. The fellow-servant rule served its purpose of subsidizing the growth of railroads during the great years of railroad expansion prior to 1860. By "the end of the century," Friedman and Ladinsky observe, "the fellow-servant rule had lost much of its reason for existence: it was no longer an efficient cost-allocating doctrine," and therefore the rule was progressively weakened by the development of exceptions.

Yet such an explanation does not account for the process by which the exceptions in fact developed. The most important exceptions, and their justifications, developed contemporaneously with the broad rule itself. . . .

Ohio's acceptance of the fellow-servant rule was so qualified, and the exception it created so broad, that it was sometimes cited as a jurisdiction which had rejected the rule outright. In fact, in the first such case to come before the Ohio Supreme Court, *Little Miami Railroad v. Stevens,* the divided court did reject the rule, the only early case to do so. The three opinions in the case present, in microcosm, the range of judicial options open to courts in the 1850s when faced for the first time with a fellow-servant question.

John Stevens was driving an eastbound train for the defendant railroad company when the train collided with a westbound train also owned by the Little Miami. As stated in the declaration, the plaintiff was "violently crushed" and was "greatly scalded, bruised, lacerated, hurt and wounded, and in consequence thereof became, and was sick, sore, lame, and disordered." The collision occurred because of a change in schedule of which the plaintiff was not informed but of which the conductor—the "commanding officer" of the train—apparently was. The jury returned a verdict for the plaintiff in the amount of $3,700.

The lawyers' arguments on appeal gave little hint of the decision that was to come. The defendant, predictably, relied on Story, *Priestley,* and *Farwell* (in that order) for the proposition that the broad fellow-servant rule was "well settled." Rather than arguing against the rule, however, the plaintiff's counsel maintained that Stevens's case ought to constitute an exception.

The majority opinion by Judge Caldwell was a model of the instrumental judicial style. Caldwell based his opinion not on precedent, but on reason and policy. He began with the general rule, "founded in reason," that a principal must be held responsible for the consequences of any power he sets "in motion for his own benefit." Such a rule was supported by policy, he observed, for "it is necessary as a preventive of mischief, and a protection to [the] community." Caldwell then confronted directly the central *Priestley* and *Farwell* legal argument that an employee's implied employment contract and his level of wages encompassed his assumption of risk of a co-worker's negligence. "If the party does contract in reference to the perils incident to the business," Caldwell noted, "he will only be presumed to contract in reference to such as necessarily attend it when conducted with ordinary care and prudence." "The employer has paid him no money for the right to break his legs," he wrote scathingly, "or, as in this case, to empty on him the contents of a boiler of scalding water."

Caldwell also rejected the policy argument employed in *Priestley* and *Farwell,* that the adoption of the fellow-servant rule would promote safety in the workplace by giving employees an incentive to be watchful of their co-workers' misconduct and carelessness. Caldwell responded to this argument by stating that "we do not think it likely that persons would be careless of their lives and persons or property" solely because they had a cause of action against the employer for provable injuries. "If men are influenced by such remote considerations to be careless of what they are likely to be most careful about," he concluded caustically, "it has never come under our observation."

Having discussed and disposed of *Priestley* and *Farwell,* Caldwell declared himself not to be bound by them. Although he could easily distinguish the case before him on its facts, he did not deign to do so. "So far as those cases decide that a recovery cannot be had in a case like the one now before the Court," he held, "we think they are contrary to the general principles of law and justice, and we cannot follow them as precedents."

Caldwell's bold style was mirrored by an equally strong dissent by Judge Spalding. Spalding, however, looked to the past for his judicial model; his opinion would not have appeared out of place in any eighteenth-century report. He carefully canvassed the precedents and concluded that "[t]he solemn adjudications of Courts of recognized authority should be followed," as "[t]here is nothing to distinguish this State from others." He direly cited Latin maxims to the effect that "even when the reason of a rule cannot well be discerned, 'the wisdom of the rule has in the end appeared, from the inconveniences that have followed the innovation.' " The irony, of course, is that the American rule which Spalding wished to protect against innovation was barely a decade old.

The third opinion in *Little Miami* was that of Chief Judge Peter Hitchcock, one of the great judges of his time. Hitchcock, far from following Caldwell's bold line, concurred in the judgment only. He was deeply disturbed by Caldwell's approach, for, he noted, "[i]f this case were, in its principal features," like *Farwell* or *Priestley,* "I should hesitate long before I would consent to disregard those decisions." "They were," he observed, "decisions made by highly respectable tribunals, and by men whose opinions are entitled to the highest considerations." Happily for Stevens, however, Hitchcock discerned a distinction that enabled him to decide for the plaintiff "without conflicting at all with the authorities."

Hitchcock began by finding that the accident was not caused by either the engineer or conductor, but rather by "the superintendent of the road." He then stated that employees of the company "engaged in making or repairing their road, or in running their cars," such as the plaintiff, are truly "servants" of the company. The *superintendent,* however, is an "*agent*" or "proper representative of the company" and is thus not an employee or servant. Hitchcock concluded that the plaintiff should therefore recover because his "injury resulted from the negligence of the company itself, or of an agent whose duty it was to give the notice" of the schedule change to the engineer and conductor, and not from the negligence of a fellow servant or employee.

Hitchcock's opinion is significant because it is an example of the preferred judicial response to an authoritative line of cases which, if followed, would yield an inappropriate result. Rather than reject the broad fellow-servant rule, as Caldwell had done, Hitchcock sought to exempt the particular case before him from its force. Courts that resisted the rule tended to do so by following Hitchcock's rather than Caldwell's approach.

Jurisprudence by Exception

. . .

As fellow-servant cases came before other jurisdictions, new "exceptions" were created, often in cases of first impression. The variety of these exceptions was limited only by the varying factual contexts in which negligent employees could kill or maim their co-workers. In several southern states, therefore, a "slave" exception developed. Courts rationalized the exception by observing that slaves were neither free to leave their employment nor to report a white co-worker's negligence and thereby avoid its consequences. Similarly, when accidents occurred between employees of a similar rank, some courts developed a "competent servant" exception, reasoning that no implied employment contract could relieve a master of his obligation to hire competent employees, nor could an employee assume the risk of a co-worker's incompetence. When accidents occurred because of defective equipment, some courts found a "warranty of safe equipment" exception, for it was the master's duty to guarantee the safety of the equipment and the employees could not be expected to assume risks of which they were not aware. When, in 1879, the Supreme Court finally adopted the fellow-servant rule, it did so in a very narrow way, listing no fewer than four "well-defined exceptions" resting "upon principles of justice, expediency, and public policy" by which the rule must be qualified.

The combined effect of these exceptions, which were adopted variously by the different jurisdictions, was to rob the rule of much of its economic effect since many railroad accidents were covered by one exception or another. Moreover, the opinions of the courts adopting the exceptions demonstrate serious dissatisfaction with the policies and premises of the fellow-servant rule. The question then becomes, why did they consistently carve out exceptions which ill served railroad interests, and which made no legal sense?

One reason for the pattern described above is intellectual. The system of precedent not only preserves the legitimacy of judicial decision-making, but it reinforces the judge's own sense of the correctness of his views. No person, even a judge, prefers to stand alone and exposed. The effect of this phenomenon on the fellow-servant rule was graphically demonstrated in Wisconsin, which, like Ohio, at first, rejected the rule. In a lengthy opinion in *Chamberlain v. Milwaukee & Mississippi Railroad,* Justice Paine criticized the entire range of fellow-servant cases, and the policies underlying them. He recognized that a great majority of them had decided that an employer should not be held liable. But he conceived that a court must not only "count but . . . weigh the cases, and by this test the majority do not always rule." He recognized that both Ohio and Indiana had couched their opposition to the rule in terms of exceptions, a tactic he believed was suggested "out of deference to authorities, from which they did not care to depart, further than the facts of the case made necessary." Wisconsin, however, would be more forthright. "We are satisfied, therefore, that the general principles of the common law sustain [an employer's] liability, and that those cases which have attempted to establish an exception [to respondeat], do not rest upon solid ground." The court thereupon reversed the lower court, which had relied upon the rule, and remanded for a new trial.

When the case came up again the next year, the court reversed itself. Its reasons were clearly stated in the short opinion by Chief Justice Dixon, who had been a member of the majority the year before. The original judgment, he noted, had been "sustained by weighty and powerful reasons," and he had not changed his views regarding its correctness. Nevertheless, he noted, in the intervening year, Ohio had moved closer to the majority rule, and Wisconsin, therefore, "stands alone." He had switched his vote, therefore, "more from that deference and respect which is always due to the enlightened and well-considered opinions of others, than from any actual change in my own views. . . . I think I am bound," he concluded, "to yield to this unbroken current of judicial opinion."

Wisconsin exemplifies the intellectual pressures weighing upon judges who were disposed to dissent from the rule, pressures which grew stronger as the number of jurisdictions adopting the rule grew. Distinguishing the case before the court on its facts, and incidentally creating new exceptions, was therefore a much more comfortable approach than outright rejection.

Such a course, moreover, was often the only one open to a court, given the questions that the lawyers had presented to them for decision. In the cases of first impression considered in this selection, no plaintiff's lawyer, save one, argued even in the alternative that a court should reject the rule. This was true not only in the late 1850s when the line of authorities had lengthened, but from the earliest cases. Rather than contesting the rule itself, lawyers invariably argued that their client's case constituted an exception. In *Farwell* itself, the plaintiff's lawyer did not question *Priestley's* holding or rationale.

The lawyers, of course, were simply doing their jobs—attempting to gain a recovery for their clients with the least risk of a contrary decision. Their instinct to distinguish, rather than argue from principle, was a product of the practical, apprenticeship system of legal training that the vast majority of nineteenth-century lawyers had received.

. . .

Conclusion

The creation of the fellow-servant rule provides a case study of the nature of legal change in nineteenth-century America. If the process by which it was created was typical, one must conclude that judicial instrumentalism, which shaped legal rules to achieve economic and social goals, coexisted with and was tempered by intellectual norms within the legal community which exerted a powerful pull toward formalism and stability. The interplay of both these forces produced the fellow servant rule.

The rule was shaped by instrumentalist judges like Shaw of Massachusetts and Caldwell of Ohio, who responded to fellow-servant cases with legal rules which, though opposite in result, drew their rationale from the economic and social realities of an industrializing society.

The spread of the rule from jurisdiction to jurisdiction, however, was largely due to judges who were motivated primarily by intellectual norms within the legal community. The history of the fellow-servant rule demonstrates that the impetus to formalism was inherent in, and inseparable from, those norms, which placed a premium on short-range precedential consistency at the expense of long-range policy goals.

. . .

More specifically, the fellow-servant cases indicate, at the least, that certain of Horwitz's conclusions should be qualified. His periodization of judicial styles, in which instrumentalist judges replace and then are replaced by formalist ones, will not account for the pattern of coexistence described above. Moreover, the theories to which the judges responded will not fit neatly into Horwitz's timeline. The cases suggest that an exaggerated respect for English precedent and a belief in common-law myths of harmony and judicial "discovery" persisted far into the nineteenth century. Similarly, the underpinnings of judicial formalism, far from surfacing in the late nineteenth century, are inherent in a precedential legal system and significantly influenced the judges who adopted the fellow-servant rule.

Finally, the development of the exceptions to the rule call into question the economic determinists' conclusions about the purposive nature of legal change. Horwitz depicted a process in which judges, armed with an instrumental conception of the power of law to aid the entrepreneurial classes, consciously and purposefully shaped private law to accomplish these ends. Yet the development of the fellow-servant rule, with its contemporaneous creation of exceptions, was anything but purposive. Judges and lawyers created this misshapen rule out of the immediate needs of the case at hand with little concern for long-range consistency.

Notes and Questions

1. It is possible that testimony elicited from fellow employees in a lawsuit pitting an employee against their mutual employer is unreliable. Employees may fear reprisals from their employer and may fear to give testimony that supports the claim against this employer. Alternatively, there may be a tendency to support one's colleagues. Are these potential biases greater than problems posed in other lawsuits? If these potential biases are serious, what legal rules would you expect to have developed? Can these dangers explain the evolution of the fellow-servant rule or of workers' compensation schemes?

2. What does the author of this selection mean by "judicial instrumentalism"? Are Shaw and Caldwell regarded as instrumentalists because of the language of their decisions or because of the results they reached? Can you give other examples of decisions in tort law that can be described as (explicitly or implicitly) instrumental?

3. Do you expect to find vicarious liability and fellow-servant doctrines (by the same or other names) in other legal systems?

Social Change and the Law of Industrial Accidents

LAWRENCE M. FRIEDMAN AND JACK LADINSKY

Sociologists recognize, in a general way, the essential role of legal institutions in the social order. They concede, as well, the responsiveness of law to social change and have made important explorations of the interrelations involved. Nevertheless, the role law plays in initiating—or reflecting—social change has never been fully explicated, either in theory or through research. The evolution of American industrial-accident law from tort principles to compensation systems is an appropriate subject for a case study on this subject. It is a topic that has been carefully treated by legal scholars, and it is also recognized by sociologists to be a significant instance of social change. This essay, using concepts drawn from both legal and sociological disciplines, aims at clarifying the concept of social change and illustrating its relationship to change in the law.

The Concept of Social Change

Social change has been defined as "any nonrepetitive alteration in the established modes of behavior in . . . society. Social change is *change in the way people relate to each other,* not change in values or in technology. Major alterations in values or technology will, of course, almost invariably be followed by changes in social relations—but they are not in themselves social change. Thus, although the change from tort law to workmen's compensation presupposed a high level of technology and certain attitudes toward the life and health of factory workers, it was not in itself a change in values or technology but in the patterning of behavior.

Social change may be revolutionary, but it normally comes about in a more or less orderly manner, out of the conscious and unconscious attempts by people to solve social problems through collective action. It is purposive and rational; although social actions have unanticipated consequences and often arise out of unconscious motivations, nonetheless social change at the conscious level involves definition of a state of affairs as a "problem" and an attempt to solve that problem by the rational use of effective means. The *problem* defined collectively in this instance was the number of injuries caused to workmen by trains, mine hazards, and factory machinery. It is clear that the number of accidents increased over the course of the century, but it is not self-evident that this objective fact necessarily gave rise to a correspondent subjective sense of a problem that had to be solved in a particular way. To understand the process of social change, one must know how and why that subjective sense evolved. This requires knowledge of how and why various segments of society perceived situations—whether they identified and defined a set of facts or a state of affairs as raising or not raising problems. It also requires an understanding of what were considered appropriate and rational means for solving that problem. The perspective of a particular period, in turn, sets limits to the way a people collectively defines problems and the means available for their solution.

This selection deals with behavior at the societal level. At that level, social change necessarily means changes in powers, duties, and rights; it will normally be reflected both in custom and law, in formal authority relations and informal ones. In mature societies, law will be an important indicator of social change; it is institutional cause and institutional effect at the same time, and a part of the broader pattern of collective perceptions and behavior in the resolution of social problems. The selection will therefore also deal with the way in which legal systems respond to their society—the social impact on law, modified and monitored by the institutional habits of the legal system.

. . .

In theory, at least, recovery for industrial accidents might have been assimilated into the existing system of tort law. The fundamental principles were broad and simple. If a factory worker was injured through the negligence of another person—including his employer—an action for damages would lie. Although as a practical matter, servants did not usually sue their master nor workers their employers, in principle they had the right to do so.

In principle, too, a worker might have had an action against his employer for any injury caused by the negligence of any other employee. The doctrine of respondeat superior was familiar and fundamental law. A principal was liable for the negligent acts of his agent.

. . .

Conceivably, then, one member of an industrial work force might sue his employer for injuries caused by the negligence of a fellow worker. A definitive body of doctrine was slow to develop, however. When it did, it rejected the broad principle of respondeat superior and took instead the form of the so-called fellow-servant rule. Under this rule, a servant (employee) could not sue his master (employer) for injuries caused by the negligence of another employee. The consequences of this doctrine were far reaching. An employee retained the right to sue the employer for injuries, provided they were caused by the employer's personal misconduct. But the factory system and corporate ownership of industry made this right virtually meaningless. The factory owner was likely to be a "soulless" legal entity; even if the owner was an individual entrepreneur, he was unlikely to concern himself physically with factory operations. In work accidents, then, legal fault would be ascribed to fellow employees, if anyone. But fellow employees were men without wealth or insurance. The fellow-servant rule was an instrument capable of relieving employers from almost all the legal consequences of industrial injuries. Moreover, the doctrine left an injured worker without any effective recourse but an empty action against his co-worker.

When labor developed a collective voice, it was bound to decry the rule as infamous, as a deliberate instrument of oppression—a sign that law served the interests of the rich and propertied, and denied the legitimate claims of the poor and the weak. The rule charged the "blood of the workingman" not to the state, the employer, or the consumer, but to the working man himself. Conventionally, then, the fellow-servant rule is explained as a deliberate or half-deliberate rejection of a well-settled principle of law in order to encourage enterprise by forcing workmen to bear the costs of industrial injury. And the overthrow of the rule is taken as a sign of a conquest by progressive forces.

It is neither possible nor desirable to avoid passing judgment on human behavior; but one's understanding of social processes can sometimes be hindered by premature moral assessments. The history of industrial accident law is much too complicated to be viewed as merely a struggle of capital against labor, with law as a handmaid of the rich, or as a struggle of good against evil. From the standpoint of social change, good and evil are social labels based on *perceptions* of conditions, not terms referring to conditions in themselves. Social change comes about when people decide that a situation is evil and must be altered, even if they were satisfied or unaware of the problem before.

. . .

Doctrinal complexity and vacillation in the upper courts, coupled with jury freedom in the lower courts, meant that by the end of the [nineteenth] century the fellow-servant rule had lost much of its reason for existence. . . . By 1911, 25 states had laws modifying or abrogating the fellow-servant doctrine for railroads. Railroad-accident law reached a state of maturity earlier than the law of industrial accidents generally; safety controls were imposed on the roads, and the common-law tort system was greatly modified by removal of the employer's most effective defense. The Interstate Commerce Commission called a conference of state regulatory authorities in 1889; the safety problem was discussed, and the commission was urged to investigate the problem and recommend legislation. In 1893, Congress required interstate railroads to equip themselves with safety appliances, and provided that any employee injured "by any locomotive, car, or train in use" without such appliances would not "be deemed . . . to have assumed the risk thereby occasioned."

The Federal Employers' Liability Act [FELA] of 1908 went much further; it abolished the fellow-servant rule for railroads and greatly reduced the strength of contributory negligence and assumption of risk as defenses. Once the employers had been stripped of these potent weapons, the relative probability of recovery by injured railroad employees was high enough so that workmen's compensation never seemed as essential for the railroads as for industry generally. The highly modified FELA tort system survives (in amended form) to this day for the railroads. It is an anachronism, but one which apparently grants some modest satisfaction to both sides. Labor and management both express discontent with FELA, but neither side has been so firmly in favor of a change to workmen's compensation as to make it a major issue.

FELA shows one of many possible outcomes of the decline in efficacy of the fellow-servant rule. Under it, the rule was eliminated, and the law turned to a "pure" tort system—pure in the sense that the proclivities of juries were not interfered with by doctrines designed to limit the chances of a worker's recovery. But the railroads were a special case. Aside from the special history of regulation, the interstate character of the major railroads made them subject to national safety standards and control by a single national authority. For other industrial employers, the FELA route was not taken; instead, workmen's compensation acts were passed. In either case, however, the fellow-servant rule was abolished, or virtually so. Either course reflects, we can assume, some kind of general agreement that the costs of the rule outweighed its benefits.

Rising Pressures for Change

The common-law doctrines were designed to preserve a certain economic balance in the community. When the courts and legislatures created numerous exceptions, the rules lost much of their efficiency as a limitation on the liability of businessmen. The rules prevented many plaintiffs from recovering, but not all; a few plaintiffs recovered large verdicts. There were costs of settlements, costs of liability insurance, costs of administration, legal fees and the salaries of staff lawyers. These costs rose steadily, at the very time when American business, especially big business, was striving to rationalize and bureaucratize its operations. It was desirable to be able to predict costs and insure against fluctuating, unpredictable risks. The costs of industrial accident liability were not easily predictable, partly because legal consequences of accidents were not predictable.

. . .

When considerations of politics were added to those of business economics and industrial peace, it was not surprising to find that businessmen gradually withdrew their veto against workmen's compensation statutes. They began to say that a reformed system was inevitable—and even desirable. A guaranteed, insurable cost—one which could be computed in advance on the basis of accident experience—would, in the long run, cost business less than the existing system. In 1910, the president of the National Association of Manufacturers (NAM) appointed a committee to study the possibility of compensating injured workmen without time-consuming and expensive litigation, and the convention that year heard a speaker tell them that no one was satisfied with the present state of the law—that the employers' liability system was "antagonistic to harmonious relations between employers and wage workers." By 1911 the NAM appeared convinced that a compensation system was inevitable and that prudence dictated that business play a positive role in shaping the design of the law—otherwise the law would be "settled for us by the demagogue, and agitator and the socialist with a vengeance." Business would benefit economically and politically from a compensation system, but only if certain conditions were present. Business, therefore, had an interest in pressing for a specific kind of program, and turned its attention to the details of the new system. For example, it was imperative that the new system be in fact as actuarially predictable as business demanded; it was important that the costs of the program be fair and equal in their impact upon particular industries, so that no competitive advantage or disadvantage flowed from the scheme. Consequently the old tort actions had to be eliminated, along with the old defenses of the company. In exchange for certainty of recovery by the worker, the companies were prepared to demand certainty and predictability of loss—that is, limitation of recovery. The jury's caprice had to be dispensed with. In short, when workmen's compensation became law, as a solution to the industrial accident problem, it did so on terms acceptable to industry. Other pressures were there to be sure, but when workmen's compensation was enacted, businessmen had come to look on it as a positive benefit rather than as a threat to their sector of the economy.

The Emergence of Workmen's Compensation Statutes

The change of the businessman's, the judge's, and the general public's attitudes toward industrial injuries was accelerated by the availability of fresh information on the extent of accidents and their cost to both management and workers. By 1900, industrial accidents and the shortcomings of the fellow-servant rule were widely perceived as *problems* that had to be solved. After 1900, state legislatures began to look for a "solution" by setting up commissions to gather statistics, to investigate possible new systems, and to recommend legislation. The commissions held public hearings and called upon employers, labor, insurance companies, and lawyers to express their opinions and propose changes. A number of commissions collected statistics on industrial accidents, costs of insurance, and amounts disbursed to injured workmen. By 1916, many states and the federal government had received more or less extensive public reports from these investigating bodies. The reports included studies of industrial-accident cases in the major industries, traced the legal history of the cases, and looked into the plight of the injured workmen and their families.

From the informaiton collected, the commissions were able to calculate the costs of workmen's compensation systems and compare them with costs under employers' liability. Most of the commissions concluded that a compensation system would be no more expensive than the existing method, and most of them recommended adoption, in one form or another, of workmen's compensation. In spite of wide variations in the systems proposed, there was agreement on one point: workmen's compensation must fix liability upon the employer regardless of fault.

Between 1910 and 1920 the method of compensating employees injured on the job was fundamentally altered in the United States. In brief, workmen's compensation statutes eliminated (or tried to eliminate) the process of fixing civil liability for industrial accidents through litigation in common-law courts. Under the statutes, compensation was based on statutory schedules, and the responsibility for initial determination of employee claims was taken from the courts and given to an administrative agency. Finally, the statutes abolished the fellow-servant rule and the defenses of assumption of risk and contributory negligence.

. . .

Compensation systems varied from state to state, but they had many features in common. The original Wisconsin law was representative of the earlier group of statutes. It set up a voluntary system—a response to the fact that New York's courts had held a compulsory scheme unconstitutional on due process grounds. Wisconsin abolished the fellow-servant rule and the defense of assumption of risk for employers of four or more employees. In turn, the compensation scheme, for employers who elected to come under it, was made the "exclusive remedy" for an employee injured accidentally on the job. The element of "fault" or "negligence" was eliminated, and the mere fact of injury at work "proximately caused by accident," and not the result of "wilful misconduct," made the employer liable to pay compensation but exempt from ordinary tort liability. The state aimed to make it expensive for employers to stay out of the system. Any employer who did so was liable to suit by injured employees and the employer was denied the common-law defenses.

The compensation plans strickly limited the employee's amount of recovery. In Wisconsin, for example, if an accident caused "partial disability," the worker was to receive 65 percent of his weekly loss in wages during the period of disability, not to exceed four times his average annual earnings. The statutes, therefore, were compensatory, not punitive, and the measure of compensation was, subject to strict limitations, the loss of earning power of the worker. In the original Wisconsin act, death benefits were also payable to dependents of the worker. If the worker who died left "no person dependent upon him for support," the death benefit was limited to "the reasonable expense of his burial, not exceeding $100." Neither death nor injury as such gave rise to a right to compensation—only the fact of economic loss to someone, either the worker himself or his family. The Wisconsin act authorized employers to buy annuities from private insurance companies to cover projected losses. Most states later made insurance or self-insurance compulsory. Some states have socialized compensation insurance, but most allow the purchase of private policies.

In essence, then, workmen's compensation was designed to replace a highly unsatisfactory system with a rational, actuarial one. It should not be viewed as the replacement of a fault-oriented compensation system with one unconcerned with fault. It should not be viewed as a victory of employees over employers. In its initial stages, the fellow-servant rule was not concerned with fault, either, but with establishing a clear-cut, workable, and predictable rule, one which substantively placed much of the risk (if not all) on the worker. Industrial accidents were not seen as a social problem—at most as an economic problem. As value perceptions changed, the rule weakened; it developed exceptions and lost its efficiency. The exceptions and counter-exceptions can be looked at as a series of brief, ad hoc, and unstable compromises between the clashing interests of labor and management. When both sides became convinced that the game was mutually unprofitable, a compensation system became possible. But this system was itself a compromise: an attempt at a new, workable, and predictable mode of handling accident liability which neatly balanced the interests of labor and management.

The Law of Industrial Accidents and Social Theory: Three Aspects of Social Change

This case study, devoted to the rise and fall of the fellow-servant rule, utilizes and supports a view of social change as a complex chain of group bargains—

economic in the sense of a continuous exchange of perceived equivalents, though not economic in the sense of crude money bargains. It also provides a useful setting for evaluating three additional popular explanations of the origin or rate of social change. First, the apparently slow development of workmen's compensation is the classic example of what Ogburn called "cultural lag." Second, since German and English statutes were enacted prior to the American laws, the establishment of compensation schemes in America can be viewed as a case of cross-cultural influence. Third, the active role of particular participants (in Wisconsin, for example, Judge Marshall and John R. Commons) may substantiate the theory which advances the causal influence of "great men" in the process of social change. . . .

The Concept of Cultural Lag

The problem of "fair and efficient incidence of industrial accident costs," in the words of Willard Hurst, "followed a fumbling course in courts and legislature for 50 years before the first broad-scale direction [leading to workmen's compensation] was applied." In a famous book written in 1922, the sociologist William Fielding Ogburn used the example of workmen's compensation and the 50-year period of fumbling to verify his "hypothesis of cultural lag." "Where one part of culture changes first," said Ogburn, "through some discovery or invention, and occasions changes in some part of culture dependent upon it, there frequently is a delay. . . . The extent of this lag will vary . . . but may exist for . . . years, during which time there may be said to be a maladjustment." In the case of workmen's compensation, the lag period was from the time when industrial accidents became numerous until the time when workmen's compensation laws were passed, "about a half-century, from 1850–70 to 1915." During this period, "the old adaptive culture, the common law of employers' liability, hung over after the material conditions had changed."

. . .

There were important reasons why 50 years elapsed before workmen's compensation became part of the law. Under the impact of industrial conditions Americans were changing their views about individual security and social welfare. Dean Pound has remarked that the twentieth century accepts the idea of insuring those unable to bear economic loss, at the expense of the nearest person at hand who can bear the loss. This conception was relatively unknown and unacceptable to judges of the nineteenth century. The fellow-servant rule could not be replaced until economic affluence, business conditions, and the state of safety technology made feasible a more social solution. Labor unions of the mid-nineteenth century did not call for a compensation plan; they were concerned with more basic (and practical) issues such as wages and hours.

. . .

What appears to some as an era of "lag" was actually a period in which issues were collectively defined and alternative solutions posed, and during which interest groups bargained for favorable formulations of law. It was a period of "false starts"—unstable compromise formulations by decision makers armed with few facts, lacking organizational machinery, and facing great, often contradictory, demands from many publics. There was no easy and suitable solution, in the light of the problem and the alignment of powers. Indeed, workmen's compensation—which today appears to be a stable solution—was only a compromise, an answer acceptable to enough people and interest groups to endure over a reasonably long period of time.

Part of what is later called "lag," then, is this period of false starts—the inadequate compromises by decision makers faced with contradictory interest groups pressing inconsistent solutions. There may not be a "solution" in light of the alignment of interests and powers with respect to the problem at any given point in time. Perhaps only a compromise "solution" is possible. What later appears to be the final answer is in fact itself a compromise—one which is stable over some significant period of time. Sociologically, that is what a "solution" to a problem is: nothing more than a stable compromise acceptable to enough people and interest groups to maintain itself over a significant period of time.

Cross-Cultural Borrowing

The adoption of workmen's compensation in America does represent an instance of what can be called conscious cross-cultural borrowing. Workmen's compensation was not an American innovation; there were numerous European antecedents. Switzerland passed a workmen's compensation act in 1881; Germany followed in 1884 with a more inclusive scheme. By 1900 compensation laws had spread to most European countries. In 1891 the United States Bureau of Labor commissioned John Graham Brooks to study and appraise the German system. His report, published in 1893, was widely distributed and successfully exposed some American opinion-leaders to the existence of the European programs. Most of the state investigating commissions also inquired into the European experience, and a number of early bills were modeled after the German and British systems.

Though workmen's compensation can therefore be viewed as an example of cross-cultural borrowing, care must be exercised in employing the concept. Successful legal solutions to social problems are often borrowed across state and national lines but this borrowing must not be confused with the actual "influence" of one legal system over another. "Influence" carries with it an implication of power or, at the least, of cultural dominance. The forces that led to a demand for workmen's compensation were entirely domestic, as this study has argued. The fact that European solutions to similar problems were studied and, to an extent, adopted here shows not dominance but an attempt to economize time, skill, and effort by borrowing an appropriate model. It would be quite wrong to detect European legal "influence" in this process. The existence of the European compensation plans was not a cause of similar American statutes. Rather, the interest shown in the foreign experiences was a response to American dissatisfaction with existing industrial accident law. Similarly, the current drive for an American *ombudsman* is not an example of the "influence" of Scandinavian law. A foreign model here sharpens discussion and provides a ready-made plan. Yet the felt need for such an officer has domestic origins.

Great Men and Social Change

Sociologists are fond of pointing out the inaccuracy of the "great-man theory of history," which holds that particular persons play irreplaceably decisive roles in determining the path of social change. The influence of single individuals, they say, is hardly as critical as historians would have us believe. The role of outstanding persons in bringing about workmen's compensation acts seems on one level quite clear. In Wisconsin, Roujet Marshall excoriated the existing system from the bench; off the bench he was a vigorous champion of the new law and, indeed, helped draft it. John R. Commons worked tirelessly for passage of the act, and served on the first Industrial Commission whose obligation it was to administer the law. His writings and teachings helped mobilize informed public opinion and virtually created a lobby of academicians for workmen's compensation. Political figures, businessmen, union leaders, and others played active roles in the passage of the law. It is quite tempting to say that the Wisconsin law would be unthinkable but for the work of Marshall, or Commons, or LaFollette and the Progressive tradition in the state, or the craftsmanship of Wisconsin's pioneering legislative reference service under the skilled leadership of Charles McCarthy. Reformers and academicians served as important middlemen in mediating between interest groups and working out compromises. Their arguments legitimated the act; their zeal enlisted support of middle-class neutrals. They were willing to do the spadework of research, drafting, and propagandizing necessary for a viable law. In the passage of many other welfare and reform laws, outstanding personalities can be found who played dominant roles in creating and leading public opinion—for example, Lawrence Veiller for the New York tenement housing law of 1901, Harvey Wiley for the Federal Food and Drug Act.

The great-man hypothesis is not susceptible of proof or disproof. But the course of events underlying workmen's compensation at least suggests that social scientists are properly suspicious of placing too much reliance on a great-man view. If the view here expressed is correct, then economic, social, political and legal forces made workmen's compensation (or some alternative, such as FELA) virtually inevitable by the end of the nineteenth century. Outstanding [individuals] may be necessary in general for the implementation of social change; someone must take the lead in creating the intellectual basis for a change in perception. Nonetheless, when a certain pattern of demand exists in society, more than one person may be capable of filling that role. Particular individuals are normally not indispensable. The need is for talent— [people] with extraordinary ability, perseverance, and personal influence who can surmount barriers and accomplish significant results. Obviously, the ab-

sence of outstanding persons interested in a particular cause can delay problem solving or lead to inept, shoddy administration. The appearance of truly
exceptional persons at the proper moment in history is undoubtedly not automatic. But talent, if not genius, may well be a constant in society; and the
social order determines whether and in what direction existing talent will be
exerted.

Thus, it would be foolish to deny that specific individuals exert great
influence upon the development of social events, and equally foolish to conclude that other persons could not have done the job as well (or better) if
given the opportunity. "Great men," however, must be in the right place,
which means that society must have properly provided for the training and
initiative of outstanding persons and for their recruitment into critical offices
when needed. In difficult times, great businessmen, political leaders, musi-
cians, or physicists will emerge. "Great men" appear "when the time is
ripe"—but only insofar as society has created the conditions for a pool of
creative manpower dedicated to the particular line of endeavor in which their
greatness lies.

Notes and Questions

1. If "social change comes about when people decide that a situation is evil and
must be altered," what precipitates the change? Is this perception described by the
authors as one possible precondition of change, as a necessary condition, or as a
sufficient condition of change?

2. In retrospect, changes we know about often seem inevitable. Viewed from the
perspective of the twentieth century, the rise of institutions such as insurance and
courts of appeal seems inevitable, as does the decline of such diverse institutions as
polygamy, public whippings, and the fellow-servant rule. But prospectively it is surely
more difficult to predict change. Do the changes described in this reading seem inevitable? Can you imagine the decline of liability insurance, for example, perhaps because
it removes a serious deterrent to tortious behavior, or is there something about this
institution or the nature of social change that guarantees its survival or expansion?
What about the rise or decline of environmental law (through the tort system or a
separate regulatory framework)? What about vicarious liability?

VI

Omissions and Commissions

There is a useful but vulnerable distinction between omissions and commissions in the law. The common law will neither penalize a passer-by who fails to rescue a stranger from drowning nor normally impose liability on someone who fails to disclose information in a negotiation or other setting. It will, however, more readily hold liable one who affirmatively begins a rescue and then behaves negligently, as well as one who affirmatively misrepresents in order to mislead another. There is, of course, a fine (or absurd) line between omissions and commissions. For example, negligent driving can sometimes be characterized as an omission, perhaps because of a failure to apply brakes. Nevertheless, our legal system regularly deters or penalizes antisocial commissions while compensating the victims of such acts, through tort, criminal, and regulatory schemes, but it rarely rewards rescuers or penalizes those who fail to effect rescues.

Inasmuch as many observers share the intuition that citizens ought to come to the aid of one another, and even be required by law to do so, the subject of rescue and the distinction between action and inaction is an interesting area in which to compare or combine interdisciplinary approaches. The first reading in this part considers the question as one of finding the optimal combination of rewards and penalties needed to get a job done. It adds a comparative approach to this functional, or economic, strategy by considering the incentive mix used in other legal systems and discusses the future evolution of the tort system.

The second selection offers an introduction to the potential links between tort and feminism theory. It highlights the argument that the legal distinction between omissions and commissions may reflect a male perspective. Although it may be difficult to imagine a world different from the one we know, the reading dares us to do so. It encourages us to think that some of the instincts we bring to the study of the law—and then discard—may be valuable windows to a different world (even if we develop these intuitions in the familiar, tainted world we know).

The final reading adopts a philosophical approach to rescue. It explores the question of whether a utilitarian argument for a duty to rescue devolves into an argument for obligatory and virtually unlimited altruism, requiring all interventions that can make others better off at reasonable cost.

There is a tendency to react to the simultaneous presentation of several approaches with a preference for one approach. And in many ways interdisciplinary approaches do compete with one another for the mind of the reader. It is surely no accident, for instance, that each of the selections in this part declines to mix philosophy, economics, and feminism. One useful question to bear in mind when reading these materials is whether this competition, if it is that, is necessary. Are these approaches mutually exclusive or is there a possibility of combining these approaches and constructing a protean, cross-bred framework?

Waiting for Rescue: An Essay on the Evolution and Incentive Structure of the Law of Affirmative Obligations

SAUL LEVMORE

. . .

The Incentive Structure for Rescue

Rewards and penalties will motivate a potential rescuer only if he is aware of these incentives and then only if in a crucial moment he is able to be influenced by this knowledge. Although there is some reason to believe that citizens are aware of the rules regarding rescue and that many rescues require prepositioning or afford time for contemplation, arguments for and against various rules must surely rest on a sense that incentives work on the margin. I do not argue . . . that one set of rules will best promote socially desirable rescues or, indeed, that these penalties and rewards have substantial behavioral—as opposed to symbolic—effects. I imagine, instead, that lawmakers will *try* to promote certain kinds of behavior with legal rules and will be especially eager to enact rules that either work on the margin or are, at worst, harmless. This is a positive and not a normative exercise.

The law could, of course, simply give a successful rescuer a reward for his deed. This "carrot" might be paid by the rescued victim, by the state, or by some third party. If the rescue is of property rather than life, as when a rescuer brings a fire under control and prevents further loss of property, then this reward could easily be varied according to the value of the property saved. Rewards for the rescue of life or property could be very substantial, sufficient to promote the equivalent of a competitive industry of rescuers, or barely enough to cover the rescuer's expenses. It is easy to quibble with this classification and insist that mere reimbursement of expenses simply erases a penalty otherwise incurred by a rescuer, but both because these expenses are not imposed by the law and because the rescuer may well regard a package of reimbursement, public acclaim, and private gratitude as a substantial carrot, it is convenient to think of a legal right to reimbursement as a reward. Apart from any other rewards, the law could exempt rescuers from liability for some or all of the injuries they happen to impose while engaged in their rescue. Finally, rescuers who are themselves injured while attempting to render assistance could be allowed to collect from the victim (whether or not actually rescued) or some other party.

Abridged and reprinted without footnotes by permission of Virginia Law Review Association and Fred B. Rothman & Company from 72 *Virginia Law Review* 879 (1986).

The available penalties are even simpler to sketch. The law must, if it seeks to penalize nonrescuers, first identify a nonrescuer and then provide some penalty, or "stick." Both because penalties are often contained within the criminal law, where doubts are usually resolved in favor of the accused, and because it would defeat the instrumental purpose of the law to punish nonrescue too seriously, the identification of nonrescuers is usually limited to persons who should realize that they could easily take some step to save life or valuable property. It is very unlikely, for instance, that any legal system would penalize someone who only called the police—even though a more direct step would have been more useful—or who genuinely believed that some other onlooker had already called for help. Once a nonrescuer is identified, the penalty that is imposed can vary over a familiar range; nonrescuers can be fined or incarcerated, held civilly liable for the injury that befalls the unassisted victim, or both.

Landes and Posner's Arguments

Any discussion of the likely behavioral effects of the carrots and sticks that might be offered to rescuers best begins with three points made by Professors Landes and Posner in their well-known work on the law and economics of rescue. They argue that: (1) an excessively large reward will sometimes lead to inefficient consequences because victims or owners will take excessive precautions to avoid paying such rewards; (2) a rescuer may be motivated by altruism or by the possibility of being regarded as a hero, and that legal inducements can impede this motivation; and (3) excessive penalties can decrease the number of rescues because potential rescuers may avoid potential "rescue spots," such as beaches, in order to avoid the reach of these penalties.

. . .

The first of Landes and Posner's points is not controversial and is easy to illustrate. If A can hire a watchman at a cost of $10,000 per year but a passerby (or even a professional rescuer) can save A's $100,000 property from fire at an average cost of $500 per event—and such an event is likely to occur once a year—then A will inefficiently (at excessive expense) hire the guard if he knows the law will require him to pay strangers who come to the rescue a commission of, say, one-third or one-half of the value of the property they save. Landes and Posner point out that under admiralty law salvors at sea are indeed not rewarded so greatly as to encourage excessive precautions by shipowners; instead, they are probably rewarded at a rate that gives them only a modest return on their investment.

. . .

It is more difficult to evaluate the argument that legal incentives may interfere with the utility of altruism as an incentive. . . . Large penalties for nonrescue, while stimulating some rescue efforts, might decrease the total number of rescue attempts if rescuers motivated by altruism realize that they will sometimes be perceived as having acted only to avoid legal sanctions. On the other hand, rewards should not generate such an effect. A rescuer can always *decline* a reward and enjoy pure hero status.

Finally, Landes and Posner's second and third arguments can usefully be combined. In considering packages of carrots and sticks that might be aimed at potential rescuers it seems sensible to reject a "large sticks and no carrots" package. Not only would a large penalty interfere with the incentive to be altruistic (or to be regarded as altruistic) but also, when unaccompanied by a large carrot, it might discourage potential rescuers from frequenting potential rescue spots if such spots can indeed be identified in advance. With these points in mind, it is now possible to generalize about those legal packages of rewards and penalties that might sensibly promote desirable rescues and then to explore further the notion that penalties may discourage potential rescuers.

Rewards

Large carrots seem unwise for a number of reasons. First, there is the potential inefficiency, just noted, of swimmers or other potential rescuees' taking excessive precautions to avoid liability. Carrots that are not financed by rescuees will, of course, not lead to precautions on their part. Moreover, private carrots may need to be somewhat more generous than publicly financed carrots because potential rescuers may fear that rescuees will be unable to pay whatever reward is required by law. This first argument against large carrots is thus limited to those that are privately financed.

Second, there is a moral hazard that potential rescuers will create the demand for their own services. This moral hazard is surely more serious in some settings than in others. It is, for example, easier to imagine that carrot-seekers might put holes in strangers' boats and then later rush out to the rescue and reward than it is to see just how such a wrongdoer might encourage swimming accidents. Even when the moral hazard seems quite plausible, it might be avoided if the carrots that are offered to rescuers are not terribly large or, more interestingly, either not in a general useful currency or accompanied by sticks. Thus, Good Samaritan statutes do not generate moral hazard problems because they concern carrots that are in a sufficiently peculiar currency (enhanced reputation) to ensure that mischievous intermeddlers cannot profit monetarily from a planned emergency. The moral hazard problem is also more serious if the carrots are publicly financed, for parties may be tempted to stage emergencies and rescues at the expense of the fisc.

There is yet another reason why it might be sensible for the law not to provide large rewards to rescuers. In developing this argument it will be useful to define a subset of "customized carrots": these are rewards, often tailor made, offered when specific rescues are needed. Thus, while successful salvors at sea normally earn uncustomized and longstanding carrots as provided by admiralty law, one who returns a lost bracelet to its owner may be entitled to a reward if the owner has advertised and thereby offered a customized carrot. That jewelry owners in general do not promise rewards to be paid in the event of future losses, and do not press for legislation that promises such rewards, may reflect the moral hazard or even the excessive precautions that such promises might generate. Standing rewards may encourage theft followed by requests for rewards and excessive precautions against loss in the first place.

But consider, now, those settings in which excessive precautions are no longer at issue; something has already been lost or is otherwise at risk and the party who is in need of "rescue," such as the owner of a lost bracelet, offers a customized carrot. That such a party may have taken suboptimal, optimal, or even excessive care is now largely irrelevant. On the other hand, the moral hazard that may be generated by carrots is quite relevant and suggests the need for legal intervention. After all, while it is understandable that one who has property in peril will assess the danger that a reward will cause more of his property to be perilized as relatively small, this party is unlikely to consider the hazard that other citizens' property is put at risk. Inasmuch as a reward might encourage potential "rescuers" to stimulate demand for their own services by other private parties who might be expected to offer and finance large rewards, the law might forbid any party in need of help to offer a carrot. It may, indeed, come to pass one day that laws are effected that prohibit victims from meeting the demands of kidnappers, blackmailers, and extortionists. On the other hand, some combination of empathy for the immediate victims of such demands, concern that kidnappers (for instance) are actually captured precisely because we allow ransoms to be paid (and followed), and suspicion that such prohibitions would be impossible to enforce may overcome any hope that the banning of carrots will lead to a decrease in the crimes that create much of the demand for rescues.

· · ·

The possibility of customized carrots is yet another reason why one should not expect to see longstanding offers of large (uncustomized) carrots. So long as citizens are permitted and the government itself is able to offer customized carrots, it is understandable that the law is less inclined to put in place standing offers of carrots to rescuers. Such carrots may generate excessive precautions where voluntary customized carrots will not. Moreover, customized carrots may create less of a moral hazard problem than carrots that are announced in advance of any specific emergency. To be sure, because potential rescuers may not always learn about the availability of customized carrots, such carrots may not always motivate rescue efforts as forcefully as would carrots that are more generally announced and available. But occasional customized carrots are surely better than no carrots at all.

In short, large carrots may best be reserved for customized offerings in settings in which the probability of successful rescue is low, the investment required of rescuers is great, or the temptation to misbehave (rather than return lost valuables, for example) may be serious. . . .

Penalties

The imposition of large penalties for the failure to rescue may also be unwise. Landes and Posner point out that rescuers may be motivated by glory (or by nothing at all) and that sanctions for nonfeasance may actually destroy this useful incentive. We do, after all, rarely regard an act as heroic when it was encouraged by the threat of punishment for inaction. Unfortunately, it is difficult to know when, if ever, legal sanctions are for this reason counterproductive. It is surely unlikely that fewer car accidents or murders would take place if we would withdraw the legal rules governing such events and rely on altruism to control behavior. On the other hand, it is conceivable that those persons who are most likely to undertake rescues respond differently to similar stimuli than those who are most likely to commit murders and torts, so that it is only the potential rescuers who behave worse when facing large penalties.

A more compelling reason to think that penalties for nonrescue may be counterproductive derives from the fact that criminal or civil liability for nonrescue can hardly extend to someone who was far from the scene of an emergency and in no position at all to act. The threat of a big stick or any stick at all may, as Landes and Posner argue, discourage potential rescuers from nearing potential rescue spots. This second argument posits an "activity-level effect"; potential rescuers must care about the low probability that an emergency will occur, imagine a significant probability that factfinders will err after the fact and find that the defendant should have known that a rescue was necessary and easy, and correctly believe that rescues are more likely to be needed in some locations than others.

. . .

Finally, note that any activity-level effect that is generated by penalties can almost surely be offset by coexisting rewards. . . .

The Balance Between Carrots and Sticks

American law does not penalize nonrescuers and seems to provide virtually no carrots for rescuers. In contrast, many civil law systems provide substantial sticks, in the form of criminal (and sometimes accompanying civil) liability, to discourage nonrescue. . . . There is, in short, some evidence of symmetry, or of balance, between carrots and sticks—the systems that penalize nonrescue offer at least reimbursement to rescuers.

. . .

The discussion in this section explores at a theoretical level the mix of carrots and sticks. I suggest that through a process of elimination one is inclined to expect at least a rough balance in the design of incentives for rescue.

The disadvantages of large carrots alone, especially in the context of a legal system that permits and presupposes customized carrots, have already been discussed. I have also argued that small sticks are unlikely to generate important activity-level effects. Very large sticks, such as long-term incarcerations, are, of course, very costly to the society that imposes them. In any event, until there is evidence that some society increases the penalties imposed on nonrescuers because of dissatisfaction with the effects of small penalties, it seems sensible to limit the present discussion to the availability of small sticks, such as fines or the threat of civil liability. These penalties need not always be imposed; occasional judgments against nonrescuers may, like a rule of partial liability, be large enough to stimulate some rescues and small enough to avoid substantial activity-level effects. And even if sticks inevitably generate counterproductive activity-level effects, it seems likely that modest carrots can offset these sticks by raising the activity level that sticks may depress. To be sure, these carrots may generate their own problems, but the point is that some balance between the dangers of carrots and the activity-level effects of sticks might be sought.

It appears, then, that large carrots alone are unwise and that sticks alone are inferior to sticks accompanied by activity-inducing carrots. By "large carrots" I mean only to point out that the disadvantages of carrots are almost surely a function of their size; indeed, I will now refer to "carrots" alone and mean to imply a presumption that they not be large. One is left with the following alternatives: (1) "sticks and carrots"; (2) "no sticks and no carrots"; and (3) "carrots and no sticks." The first package is symmetrical, or balanced, and is close to what is found in most civil law countries. The second package is also balanced and fits the usual description of American law. One might dispute this description and argue that because American culture exalts rescuers, and because state statutes in the United States generally exempt some rescuers from tort liability, it is the third package that most accurately characterizes American law. But because I mean to focus on carrots that are larger than those provided through Good Samaritan statutes and cultural attitudes toward heroism, I ask the reader who believes that Good Samaritan statutes contain important or substantial carrots to rephrase the third stick-and-carrot package in a way that permits one to pose what I regard as an important unanswered question: Why do we appear to find (1) "sticks and carrots" and (2) "no sticks and *virtually* no carrots" when it would seem that (3) "carrots and no sticks" is a package that is to be preferred to package (2)? Again, the power of altruism as a motivation to rescue is not something that argues strongly in favor of package (2) over package (3) because so long as carrots can be declined they should not defeat this power. Only one unbalanced package, (3), that of "carrots and no sticks," remains to be promoted or explained away; if it generates no counterproductive behavior and seems superior to package (2), then perhaps it should be hailed as an appropriate goal of law reform.

One might insist that the carrots contained in package (3), even though they are not large, would stimulate excessive precautions by potential rescuees. But it is difficult to imagine the excessive precautions that might be undertaken by swimmers or homeowners who knew that those who one day rescued them from water or fire would be entitled to a reward of, say, two hundred dollars. And if excessive precautions by such potential rescuees were imaginable, then it would even seem worthwhile for the state to absorb the cost of carrots in package (3) so that potential rescuees would not so fear the cost of being rescued. Alternatively, it might be argued that the carrots in package (3) would, as discussed earlier, generate the moral hazard of intermeddlers' creating the demand for their own services. But this argument should at most call for a modification of package (3) to contain "no sticks and *skeleton* carrots"—that is, carrots that will make rescue somewhat more attractive but not induce the strategic creation of emergencies. The legal right to reimbursement for the actual expense of a rescue would, of course, be just such a skeleton carrot. To be sure, such compensatory or skeleton carrots might be difficult and relatively expensive to administer. One might, therefore, expect to see some mix of package (3) with a balanced package depending on the valuation and administrative problems of particular settings.

· · ·

In sum, there are reasons to expect a rough balance between the rewards and penalties for rescues. Greater sticks must almost surely be accompanied by greater carrots, unless there is no activity-level effect at all, and large carrots may be counterproductive even when accompanied by sticks. But a slightly unbalanced package ("no sticks and skeleton carrots") seems, at least theoretically, to be superior to the balanced package of "no sticks and no carrots." That the latter package appears to describe the strategy expressed in American law is, therefore, somewhat puzzling. . . .

Variety and Uniformity in the Treatment of Rescuers

I argued [earlier] that "sticks and (not too large) carrots" and "no sticks and skeleton carrots" would be sensible incentive packages for the purpose of promoting desirable rescues. I now turn to the incentives actually offered to potential rescuers and develop two points. First, American law may actually offer "no sticks and skeleton carrots" and, second, although different legal systems may provide dissimilar solutions to given problems—such as the promotion of desirable rescues—their treatments may be constrained by expectations about behavioral responses that may be quite alike in various societies.

· · ·

This second point is part of a larger positivist argument that I have developed elsewhere about "uniformity" and "variety" among legal systems. My suggestion is that different legal systems can be expected to display uniformity when a particularly necessary behavioral effect can be accomplished only with a certain rule. Variety, on the other hand, is to be expected either over a range in which rules do not greatly affect behavior or when reasonable people could disagree over which rule best accomplishes a desired effect. . . .

Incentives for Rescuers in American Law

Life Rescuers. Medals and media attention aside, "no sticks and no carrots" is said to be the Anglo-American norm for those not contractually bound to undertake emergency rescue efforts. There are some well-known exceptions to this norm. Physicians and hospitals may be liable for ignoring emergencies and can certainly collect competitive carrots for any rescue efforts that they do undertake. Such exceptions are not at all surprising for . . . "sticks and carrots" is in some settings quite a sensible package. In particular, it is not likely that serious moral hazard and excessive precaution problems would accompany the use of carrots for emergency professional medical services. A second well-known exception to the "no sticks and no carrots" rule is the statutory development of small sticks for nonrescue in some jurisdictions. Vermont and Minnesota have both legislated a "duty to rescue" when such rescue is riskless and have provided for a fine up to $100 for violations of this duty; Massachusetts and Rhode Island have legislated a duty to report a crime and have provided for substantial penalties for failure to report. Inasmuch as the activity-level effect of penalties is quite small, this development is hardly surprising. Still, I would expect these states to develop at least skeleton carrots to accompany these small sticks. Meanwhile, it is noteworthy that there have been no reported prosecutions under these statutes.

The most curious thing about American law—even outside of Vermont, Minnesota, Massachusetts, and Rhode Island—is the apparent absence of even skeleton carrots, for such carrots may promote desirable rescues at low cost. This curiosity might be dealt with in a number of ways. First, it could be argued that the costs of legal intervention must figure prominently in uniformity-variety predictions; American law may simply opt for little intervention. Inasmuch as American law encourages physicians to provide emergency medical services and thus intervenes in at least some settings, a more sophisticated form of this argument would be that the costs of legal intervention may, in the eyes of American lawmakers, overcome whatever advantages carrots generate. Similarly, it might be argued that the rule does not much matter so that no level of carrots would be particularly surprising. Large carrots may generate large problems but skeleton carrots may offer such small incentives that variety ought to be expected between "no carrots" and "skeleton carrots." But one need not choose between these explanations of American law, for the law and the curiosity just noted are generally misstated. In fact, American law does often provide at least skeleton carrots. The discussion in this section considers the ways in which American law rewards not only physicians but also other persons who voluntarily rescue life and property.

· · ·

An important, if skeleton, carrot in American law consists of a promise to potential rescuers that they or their survivors will be able to recover for injuries suffered in a rescue attempt. It is not particularly interesting to find that such recovery is generally available from a negligent third party who creates the need for the rescue so long as the rescuer was not also a cause of the emergency. Courts may struggle a bit with causality doctrines, but it is rather clear that if without a rescue attempt of A by B, C would be liable for injuries suffered by A, C should and will be liable for injuries suffered by B while undertaking a reasonable rescue. It is , however, much more interesting to note that a rescuer stands an excellent chance of recovering for his injuries when the *victim's* negligence created the need for rescue. It is difficult to see why the presence or absence of negligence by the rescuee himself should affect the rescuer's changes of recovery. The rescuee himself should affect the rescuer's chances of recovery. The rescuer acts to save the rescuee the burden of his own injury—but this is so whether or not the emergency arose because of the rescuee's negligence. It is thus arguable that recovery by the rescuer in these cases not only represents an important subset of rescue cases in which carrots are available but also anticipates or opens up the possibility of recovery by rescuers in future cases in which there is no negligence at all. It is possible that judges exploit the presence of negligence, however unconnected it is to arguments for recovery, as a means of protection against criticism that there will be no limit to their intermeddling. Over time, however, it may become easier to build on the many precedents in favor of recovery nad argue that there is no reason why the absence of negligence should require a decision against the rescuer. Indeed, the dicta of the cases involving negligence by the rescuee himself have long contained the seeds of such evolution, for they speak of the rescuee as having created the need for rescue without insisting that the creation of the rescue-inviting situation was negligent.

Another relevant development in American law may point to the development of sticks more than it does to that of carrots. Although the general rule of no sticks, or no duty to rescue, is solidly entrenched in most jurisdictions, many exceptions to this rule have been discovered in the form of "special relationships" out of which affirmative duties are said to arise. These exceptions have obviously not appeared in all jurisdictions but the overall appearance of these exceptions signals expansion. Although a finding of a special relationship normally precedes a finding of tort liability for nonrescue, it is not unreasonable to expect that carrots may grow next to these new sticks. After all, just as a physician's ability to recover for noncontractual emergency services seems to accompany the physician's duty to assist those in need of his services, perhaps because sticks alone would create an undesirable activity-level effect, so too might carrots accompany the sticks that are extended to "special relationships." The expanding set of exceptional special relationships to the "no duty" rule includes common carrier-passenger; innkeeper-guest; innkeeper-stranger (a duty to protect a stranger from injury by a guest); employer-employee; ship-crewman; shopkeeper-business visitor; host-social guest; jailer-prisoner; school-pupil; drinking companions; landlord-trapped trespasser; safety engineer-laborer; physician-patient; psychologist-stranger (a duty to protect a stranger from harm at the hands of the psychologist's patient); manufacturer-consumer; landlord-tenant; parole board-stranger (a duty to protect strangers from a released prisoner); husband-wife; parent-child; and tavern keeper-patron. In all these settings it appears increasingly likely that duties will again be pronounced and that liability for nonrescue will follow.

. . .

Apart from the familiar case of the physician rendering emergency services, four of the special relationships just cited do not involve preexisting relationships, although this is a concept that is often said to be related to liability for nonrescue. The liability of an innkeeper for not preventing injuries to *strangers* does not, for example, seem far from the question of liability for not rescuing one's neighbor. And, once again, carrots for rescuing one's neighbor seem related to sticks for not rescuing one's neighbor.

. . .

These developments and the preceding arguments lead one to believe that American law *already* grants some rewards to rescuers and may not be far from granting many more. I suspect that a rescuer who seeks to recover out-of-pocket expenses or other costs that are easily valued is more likely than other rescuers to secure recovery. But it is difficult to assess the precise degree to which carrots are already available. If, for example, judges feel bound to the doctrinal steppingstone of negligence as a requirement for recovery by an injured rescuer, then it may take some time before suficient precedents are created by more aggressive judges who overlook this apparent doctrinal requirement or create new "special relationships" that trigger both sticks and carrots. . . .

European Civil Law Systems

. . .

The most striking aspect of the treatment of rescue in modern European civil law systems is that sticks are often available for and applied to nonrescuers. These sticks usually take the form of criminal penalties, consisting of fines, imprisonment, or both, and may be accompanied or extended by exposure to civil liability. French law, for example, provides a fine and jail sentence for someone who does not undertake a riskless rescue of an endangered person. The statute applies only when a person—and not property alone—is in danger. The nonrescuer is also subject to civil liability, and such liability is triggered by both intentional and negligent omissions.

In practice, French law provides skeleton carrots to go along with the sticks just described. If a rescuer has suffered damage or incurred expenses, recovery will usually be possible from the rescuee who will be said to have impliedly contracted for the rescue effort, from a tortfeasor who will be said to have caused the rescue attempt, from the rescuee or his insurer who may be identified as unjustly enriched, or simply from the rescuee by an action of *negotiorum gestio,* under which one who acts for the benefit of someone who is unable to take care of himself is able to recover, especially when the unilaterally created "agency" turns out to be useful to the principal. . . . The "sticks and skeleton carrots" package offered in French law is an illustration of the sort of variety one might expect after considering the likely behavioral effects of penalties and rewards of various size. . . .

Other European legal systems also provide "sticks and skeleton carrots."

Nonrescue is a criminal offense in most of Europe; penalties range from a fine only to imprisonment of up to five years and fine. . . .

The Future of Rescue

There is no reason why the law regarding rescue could not remain just as it now is. There is no sudden outbreak of conflicting decisions or hostile public opinion on the matter. Concern about activity-level effects, excessive precautions, unwanted intermeddling, unnecessary judicial intervention in the private affairs of citizens, and the monetization of moral norms may forestall any changes in rescue law or even ensure that the clock is rolled back a bit and that "no sticks and no carrots" is practiced as well as advertised. On the other hand, courts may continue to designate special relationships and may often decide that rescuers shall recover for their expenses when rescuees were negligent; if so, the general rule will be "occasional sticks and infrequent carrots," rather than virtually none of each. Indeed, courts may even drop these preconditions to their own intervention. My intuition is that this second scenario, a steady evolution toward the European model of "sticks and carrots," is more likely.

. . .

The aim of the discussion in this section is not, however, to demonstrate that the law of rescue is especially ripe for change. I argue instead that an important theme that runs through the rescue cases also appears in many important modern tort cases—and that the evolution of rescue law may track that of tort law. Tort law can be described in a manner that may first seem paradoxical as both imposing liability on the *single* party that is the least-cost avoider and as increasingly imposing some form of joint liability, rather than no liability, when no such single avoider exists. And rescue law can be described as expanding the obligations of potential rescuers precisely where a single party is the best or most obvious rescuer. Finally, it is possible that rescue law will not follow tort law and develop means with which to grapple with cases where no single potential rescuer stands out. The recent evolution of tort law may therefore give us a sense of the future of rescue law.

Single and Multiple Accident-Avoiders in Tort Law

The notion that much of tort law can be viewed as imposing liability on the party, or single avoider, best able to accomplish a socially desirable or "would-be-bargain" result needs no rehearsal. The search for a single and best avoider is easy to understand in terms of transaction costs. If no liability is imposed, then the victim, assuming he is not the best avoider, must try to contract with potential tortfeasors in order to avoid injury. Such bargains with many unknown parties will obviously be difficult to accomplish. If the threat of liability hangs over multiple potential avoiders, then they may undertake duplicative precautions or may each hope that someone else avoids the accident so that no one in fact takes the necessary precaution. In short, when transaction costs are substantial, not only is the need for legal intervention clear but also the advantages of aiming that intervention at a single avoider are apparent. The single-avoider notion is somewhat contradicted by the evolution of products liability law where a plaintiff can now sue both the manufacturer and the dealer of the product he regards as defective; but it is fairly clear that in these cases the contractual ties between the potential defendants must usually be thick enough to overcome the potential for duplicative or inadequate precautions that normally exists with multiple potential avoiders.

In more complex settings the problems posed by multiple parties are not so easily solved by imposing liability on a single avoider. There is, I think, no paradox in suggesting that when single-avoider solutions are feasible tort law evolves toward such solutions, but that when multiple party problems are endemic tort law will simply try to solve the deterrence problem as well as possible. Broadly speaking, the multiple party problems are of two types, interactive and horizontal. By interactive I refer to circumstances in which harms can be prevented by multiple parties acting jointly or alternatively. The manufacturer who can add more safety devices to a machine and the consumer who uses this machine are examples of such interactive parties. In such settings, the ideal behavior of one party depends on the behavior of another, whose own ideal behavior depends in turn on expectations about the first party's behavior. Unless the parties that are invovled and the factfinder that reviews the situation later are extremely well-informed, these interactive (multiple party) problems are confounding. Tort law has experimented with different kinds of liability rules including various contributory and comparative negligence rules, but no rule is quite the match for the problem. I think it fair to say that the direction of change has been toward partial recovery; comparative negligence rules, under which liability is spread among parties who might have taken accident-avoiding steps, have become increasingly popular.

"Horizontal" problems involve multiple misbehaving or potentially misbehaving parties (whose ideal behavior depends very little on that of other parties) and uncertainty regarding which of these parties caused a particular harm. The law has had great difficulty in settling on rules that best "match" injured plaintiffs with those negligent defendants who truly caused their injuries. The law dealing with these horizontal problems is hardly settled, but it is at least arguable that recoveries are unlikely when there is a horizontal matching problem with limited misbehavior, as when plaintiff only knows that he was hit by a bus and more than one party might have owned the bus that hit him, but that probabilistic recoveries are increasingly likely when horizontally aligned defendants have all misbehaved, as when plaintiff's injury is caused by a defective drug manufactured by one untraceable member of a group of unrelated companies making the same product. This review of single and multiple tortfeasors, or potential accident-avoiders, sets the stage for a discussion of single and multiple omitters, or potential rescuers. As we will see, the increasing willingness of courts to accept probabilistic causation arguments in

cases involving multiple tortfeasors may signal—at least as a logical matter—
a willingness to accept similar causation arguments that must underlie claims
that assert a duty to rescue.

Single and Multiple Potential Rescuers

. . .

As a descriptive matter tort law can be characterized as searching for a single
best avoider, but increasingly imposing liability when there are multiple avoiders and no single best avoider. Similarly, the discussion in this section now
links the growth in liability for potential rescuers who have special relationships with victims to the notion of the single rescuer. Although this link is
promoted in this article as descriptively useful, it is worthwhile—if only because positive arguments are more appealing when they are coextensive with
arguments that are normatively plausibble—to explore the reasons why lawmakers might limit sticks (or sticks and carrots) to circumstances in which
there is a clear single potential rescuer.

. . .

When a victim's call for help goes unanswered, there may be more than
one potential rescuer within hearing range. Indeed, it must often be the case
that when B could have rescued A, B's presence at the scene only comes to
the attention of the law because other potential (and often unhelpful) rescuers, C and D, were also at the scene of A's trouble. Such multiple potential
rescuers pose a doctrinal problem; if no rescue is attempted, it will be unclear
whether B, C, or D "caused" A's injury—and misbehavior without causation
of an injury is traditionally an insufficient basis for liability. The problem may
not seem any different from one in which two hunters negligently shoot in the
direction of a third person in their party and one (but it is unclear which) hits
this unintended mark. The hunters will, of course, be held jointly and severally liable. Apart from insisting that three potential causal agents present a
very different problem than do two, there are at least three ways to distinguish
the multiple nonrescuers from the multiple shooters. First, one can simply
follow the traditional line and insist that the law distinguishes omissions from
commissions. This route will not be a terribly successful one both because
some omissions, such as draft evasion and tax evasion, are seriously and
understandably penalized and because, as reviewed presently, rescue *is* often
obligatory.

Second, it is arguable that the causation problem in rescue goes to the root
of negligence so that rescue cases are more like the bus case, in which only
one actor is negligent and none is held liable, than like the shooting or defective drug cases, in which all actors are negligent and at least partially liable. In
rescue cases, after all, if B rescues A, C and D's inaction is perfectly acceptable in both legal and moral terms. As such, because in many cases each
potential rescuer is probably unaware of the actions of other potential rescuers, it is conceivable, even if horrible, that each behaves *nonnegligently* because each thinks that there is no need for his own efforts. On the other hand,

no hunter should shoot in the direction of a camouflaged buddy and no drug manufacturer should turn out defective drugs regardless of the actions of fellow hunters and manufacturers. I must add that my own normative view is that this distinction is unimportant. It may be sensible to deny recovery to a plaintiff who offers only probabilistic evidence regarding which of a number of bus companies caused his injury because in many cases this rule might lead plaintiffs to investigate further and identify their tortfeasors more precisely. In contrast, the probabilistic basis for believing that a nonrescuer may not have been negligent is not capable of being altered after the fact. The problems of probabilistic proof and probabilistic need for rescue are thus different. Still, I advance this second argument for distinguishing multiple nonrescue from multiple tortious behavior because it is certainly possible that it is this sort of argument that has caused judges to hold multiple shooters and not multiple nonrescuers liable.

A third difference between multiple nonrescue and negligence may be one of time and not substance. Tort law has only recently evolved to the point of assigning liability in probabilistic fashion among numerous defendants. Two negligent hunters seemed to present an easy case, perhaps because the probability of a "match" was just 1 percent below that required in the usual more-likely-than-not test. But multiple drug manufacturers would almost surely not have been liable 20 or 50 years ago unless the probability of a match between a plaintiff and a given manufacturer was also in the 50 percent range. It is thus arguable that multiple nonrescue and multiple affirmative negligence were treated alike for many years and that evolution or experimentation has recently affected the treatment of the latter only so that we are now in a developmental period in which multiple negligent parties but not multiple nonrescuers incur liability. This argument suggests that the law may change in either area once again and multiple nonrescuers and multiple negligent tortfeasors will then be similarly treated.

A single identifiable nonrescuer, however, presents no matching or deterrence difficulties and is very much like a paradigmatic single tortfeasor. To be sure, it is possible to insist that nonrescue always presents the problem of multiple parties. When, for example, A's surviving family learns that B could have rescued A, it is always possible that C and D could also have effected a rescue but that their identities or presence at the scene of A's tragedy remains unknown. But this uncertainty alone should not lead one to expect that single nonrescuers will be excused from liability, because this is the sort of uncertainty that the tort system frequently indulges. When Z's negligently started fire sweeps across X's property, Z may be liable even though Y may also have started a fire that was on its way but is now unknown—perhaps because Z's fire consumed the evidence of its origins and path. The case presents no difficulty; X bears the burden of proving that there was at least one negligently set fire, and Z can only escape full liability by identifying Y and then passing on the blame, sharing it, or seeking contribution. The remote possibility that a negligent Y was involved but that Z is unable to identify Y and prove Y's involvement is hardly allowed to interfere with Z's liability to X. Similarly, the mere possibility of multiple nonrescuers is unlikely to interfere with the liability of a single identifiable nonrescuer.

But the most important reason to think that the failure to rescue is really not so different from the commission of a negligent act is that, as indicated earlier, courts have discovered a surprising number of special relationships as bases for the imposition of duties to rescue. Most significantly, these special relationships have one thing in common: when there is a special relationship there is no multiple nonrescuer problem, for such a relationship is pronounced only in circumstances in which there is one identifiable or best-situated nonrescuer. This is not to say that whenever there is but one nonrescuer, courts will insist that he and the victim had a special relationship out of which a duty to rescue arose. The presence of a single nonrescuer is at present a necessary but not a sufficient condition for liability; it is the growing number of special relationships that indicates an increasing likelihood of liability on the single nonrescuer. As indicated earlier, the most remarkable thing about these cases in which special relationships are found is that some involve a single nonrescuer and a *stranger.* Liability has been found appropriate for an innkeeper who could have protected a stranger from injury by one of the innkeeper's guests, a safety engineer who could have prevented an injury to a laborer he did not employ, a psychologist who might have warned an identifiable stranger his patient was intent on harming, and, similarly, a parole board, acting as a single entity, that might have warned someone who was the target of a released convict. In these cases, there is of course no "relationship" at all. Instead, these cases contain three elements. First, there is a single nonrescuer. Second, this nonrescuer could with little effort have prevented a serious loss. Third, this nonrescuer had no reason to think that someone else would save the day. In short, although I think that a single nonrescuer is not very different from a single negligent tortfeasor, it is hardly arguable that the two are treated the same under current law. But it is arguable that nonrescuers are increasingly incurring liability and that through the concept of special relationships this liability is most often imposed precisely where there is a clear, single nonrescuer.

There is another clue that suggests in a striking way that the major obstacle to holding nonrescuers liable is the historical disinclination to impose liability where multiple parties are at fault. It has long been the case that if B begins to rescue A but then abandons the rescue effort, then B will often be held liable for nonrescue even though he is said to have had no duty to rescue in the first place. Occasionally, B may have made A worse off by leaving him in a place where other rescuers were less likely to help or simply by using up time in which other rescuers might have appeared. But it is rather clear that courts do not look carefully for evidence of such causation but instead often hold B liable for what might be regarded as active, or conscious, nonrescue. Clearly, B's initial efforts single him out and "solve" any multiple party problem. B's false start allows the law to identify him—as opposed to other potential rescuers—as the nonrescuer and to see him in a light that is so similar to that which falls on a typical tortfeasor. The inclination to hold the "withdrawing rescuer," B, liable is thus a further indication that it is the problem of multiple nonrescuers that is at the heart of any difference between omissions and commissions. Similarly, the admiralty law rule that a vessel involved in a collision must "stand by" to assist the other vessel or be rebuttably presumed to have been at fault is, like attitudes towards a hit-and-run driver on land, an example of the imposition of an affirmitive obligation where an identifiable, single and obviously least-cost rescuer is on hand.

These arguments—that there is not a convincing analytic difference between the case involving two negligent hunters and that involving two nonrescuers and that the expansion of special relationships is no more or less than the gradual (if slow) imposition of a duty to rescue on single identifiable potential rescuers—lead to fairly specific predictions about the future of the use of sticks in rescue law. First, there is every reason to think that nonrescuers will increasingly be held liable, even in the absence of further statutory innovation. Second, multiple nonrescuers may also be held liable in the future, for now that tort law has grappled with multiple parties there is less reason to think that rescue law will not do the same. Finally, although this section has focused on sticks and not on carrots, it is useful to repeat the thesis developed [earlier]: American law may already provide at least skeleton carrots to many rescuers. Whether or not the expansion of special relationships will mean not only that an identifiable A might be liable for not rescuing B but also that A could collect his expenses (or more) from B is a difficult question. Many of the special relationships that, at least at first, link A and B are contractual, so that A could collect "negotiated carrots" from B. Landlords and psychologists, for example, could surely raise rents and fees to finance the cost of their increased precautions (or liability). It is, in short, possible that American law will raise the rewards offered to potential rescuers but also possible that it will, instead, simply leave these rewards to market forces. If liability for nonrescue extends one day to all single or even multiple nonrescuers, then it is most likely that skeleton or statutory carrots will continue to sprout.

Notes and Questions

1. The analysis in this reading begins with some of the ideas introduced in William Landes and Richard Posner, "Salvors, Finders, Good Samaritans, and Other Rescuers: An Economic Study of Law and Altruism," 7 *Journal of Legal Studies* 83 (1978). Consider first the Landes and Posner claim that excessive penalties for failure to rescue can actually make potential rescuees worse off. If potential rescuers avoid situations or locations, such as beaches, where the "duty" to rescue may arise, there will be fewer rescues than under a no-duty rule. Can you think of many examples of such an avoidance strategy? If you think this is an unimportant effect, then how does the remainder of the analysis in the reading change?

2. Consider the argument that legal inducements can interfere with altruistic motivations. Is the same true of tort law more generally or even of criminal law? Do we sometimes decline to regard behavior as criminal or tortious because we think that an

appeal to pure altruism is more likely to succeed than the threat of penalty? Might we have less embezzlement or drug trafficking if there were no laws against these activities?

3. Do you expect that rules regarding the rescue of property (as in the treatment of one who finds lost property), as opposed to the preservation of life, will be similarly aligned along the omission-commission axis?

4. Imagine a legal system that drew no distinction between omissions and commissions. In such a system there would be no immediate doctrinal impediment to bringing suit against a passer-by who had not responded to pleas for assistance by a victim-plaintiff. On the other hand, the defendant in such a nonrescue case could presumably argue that his or her behavior was not negligent because the risks of intervention were greater than the likely benefits. In turn, a plaintiff who suffered an assault, for example, might be expected to win if the defendant had not even taken the trouble to notify the police of the problem at hand. How might such a legal system treat cases in which numerous passers-by failed to call the police? What if each one argues that there was every reason to assume that the police had already been notified by other passers-by?

A Lawyer's Primer
on Feminist Theory and Tort
LESLIE BENDER

. . .

Tort law needs to be more of a system of response and caring than it is now. Its focus should be on interdependence and collective responsibility rather than on individuality, and on safety and help for the injured rather than on "reasonableness" and economic efficiency. Feminism raises questions about both the external structure of our negligence analysis—how we frame our understanding of negligence problems—and its internal categories, such as duty and the standard of care. My reflections on negligence law are interwoven with a primer on feminist theory in order to suggest how feminist theory can help us think about the traditional structures of our laws, legal analyses, and legal system.

. . .

Feminism integrates practice and theory. It is a woman-centered methodology of critically questioning our ideological premises and reimagining the world. Feminists applaud women's accomplishments that have been unmentioned or slighted in others' tellings. We study women's oppression in order to understand what it is, how it happened, the subtle ways it works, and how oppression, exploitation, and exclusion affect different aspects of our lives and thinking. Whatever the focus of our particular work, all feminist efforts

Abridged and reprinted without footnotes by permission from 38 *Journal of Legal Education* 3 (1988). Copyright © 1988 American Association of Law Schools.

are combined in struggle to eradicate women's subordinate status. This is no small task, because male dominance pervades all aspects of our lives—from our knowledge formation to our reproductive control, from our political and social organization to our self-perceptions. Feminism is political, methodological, philosophical, and intent upon social transformation. It is about women, but it is not for women only. Feminism seeks to be inclusive, not exclusive; its teachings illustrate the harm that flows from exclusion. To achieve transformation it requires large-scale participation and a revision of our society.

. . .

Men have created and named a world in which men have power over women—physical power, political power, opportunity power, silencing power. We must learn how our social and political organizations have been constructed by men in their own image and explore how a world constructed by women and men for women and men would be different. Do women have a separate and complementary value system and perspective? For if institutions are designed by men to reflect male values, characteristics, conceptions of reality, and needs, and if women are acculturated differently or have alternative perspectives, then even women who have access to male institutions will have to relinquish their female training and identity in order to succeed.

Consider an example from the legal profession. Men have constructed an adversary system, with its competitive, sparring style, for the resolution of legal problems. In many ways it is an intellectualized substitute for duelling or medieval jousting. Much of legal practice is a win-lose performance, full of one-upmanship and bravado. If it were to turn out that competitive sparring is not the way a majority of women function most effectively, then within patriarchy's terms it could be concluded that women are not well suited for legal practice. But rather than regarding legal practice as fixed, we can question whether a competitive, win-lose approach is necessary and examine how it has been modelled by men in their own image. When we look anew for methods for resolving conflicts, we may decide that win-lose, adversary methods are not the only, or not the best, or even not a preferable method for dispute resolution. Perhaps we could design alternative models that incorporate the perspectives of women and men.

The role of patriarchy within the construction of substantive law as it affects all women is even more deleterious than its effect upon women practicing law. For centuries a woman lost her legal identity when she married. Until recently rape has been defined and "remedied" from a male perspective, and such experiences as sexual harassment have been excluded from the legal system. Intraspousal and intrafamilial tort immunities shrouded domestic violence against women and children from legal redress. The Supreme Court has even held that discrimination on the basis of pregnancy is not sex discrimination. Because our legal system has developed from an unstated male norm, it has never focussed adequately on harms to women.

. . .

The primary task of feminist scholars is to awaken women and men to the insidious ways in which patriarchy distorts all of our lives. Patriarchy deprives us of the richness of women's voice in public discourse and curtails the range of women's contributions to shaping our world. It takes extraordinary effort to release ourselves from the overpowering ideological yoke of patriarchy. Rather than letting patriarchy overwhelm us, feminism carefully examines its contours. Feminism rejects patriarchy in order to construct a world in which every person is empowered rather than one in which some people are overpowered.

Unearthing each shard of patriarchy is especially difficult because of the powerful assumptions embedded in our langauge and logic. Western culture teaches us that the patriarchal description of reality is not biased but neutral; that our knowledge and truths are not subjective, intersubjective, relative, or constructed from narrow perspectives but objective, scientifically based, and universal; that our human nature is autonomous, self-interested, and even aggressively competitive, not interdependent, collective, cooperative, and caring. How do we move beyond traditional ways of knowing, understanding our natures, and theorizing about our social and political structures when we are a product of them? Can we step out of the frame that limits our perceptions?

· · ·

We need to raise our consciousnesses about a system that has ignored or misapprehended the concerns and participation of over half the population while priding itself on its commitment to ideas such as "democracy" and "equality." Democracy and equality for whom? These shibboleths have primarily served only those who are empowered. Because the legal system is one of the most effective tools for distributing and equalizing power, it must be a major focus of feminist struggle. Feminism helps us look behind such reified concepts as "equality" to the real experiences of women, minorities, and the unempowered. If feminism and its consciousness-raising methodology can help us break through our legal system's myths, perhaps we can design transformative social, political, and legal systems that emphasize feminist values and concerns—values and concerns that emerge, for instance, from viewing human nature as interdependent instead of isolated and autonomous, and from attending to contexts and particularity instead of relying on universals and abstractions. Feminists value a collective and dialectical creation of knowledge through sharings of multiple perspectives and reject the false notion of neutral, objective, unsituated knowledge. We can create a system committed to a continuous reexamination of our premises and questioning of our assumptions. Feminism values openness, inclusivity, and equal respect, not exclusivity and hierarchy; diversity and difference, not uniformity and sameness. It prefers caring and responsibility over rights and property; health and safety over profit and self-interest. A system with feminist underpinnings would value emotion and instinct, as well as reason and intellect, and reject a dualistic either-or, self-other approach to understanding.

· · ·

Part of the problem with having our world named and language created by men is that although men consistently use themselves as norms, they do not do so explicitly. It takes careful sifting to discover where and how the implicit male norm has skewed our understanding. Carol Gilligan's *In a Different*

Voice is a scholarly effort to expose biases in our knowledge due to unstated male norms.

One cannot stress enough how deep the critique of the male power of naming goes. All of our norms and standards have been male. If we extract the male biases from our language, methods, and structures, we will have nothing—no words, no concepts, no science, no methods, no law. The biases are coextensive with the concepts around which we have structured our political and social systems. . . .

Negligence Law: The "Reasonable Person" Standard as an Example of Male Naming and the Implicit Male Norm

That implicit male norms have been used to skew legal analysis can be seen in tort negligence law. To assess whether a defendant's conduct is negligent, and hence subject to liability, we ask whether the defendant has a duty to the plaintiff and whether she has met the legally required standard of conduct or care. "Standard of care" is a term of art in the law. It is alternatively described as the care required of a reasonably prudent person under the same or similar circumstances, or of a reasonable person of ordinary prudence, an ordinarily prudent man, or a man of average prudence. Prosser and Keeton explain the standard as some "blend of reason and caution." A "reasonable person" standard is an attempt to establish a universally applicable measure for conduct. This reasonable person is a hypothetical construct, not a real person, and is allegedly objective rather than subjective.

Not surprisingly, the standard was first articulated as a reasonable *man* or *man* of ordinary prudence. Recognizing the original standard's overt sexism, many courts and legal scholars now use a "reasonable person" standard. My concern with the "reasonable person" standard is twofold. Does converting a "reasonable man" to a "reasonable person" in an attempt to eradicate the term's sexism actually exorcise the sexism or instead embed it? My second concern is related. Should our standard of care focus on "reason and caution" or something else?

It was originally believed that the "reasonable man" standard was gender neutral. "Man" was used in the generic sense to mean person or human being. But man is not generic except to other men. Would men regard a "prudent woman" standard as an appropriate measure of their due care? As our social sensitivity to sexism developed, our legal institutions did the "gentlemanly" thing and substituted the neutral word "person" for "man." Because "reasonable man" was intended to be a universal term, the change to "reasonable person" was thought to continue the same universal standard without utilizing the gendered term "man." The language of tort law was neutered, made "politically correct," and sensitized. Although tort law protected itself from allegations of sexism, it did not change its content and character.

. . .

Changing the word without changing the underlying model does not work. Specifically addressing the "reasonable person" tort standard, Guido Calabresi challenges whether the "reasonable person" is in any way meant to include women or, for that matter, people of non-WASP beliefs or attitudes. Calabresi explains that use of a universal standard is intended to cause those who are "different" from that standard to adopt the dominant ideological stance. Like the notion of America as a melting pot, the reasonable person standard encourages conformism and the suppression of different voices.

Not only does "reasonable person" still mean "reasonable man"— "reason" and "reasonableness" are gendered concepts as well. Gender distinctions have often been reinforced by dualistic attributions of reason and rationality to men, emotion and intuition (or instinct) to women. Much of Western philosophy is built on that distinction. Aristotle describes the female body as "a deformity, though one which occurs in the ordinary course of nature," and regards women as inferior beings whose reasoning capacity is defective. Immanuel Kant observes that women are devoid of characteristics necessary for moral action because they act on feelings, not reason.

. . .

We would be hard pressed today to find many people who would openly assert that women cannot be reasonable. Today we are taught to consider women reasonable when they act as men would under the same circumstances, and unreasonable when they act more as they themselves or as other women act. If it is true that somewhere, at some subconscious level, we believe men's behavior is more reasonable and objective than women's, then changing the phrase "reasonable man" to "reasonable person" does not really change the hypothetical character against whom we measure the actors in torts problems. By appending the very term "reasonable," we attach connotations and characteristics of maleness to the standard of conduct. If we are wedded to the idea of an objective measure, would it not be better to measure the conduct of a tortfeasor by the care that would be taken by a "neighbor" or "social acquaintance" or "responsible person with conscious care and concern for another's safety"?

Perhaps we have gone astray in tort-law analysis because we use "reason" and caution as our standard of care, rather than focusing on care and concern. Further study of feminist theory may help to suggest how a feminist ethic can affect our understanding of standards of care in negligence law.

. . .

The concept of an ethic based on care and responsibility informs a great deal of feminist scholarship. Carol Gilligan suggests that women's moral development reflects a focus on responsibility and contextuality, as opposed to men's, which relies more heavily on rights and abstract justice. After studying responses to interview questions in three studies (a college student study, an abortion-decision study, and a rights-and-responsibilities study), Gilligan recognized that there are two thematic approaches to problem solving that generally correlate with gender, although she makes no claims about the origin of

the difference. Traditional psychological and moral-development theory recognizes and rewards one approach but undervalues or fails to define the other,
the approach Gilligan calls "a different voice." When she asked what characterizes the different methods for resolving and analyzing moral dilemmas,
Gilligan found that the "right" answers (according to the traditionally formulated stages of moral development) involve abstract, objective, rule-based
decisions supported by notions of individual autonomy, individual rights, the
separation of self from others, equality, and fairness. Often the answers provided by women focus on the particular contexts of the problems, relationships, caring (compassion and need), equity, and responsibility. For this different voice "responsibility" means "response to" rather than "obligation for."
The first voice understands relationships in terms of hierarchies or "ladders,"
whereas the "feminine" voice communicates about relationships as "webs of
interconnectedness."

· · ·

Our legal system must learn from Gilligan's study; it must attend to the
relationships between people, our interdependencies and interconnectedness,
to our responsibilities for and toward one another, and to the need to be
responsive and caring. It must also recognize that it has been formualted in
one voice, the masculine voice, and that it must listen to the meanings of
different voices and reconstitute itself accordingly. If we are willing to accept
the implications of Gilligan's study, that is, that women have developed different ethical priorities and approaches to experience, then how does a feminist
perspective help us think about negligence law?

Our traditional negligence analysis asks whether the defendant met the
requisite standard of care to avoid liability. As I have suggested, if we continue to feel obliged to apply a "universal" standard, we might be better off if
we at least expect people to act not as "reasonable persons" but as "responsible neighbors" or with the care and concern of "social acquaintances." But
figuring out against whom we should measure our conduct is only a partial
solution to the problem. We also need to determine what standard of care to
apply and what conduct is negligent or falls below that standard.

In tort law we generally use the phrase "standard of care" to mean "level
of caution." How careful should the person have been? What precautions do
we expect people to take to avoid accidents? We look to the carefulness a
reasonable person would exercise to avoid impairing another's rights or interests. If a defendant did not act carefully, reasonably, or prudently by guarding
against foreseeable harm, she would be liable. The idea of care and prudence
in this context is translated into reasonableness, which is frequently measured
instrumentally in terms of utility or economic efficiency.

When the standard of care is equated with economic efficiency or levels of
caution, decisions that assign dollar values to harms to human life and health
and then balance those dollars against profit dollars and other evidences of
benefit become commonplace. Such cost-benefit and risk-utility analyses turn
losses, whether to property or to persons, into commodities in fungible dollar
amounts. The standard of care is converted into a floor of unprofitability or

inefficiency. People are abstracted from their suffering; they are dehumanized. The risk of their pain and loss becomes a potential debit to be weighed against the benefits or profits to others. The result has little to do with care or even with caution, if caution is understood as concern for safety.

There is another possible understanding of "standard of care" that conforms more closely to Gilligan's "different voice," an alternative perspective rooted in notions of interconnectedness, responsibility, and caring. What would happen if we understood the "reasonableness" of the standard of care to mean "responsibility" and the "standard of care" to mean the "standard of caring" or "consideration of another's safety and interests"? What if, instead of measuring carefulness or caution, we measured concern and responsibility for the well-being of others and their protection from harm? Negligence law could begin with Gilligan's articulation of the feminine voice's ethic of care—a premise that no one should be hurt. We could convert the present standard of "care of a reasonable person under the same or similar circumstances" to a standard of "conscious care and concern of a responsible neighbor or social acquaintance for another under the same or similar circumstances."

The legal standard of care may serve as the minimally acceptable standard of behavior, failing which one becomes liable. But the standard need not be set at the minimum—we do not need to follow Justice Holmes' advice and write laws for the "bad man." Have we gained anything from legally condoning behavior that causes enormous physical and mental distress and yet is economically efficient? The law can be a positive force in encouraging and improving our social relations, rather than reinforcing our divisions, disparities of power, and isolation.

The recognition that we are all interdependent and connected and that we are by nature social beings who must interact with one another should lead us to judge conduct as tortious when it does not evidence responsible care or concern for another's safety, welfare, or health. Tort law should begin with a premise of responsibility rather than rights, or interconnectedness rather than separation, and a priority of safety rather than profit or efficiency. The masculine voice of rights, autonomy, and abstraction has led to a standard that protects efficiency and profit; the feminine voice can design a tort system that encourages behavior that is caring about others' safety and responsive to others' needs or hurts, and that attends to human contexts and consequences.

. . .

Through a feminist focus on caring, context, and interconnectedness, we can move beyond measuring appropriate behavior by algebraic formulas to assessing behavior by its promotion of human safety and welfare. This approach will clearly lead to liability for some behaviors for which there was none before. If we do not act responsibly with care and concern for others, then we will be deemed negligent. Just as we can now evaluate behavior as negligent if its utility fails to outweigh its risks of harm, we could evaluate behavior as negligent if its care or concern for another's safety or health fails to outweigh its risks of harm. From a feminist perspective the duty of care required by negligence law might mean "acting responsibly towards others to avoid harm, with a concern about the human consequences of our acts or failure to act."

. . .

One of the most difficult areas in which questions of duty and the standard of care arise is the "no duty to rescue" case. The problem is traditionally illustrated by the drowning-stranger hypothetical.

. . .

Each year that I teach torts I watch again as a majority of my students initially find this legal "no duty" rule reprehensible. After the rationale is explained and the students become immersed in the "reasoned" analysis, and after they take a distanced, objective posture informed by liberalism's concerns for autonomy and liberty, many come to accept the legal rule that intuitively had seemed so wrong to them. They are taught to reject their emotions, instincts, and ethics, and to view accidents and tragedies abstractly, removed from their social and particularized contexts, and to apply instead rationally-derived universal principles and a vision of human nature as atomistic, self-interested, and as free from constraint as possible. They are also taught that there are legally relevant distinctions between acts and omissions.

How would this drowning-strager hypothetical look from a new legal perspective informed by a feminist ethic based upon notions of caring, responsibility, interconnectedness, and cooperation? If we put abstract reasoning and autonomy aside momentarily, we can see what else matters. In defining duty, what matters is that someone, a human being, a part of us, is drowning and will die without some affirmative action. That seems more urgent, more imperative, more important than any possible infringement of individual autonomy by the imposition of an affirmative duty. If we think about the stranger as a human being for a moment, we may realize that much more is involved than balancing one person's interest in having his life saved and another's interest in not having affirmative duties imposed upon him in the absence of a special relationship, although even then the balance seems to me to weigh in favor of imposing a duty or standard of care that requires action. The drowning stranger is not the only person affected by the lack of care. He is not detached from everyone else. He no doubt has people who care about him—parents, spouse, children, friends, colleagues; groups he participates in—religious, social, athletic, artistic, political, educational, work-related; he may even have people who depend upon him for emotional or financial support. He is interconnected with others. If the stranger drowns, many will be harmed. It is not an isolated event with one person's interests balanced against another's. When our legal system trains us to understand the drowning-stranger story as a limited event between two people, both of whom have interests at least equally worth protecting, and when the social ramifications we credit most are the impositions on personal liberty of action, we take a human situation and translate it into a cold, dehumanized algebraic equation. We forget that we are talking about human death or grave physical harms and their reverberating consequences when we equate the consequences with such things as one person's momentary freedom not to act. People are decontextualized for the analysis, yet no one really lives an acontextual life. What gives us the authority to take contextual, actual problems and encode them in a language of numbers, letters, and symbols that represents no reality in any actual person's life?

If instead we impose a duty of acting responsibly with the same self-conscious care for the safety of others that we would give our neighbors or people we know, we require the actor to consider the human consequences of her failure to rescue. Even though it is easier to understand the problem if we hone it down to "relevant facts," which may include abstracting the parties into letter symbols (either A and B or P and D) or roles (driver and passenger), why is it that "relevant facts" do not include the web of relationships and connected people affected by a failure to act responsibly with care for that person's safety? Why is it that our legal training forces us to exclude that information when we solve problems and make rules governing social behavior or for compensating some victims of accidents? Why should our autonomy or freedom not to rescue weigh more heavily in law than a stranger's harms and the consequent harms to people with whom she is interconnected?

The "no duty" rule is a conseqeunce of a legal system devoid of care and responsiveness to the safety of others. We certainly could create a duty to aid generated from a legal recognition of our interconnectedness, an elevated sense of the importance of physical health and safety, a rejection of the act/omission dualism, and a strong legal value placed on care and concern for others rather than on economic efficiency or individual liberty. The duty to act with care for another's safety, which under appropriate circumstances would include an affirmative duty to act to protect or prevent harm to another, would be shaped by the particular context. One's ability to aid and one's proximity to the need would be relevant considerations. Whether one met that duty would not be determined by how a reasonable person would have acted under the circumstances but by whether one acted out of a conscious care and concern for the safety, health, and well-being of the victim in the way one would act out of care for a neighbor or friend. If someone is clearly in need of medical help or police protection and your own abilities or limitations make you incapable of providing it, a duty to aid arising from care and concern for another's safety may require you to call for help expeditiously. In circumstances that do not call for rescue, the duty of care would require that one's behavior be governed by a conscious regard for another's safety. This seemingly minor change would transform the core of negligence law to a human, responsive system.

. . .

A feminist critique of tort law questions how we structure our legal understanding of accidents, the definitions we use, the content of our classifications, the perspectives from which they are made, and their internal components. Beyond the standard of care, other aspects of negligence—duty, breach of duty, causation (legal and factual), and damages—pose equally serious problems. There are also complicated problems with the patriarchal biases in terms of art, for instance, in "reasonableness," "foreseeability," and "assumption of the risk."

Other questions are more systemic and focus on the legitimacy of the entire analytical framework, not just on the meanings of the questions we ask and the terms we use, but on which questions we choose to ask and in what order; on whose and what kinds of interests are protected or recognized, and whose or what kinds are not "legally cognizable"; on permissible and impermissible problem-solving processes; on who counts as a "party" to situations causing accidental harms, and therefore who may be asked to contribute to solutions. Why, for instance, do tort damages recognize financial loss and yet remain reluctant to recognize relational loss, such as loss of the companionship of a child, or intangible harms, such as an increased risk of cancer or loss of a less-than-even chance of survival? Why are tort remedies all translated into money values instead of other forms of compensation? Why do we settle for the ease of monetary payment (particularly insurance premiums) instead of requiring tortfeasors to take fuller and more personally active responsibility for the harms they cause? How does tort analysis serve to perpetuate existing power hierarchies? Feminist critiques challenge the implicit assumptions in the very structure of the analysis we use.

Notes and Questions

1. Is there a sense in which the choice between negligence and strict liability might be said to be distorted by the suppression of the female voice? Does this distortion also operate for comparative and contributory negligence?

2. A majority of the author's students found the no-duty-to-rescue rule reprehensible, and many then "learned" to reject these previously held intuitions. What other doctrines in tort law seem counterintuitive to many observers? Do these doctrines also reflect a view of the world that can be described as generated by a dominant religion, culture, or gender?

3. According to the author, "Tort law should begin with a premise of responsibility rather than rights or interconnectedness rather than separation, and a priority of safety rather than profit or efficiency." If tort law began this way, what other rules (besides rescue) might be different? Would you expect most people to vote for, or otherwise prefer, such changes? What are the obstacles to change? What are the obstacles to voluntary, incremental change—that is, to a personal commitment to behave differently even though such behavior is not encouraged by law?

4. How do we know when something is distorted by male (or other) dominance? Consider the example of classroom dynamics. If we were to observe that female students tend to be supportive of their classmates while male students tend to be competitive, what might we conclude? What if the situation were reversed, or the behavior seemed indistinguishable by gender? Whose observations should we rely on to judge supportiveness and competitiveness? If support and competition do not seem sufficiently distinct from one another, what everyday behavior might serve as a proxy for cooperation, rescue, or efficiency?

The Case for a Duty to Rescue

ERNEST J. WEINRIB

. . .

Consideration of the utilitarian approach towards rescue must begin with Jeremy Bentham's thought on the problem. "[I]n cases where the person is in danger," he asked, "why should it not be made the duty of every man to save another from mischief, when it can be done without prejudicing himself . . . ? Bentham supported the implicit answer to this question with several illustrations: using water at hand to quench a fire in a woman's headdress; moving a sleeping drunk whose face is in a puddle; warning a person about to carry a lighted candle into a room strewn with gunpowder. Bentham clearly had in mind a legal duty that would be triggered by the combination of the victim's emergency and the absence of inconvenience to the rescuer—that is, by the features of most of the proposed reforms requiring rescue. Unfortunately, the rhetorical question was the whole of Bentham's argument for his position.

. . .

Can one supply the Benthamite justification that Bentham himself omitted? Because the avoidance of injury or death obviously contributes to the greatest happiness of the greatest number, the difficulties revolve not around the basic requirement of rescue but around the limitations placed upon that requirement by the notions of emergency and absence of inconvenience.

. . .

The utilitarian's only concern is that an individual bring about a situation that results in a higher surplus of pleasure over pain than would any of the alternative situations that his actions could produce. Consequences are important; how they are reached is not. The distinction between nonfeasance and misfeasance has no place in this theory, and neither would the rescue duty's emergency or convenience limitations, which apply only after that distinction is made.

One solution to the apparent inconsistency between the rescue limitations and Benthamite theory's regard only for consequences is to drop the conditions of emergency and convenience as limitations on the duty to rescue. The position could be taken that there is an obligation to rescue whenever rescuing would result in greater net happiness than not rescuing. This principle, it is important to observe, cannot really be a principle about rescuing as that concept is generally understood. As a matter of common usage, a rescue presupposes the existence of an emergency, of a predicament that poses danger of greater magnitude and imminence than one ordinarily encounters. The

Abridged and reprinted without footnotes by permission of The Yale Law Journal Company and Fred B. Rothman & Company from *The Yale Law Journal*, vol. 90, pp. 247–293.

proposed principle, however, requires no emergency to trigger a duty to act. The principle, in fact, is one of beneficence, not rescue, and should be formulated more generally to require providing aid whenever it will yield greater net happiness than not providing aid.

Eliminating the limitations regarding emergency and convenience might transform a requirement of rescue conceived along utilitarian lines into a requirement of perfect and general altruism. This demand of prefect altruism would be undesirable for several reasons. First, it would encourage the obnoxious character known to the law as the officious intermeddler. Also, its imposition of a duty of continual saintliness and heroism is unrealistic. Moreover, it would overwhelm the relationships founded on friendship and love as well as the distinction between the praiseworthy and the required; it would thereby obscure some efficient ways, in the utilitarian's eyes, of organizing and stimulating beneficence. Finally, and most fundamentally, it would be self-defeating. The requirement of aid assumes that there is some other person who has at least a minimal core of personhood as well as projects of his own that the altruist can further. In a society of perfect and general altruism, however, any potential recipient of aid would himself be an altruist, who must, accordingly, subordinate the pursuit of his own projects to the rendering of aid to others. No one could claim for his own projects the priority that would provide others with a stable object of their altruistic ministrations. Each person would continually find himself obligated to attempt to embrace a phantom.

Although the utilitarian principle that requires the provision of aid whenever it will result in greater net happiness than failure to aid easily slips into the pure-altruism duty, it need not lead to so extreme a position. The obvious alternative interpretation of the principle is that aid is not obligatory whenever the costs to one's own projects outweigh the benefits to the recipient's. This interpretation avoids the embracing-of-phantoms objection to pure altruism, but it is subject to all the other criticisms of the purer theory. Because the cost-benefit calculus is so difficult to perform in particular instances, the duty would remain ill-defined. In many cases, therefore, it would encourage the officious intermeddler, seem unrealistically to require saintliness, voerwhelm friendship and love, and obliterate the distinction between the praiseworthy and the required. Moreover, the vagueness of the duty would lead many individuals unhappily and inefficiently to drop their own projects in preference for those of others.

A different formulation of the rescue duty is needed to harness and temper the utilitarian impulses toward altruism and to direct them more precisely toward an intelligible goal. One important weakness of a too-generally beneficent utilitarianism is that it tempts one to consider only the immediate consequences of particular acts, and not the longer-term consequences, the most important of which are the expectations generated that such acts will continue. If, as the classical utilitarians believed, the general happiness is advanced when people engage in productive activities that are of value to others, the harm done by a duty of general beneficence, in either version discussed above, would override its specific benefits. The deadening of industry resulting from both reliance on beneficence and devotion to beneficence would in the long run be an evil greater than the countenancing of individual instances of unfulfilled needs or wants.

. . .

Utilitarianism can use the notion of reliance to restrict the requirement of beneficence. If an act of beneficence would tend to induce reliance on similar acts, it should be avoided. If the act of beneficence does not have this tendency, it should be performed as long as the benefit produced is greater than the cost of performance. In the latter case, there are no harmful effects on industry flowing from excessive reliance to outweigh the specific benefits. This rule can account for Bentham's restriction of the duty to rescue to situations of emergency. People do not regularly expose themselves to extraordinary dangers in reliance on the relief that may be available if the emergency materializes, and only a fool would deliberately court a peril because he or others had previously been rescued from a similar one.

. . .

Bentham's intuitive restriction of beneficence to situations of emergency can thus be supported on utilitarian grounds. Is the same true of the inconvenience limitation? As with the emergency restriction, finding utilitarian support requires looking behind the specific action to its social and legal context. For the utilitarian, the enforcement of a duty through legal sanctions is always an evil, which can be justified only to avoid a greater evil. If the sanction is applied, the offender suffers the pain of punishment. If the prospect of the sanction is sufficient to deter conduct, those deterred suffer the detriment of frustrated preferences. Moreover, the apparatus of enforcement siphons off social resources from other projects promoting the general happiness.

Accordingly, a utilitarian will be restrained and circumspect in the elaboration of legal duties. In particular, he will not pitch a standard of behavior at too high a level: the higher the standard, the more onerous it will be to the person subjected to it, the greater the pleasure that he must forego in adhering to it, and the greater his resistance to its demands. A high standard entails both more severe punishment and a more elaborate apparatus of detection and enforcement. Applied to the rescue situation, this reasoning implies that some convenience restriction should be adopted as part of the duty. Compelling the rescuer to place himself in physical danger, for instance, would be inefficacious, to use Bentham's terminology, because such coercion cannot influence the will: "the evil, which he sees himself about to undergo . . . is so great that the evil denounced by the penal clause . . . cannot appear greater." Limiting the duty of rescue to emergency situations where the rescue will not inconvenience the rescuer—as judicial decisions would elaborate that limitation and thus give direction to individuals—minimizes both the interference with the rescuer's own preferences and the difficulties of enforcement that would result form recalcitrance. Bentham's second limitation can thus also be supported on a utilitarian basis.

Notes and Questions

1. The argument in the reading seems to converge on the moral intuition that many observers, and perhaps the author, begin with: easy ("convenient") rescues should be required in emergency situations. Can you think of other areas of tort law where moral intuitions do not match legal rules? If in most instances the two converge, why is this so? And then why is it not always the case?

2. Compare this selection with the preceding two. Did the language of these readings seem different or at odds with one another? What is it that makes an approach philosophical? Is there room to combine the philosophical, feminist, and economic approaches to rescue law?

3. As reported in the first reading, if A is in danger as a result of her own negligence, and B, though not required to do so by law, goes to A's assistance but is injured in the process, a fair number of cases allow B compensation from A. A separate doctrinal development allows A to recover from B if B begins rescuing A but then abandons the effort. Does any approach advanced in this part help explain these rules (against the background rule which allows B to do nothing)? Does any combination of approaches cast light on these doctrines?

VII

Tort Liability and Other Means of Social Control

The previous parts in this anthology have dealt with fundamental questions that arise in the tort system. The present part seeks to step outside the boundaries of tort law in order to examine questions about, rather than within, tort law. The first reading may well be the best-known article written about tort law. It asks when and why we use liability rules, as opposed to rules of property or inalienability (that is, wholesale bans on certain transactions) to protect "entitlements." Entitlements refer to the initial positions or licenses granted within the legal system. Thus if B sues factory A and shows that A's smokestack emissions injure or annoy B, and a court rules that A can continue as before with no liability to B, A is said to have an entitlement to pollute; such an entitlement is a property right in the sense that B could buy it from A. B might buy the factory or pay A to operate the factory differently, just as most property rights can be transferred by a willing seller. The reading also generalizes about the initial assignment of these entitlements (sometimes to factories and sometimes to the B's of the world). One way to think of this reading is as an attempt to understand why and when tort liability is used as compared to other legal tools.

If this first reading is understood to concern choices among legal tools available to the government and its courts in regulating behavior, then the second reading can be said to begin with the claim that behavior is controlled by much more than legal tools. In a world in which people are often ignorant of the content of legal rules, and in which they often behave in ways not required or encouraged by the rules, it is likely that there is much to be gained by focusing on customs and other social patterns of control that arise apart from (or even despite) the government's role in making and enforcing rules. These patterns have often been ignored by traditional lawyers, historians, and economists, but they are the focus of the work of sociologists. This comments on the progress and promise of a law-and-sociology or "law-and-society" approach. It steps outside of tort law not to ask how and when other law can be used but rather to ask how something other than law affects the very same behavior that tort law attempts to regulate.

The final section steps outside the boundaries of tort law, and of common-law subjects in general, to argue that it is easy to overstate the impact of legal decisions in these areas of law on economic and social change. Thus it may be common to regard the assignment of legal entitlements and the development of tort liability rules as having a profound effect on the distribution of wealth in society, but the reading suggests that in fact these rules have a relatively small effect on wealth. One can think of this reading as supporting the claim that nonlegal, noncentralized decisions have a profound impact on behavior, but one can also regard this reading as an opportunity to revisit part II. It is a reminder of both the historical controversy regarding the relationship between tort law and the industrial revolution and the literature on the choice between negligence and strict liability.

Property Rules, Liability Rules, and Inalienability: One View of the Cathedral

GUIDO CALABRESI AND A. DOUGLAS MELAMED

Only rarely are property and torts approached from a unified perspective. Recent writings by lawyers concerned with economics and by economists concerned with law suggest, however, that an attempt at integrating the various legal relationships treated by these subjects would be useful both for the beginning student and the sophisticated scholar. By articulating a concept of "entitlements" which are protected by property, liability, or inalienability rules, we present one framework for such an approach.

· · ·

The first issue which must be faced by any legal system is one we call the problem of "entitlement." Whenever a state is presented with the conflicting interests of two or more people, or two or more groups of people, it must decide which side to favor. Absent such a decision, access to goods, services, and life itself will be decided on the basis of "might makes right"—whoever is stronger or shrewder will win. Hence the fundamental thing that law does is to decide which of the conflicting parties will be entitled to prevail. The entitlement to make noise versus the entitlement to have silence, the entitlement to pollute versus the entitlement to breathe clear air, the entitlement to have children versus the entitlement to forbid them—these are the first order of legal decisions.

Having made its initial choice, society must enforce that choice. Simply setting the entitlement does not avoid the problem of "might makes right"; a minimum of state intervention is always necessary.

· · ·

These decisions go to the manner in which entitlements are protected and to whether an individual is allowed to sell or trade the entitlement. In any given dispute, for example, the state must decide not only which side wins but also the kind of protection to grant. It is with the latter decisions, decisions which shape the subsequent relationship between the winner and the loser, that this article is primarily concerned. We shall consider three types of entitlements—entitlements protected by property rules, entitlements protected by liability rules, and inalienable entitlements.

· · ·

An entitlement is protected by a property rule to the extent that someone who wishes to remove the entitlement from its holder must buy it from him in a voluntary transaction in which the value of the entitlement is agreed upon by

Abridged and reprinted without footnotes by permission from 85 *Harvard Law Review* 1089 (1972).

the seller. It is the form of entitlement which gives rise to the least amount of state intervention: once the original entitlement is decided upon, the state does not try to decide its value.

. . .

Whenever someone may destroy the initial entitlement if he is willing to pay an objectively determined value for it, an entitlement is protected by a liability rule. This value may be what it is thought the original holder of the entitlement would have sold it for. But the holder's complaint that he would have demanded more will not avail him once the objectively determined value is set. Obviously, liability rules involve an additional stage of state intervention: not only are entitlements protected, but their transfer or destruction is allowed on the basis of a value determined by some organ of the state rather than by the parties themselves.

An entitlement is inalienable to the extent that its transfer is not permitted between a willing buyer and a willing seller. The state intervenes not only to determine who is initially entitled and to determine the compensation that must be paid if the entitlement is taken or destroyed, but also to forbid its sale under some or all circumstances. Inalienability rules are thus quite different from property and liability rules. Unlike those rules, rules of inalienability not only "protect" the entitlement; they may also be viewed as limiting or regulating the grant of the entitlement itself.

. . .

This article will explore two primary questions: (1) In what circumstances should we grant a particular entitlement? and (2) in what circumstances should we decide to protect that entitlement by using a property, liability, or inalienability rule?

The Setting of Entitlements

What are the reasons for deciding to entitle people to pollute or to entitle people to forbid pollution, to have children freely or to limit procreation, to own property or to share property? They can be grouped under three headings: economics efficiency, distributional preferences, and other justice considerations.

Economic Efficiency

Perhaps the simplest reason for a particular entitlement is to minimize the administrative costs of enforcement. . . . By itself this reason will never justify any result except that of letting the stronger win, for obviously that result minimizes enforcement costs. Nevertheless, administrative efficiency may be relevant to choosing entitlements when other reasons are taken into account.

. . .

But administrative efficiency is just one aspect of the broader concept of economic efficiency. Economic efficiency asks that we choose the set of entitlements which would lead to that allocation of resources which could not be improved in the sense that a further change would not so improve the condition of those who gained by it that they could compensate those who lost from it and still be better off than before. This is often called Pareto optimality.

. . .

Recently it has been argued that on certain assumptions, usually termed the absence of transaction costs, Pareto optimality or economic efficiency will occur regardless of the initial entitlement. For this to hold, "no transaction costs" must be understood extremely broadly as involving both perfect knowledge and the absence of any impediments or costs of negotiating. Negotiation costs include, for example, the cost of excluding would-be freeloaders from the fruits of market bargains. In such a frictionless society, transactions would occur until no one could be made better off as a result of further transactions without making someone else worse off. This, we would suggest, is a necessary, indeed a tautological, result of the definitions of Pareto optimality and of transaction costs which we have given.

Such a result would not mean, however, that the *same* allocation of resources would exist regardless of the initial set of entitlements. Taney's willingness to pay for the right to make noise may depend on how rich he is; Marshall's willingness to pay for silence may depend on his wealth. In a society which entitles Taney to make noise and which forces Marshall to buy silence from Taney, Taney is wealthier and Marshall poorer than each would be in a society which had the converse set of entitlements. Depending on how Marshall's desire for silence and Taney's for noise vary with their wealth, an entitlement to noise will result in negotiations which will lead to a different quantum of noise than would an entitlement to silence. This variation in the quantity of noise and silence can be viewed as no more than an instance of the well-accepted proposition that what is a Pareto optimal, or economically efficient, solution varies with the starting distribution of wealth. Pareto optimality is optimal *given* a distribution of wealth, but different distributions of wealth imply their own Pareto optimal allocation of resources.

All this suggests why distributions of wealth may affect a society's choice of entitlements. It does not suggest why *economic efficiency* should affect the choice, if we assume an absence of any transaction costs. But no one makes an assumption of no transaction costs in practice. . . . [T]he assumption of no transaction costs may be a useful starting point, a device which helps us see how, as different elements which may be termed transaction costs become important, the goal of economic efficiency starts to prefer one allocation of entitlements over another.

. . .

[I]t is enough to say here: (1) that economic efficiency standing alone would dictate that set of entitlements which favors knowledgeable choices between social benefits and the social costs of obtaining them, and between social costs and the social costs of avoiding them; (2) that this implies, in the absence of certainty as to whether a benefit is worth its costs to society, that the cost should be put on the party or activity best located to make such a cost-benefit analysis; (3) that in particular contexts like accidents or pollution this suggests putting costs on the party or activity which can most cheaply avoid them; (4) that in the absence of certainty as to who that party or activity is, the costs should be put on the party or activity which can with the lowest transaction costs act in the market to correct an error in entitlements by inducing the party who can avoid social costs most cheaply to do so; and (5) that since we are in an area where by hypothesis markets do not work perfectly—there are transaction costs—a decision will often have to be made on whether market transactions or collective fiat is most likely to bring us closer to the Pareto optimal result the "perfect" market would reach. . . .

Distribution Goals

There are, we would suggest, at least two types of distributional concerns which may affect the choice of entitlements. These involve distribution of wealth itself and distribution of certain specific goods, which have sometimes been called merit goods.

. . .

Difficult as wealth distribution preferences are to analyze, it should be obvious that they play a crucial role in the setting of entitlements. For the placement of entitlements has a fundamental effect on a society's distribution of wealth. It is not enough, if a society wishes absolute equality, to start everyone off with the same amount of money. A financially egalitarian society which gives individuals the right to make noise immediately makes the would-be noisemaker richer than the silence-loving hermit. Similarly, a society which entitles the person with brains to keep what his shrewdness gains him implies a different distribution of wealth from a society which demands from each according to his relative ability but gives to each according to his relative desire. . . . The consequence of this is that it is very difficult to imagine a society in which there is complete equality of wealth.

. . .

If perfect equality is impossible, a society must choose what entitlements it wishes to have on the basis of criteria other than perfect equality. In doing this, a society often has a choice of methods, and the method chosen will have important distributional implications. Society can, for instance, give an entitlement away free and then, by paying the holders of the entitlement to limit their use of it, protect those who are injured by the free entitlement. Conversely, it can allow people to do a given thing only if they buy the right from the government. Thus a society can decide whether to entitle people to have children and then induce them to exercise control in procreating, or to require people to buy the right to have children in the first place. A society can also decide whether to entitle people to be free of military service and then induce them to join up, or to require all to serve but enable each to buy his way out. Which entitlement a society decides to sell, and which it decides to give away, will likely depend in part on which determination promotes the wealth distribution that society favors.

If the choice of entitlements affects wealth distribution generally, it also affects the chances that people will obtain what have sometimes been called merit goods. Whenever a society wishes to maximize the chances that individuals will have at least a minimum endowment of certain particular goods— education, clothes, bodily integrity—the society is likely to begin by giving the individuals an entitlement to them. If the society deems such an endowment to be essential regardless of individual desires, it will, of course, make the entitlement inalienable. Why, however, would a society entitle individuals to specific goods rather than to money with which they can buy what they wish, unless it deems that it can decide better than the individuals what benefits them and society; unless, in other words, it wishes to make the entitlement inalienable?

We have seen that an entitlement to a good or to its converse is essentially inevitable. We either are entitled to have silence or entitled to make noise in a given set of circumstances. We either have the right to our own property or body or the right to share others' property or bodies. We may buy or sell ourselves into the opposite position, but we must start somewhere. Under these circumstances, a society which prefers people to have silence, or own property, or have bodily integrity, but which does not hold the grounds for its preference to be sufficiently strong to justify overriding contrary preferences by individuals, will give such entitlements according to the collective preference, even though it will allow them to be sold thereafter.

Whenever transactions to sell or buy entitlements are very expensive, such an initial entitlement decision will be nearly as effective in assuring that individuals will have the merit good as would be making the entitlement inalienable. . . . In such circumstances society will pick the entitlement it deems favorable to the general welfare and not worry about coercion or alienability: it has increased the chances that individuals will have a particular good without increasing the degree of coercion imposed on individuals.

. . .

Whenever society chooses an initial entitlement it must also determine whether to protect the entitlement by property rules, by liability rules, or by rules of inalienability. In our framework, much of what is generally called private property can be viewed as an entitlement which is protected by a property rule. No one can take the entitlement to private property from the holder unless the holder sells it willingly and at the price at which he subjectively values the property. Yet a nuisance with sufficient public utility to avoid injunction has, in effect, the right to take property with compensation. In such a circumstance the entitlement to the property is protected only by what we call a liability rule: an external, objective standard of value is used to facilitate the transfer of the entitlement from the holder to the nuisance. Finally, in some instances we will not allow the sale of the property at all, that is, we will occasionally make the entitlement inalienable.

. . .

Why cannot a society simply decide on the basis of the already mentioned criteria who should receive any given entitlement, and then let its transfer occur only through a voluntary negotiation? Why, in other words, cannot society limit itself to the property rule?

. . .

In terms of economic efficiency the reason is easy enough to see. Often the cost of establishing the value of an initial entitlement by negotiation is so great that even though a transfer of the entitlement would benefit all concerned, such a transfer will not occur. If a collective determination of the value were available instead, the beneficial transfer would quickly come about.

Eminent domain is a good example. A park where Guidacres, a tract of land owned by 1,000 owners in 1,000 parcels, now sits would, let us assume, benefit a neighboring town enough so that the 100,000 citizens of the town would each be willing to pay an average of $100 to have it. The park is Pareto desirable if the owners of the tracts of land in Guidacres actually value their entitlements at less than $10,000,000 or an average of $10,000 a tract. Let us assume that in fact the parcels are all the same and all the owners value them at $8,000. On this assumption, the park is, in economic efficiency terms, desirable—in values foregone it costs $8,000,000 and is worth $10,000,000 to the buyers. And yet it may well not be established. If enough of the owners hold out for more than $10,000 in order to get a share of the $2,000,000 that they guess the buyers are willing to pay over the value which the sellers in actuality attach, the price demanded will be more than $10,000,000 and no park will result. The sellers have an incentive to hide their true valuation and the market will not succeed in establishing it.

An equally valid example could be made on the buying side. Suppose the sellers of Guidacres have agreed to a sales price of $8,000,000 (they are all relatives and at a family banquet decided that trying to hold out would leave them all losers). It does not follow that the buyers can raise that much even though each of 100,000 citizens *in fact* values the park at $100. Some citizens may try to free-load and say the park is only worth $50 or even nothing to them, hoping that enough others will admit to a higher desire and make up the $8,000,000 price. Again there is no reason to believe that a market, a decentralized system of valuing, will cause people to express their true valuations and hence yield results which all would *in fact* agree are desirable.

Whenever this is the case an argument can readily be made for moving from a property rule to a liability rule. If society can remove from the market the valuation of each tract of land, decide the value collectively, and impose it, then the holdout problem is gone. Similarly, if society can value collectively each individual citizen's desire to have a park and charge him a "benefits" tax based upon it, the freeloader problem is gone.

. . .

Even if holdout and freeloader problems can be met feasibly by the market, an argument may remain for employing a liability rule. Assume that in our hypothetical, freeloaders can be excluded at the cost of $1,000,000 and that all owners of tracts in Guidacres can be convinced, by the use of $500,000 worth of advertising and cocktail parties, that a sale will only occur if they reveal their true land valuations. Since $8,000,000 plus $1,500,000 is less than

$10,000,000, the park will be established. But if collective valuation of the tracts and of the benefits of the prospective park would have cost less than $1,500,000, it would have been inefficient to establish the park through the market—a market which was not worth having would have been paid for.

Of course, the problems with liability rules are equally real. We cannot be at all sure that landowner Taney is lying or holding out when he says his land is worth $12,000 to him. The fact that several neighbors sold identical tracts for $10,000 does not help us very much; Taney may be sentimentally attached to his land. As a result, eminent domain may grossly undervalue what Taney would actually sell for, even if it sought to give him his true valuation of his tract. In practice, it is so hard to determine Taney's true valuation that eminent domain simply gives him what the land is worth "objectively," in the full knowledge that this may result in over or undercompensation.

. . .

The example of eminent domain is simply one of numerous instances in which society uses liability rules. Accidents is another. If we were to give victims a property entitlement not to be accidentally injured we would have to require all who engage in activities that may injure individuals to negotiate with them before an accident, and to buy the right to knock off an arm or a leg. Such preaccident negotiations would be extremely expensive, often prohibitively so. To require them would thus preclude many activities that might, in fact, be worth having. And, after an accident, the loser of the arm or leg can always very plausibly deny that he would have sold it at the price the buyer would have offered.

. . .

It is enough for our purposes to note that a very common reason, perhaps the most common one, for employing a liability rule rather than a property rule to protect an entitlement is that market valuation of the entitlement is deemed inefficient, that is, it is either unavailable or too expensive compared to a collective valuation.

We should also recognize that efficiency is not the sole ground for employing liability rules rather than property rules. Just as the initial entitlement is often decided upon for distributional reasons, so too the choice of a liability rule is often made because it facilitates a combination of efficiency and distributive results which would be difficult to achieve under a property rule.

. . .

Thus far we have focused on the questions of when society should protect an entitlement by property or liability rules. However, there remain many entitlements which involve a still greater degree of societal intervention: the law not only decides who is to own something and what price is to be paid for it if it is taken or destroyed, but also regulates its sale—by, for example, prescribing preconditions for a valid sale or forbidding a sale altogether.

. . .

For instance, if Taney were allowed to sell his land to Chase, a polluter, he would injure his neighbor Marshall by lowering the value of Marshall's land. Conceivably, Marshall could pay Taney not to sell his land; but, because there are many injured Marshalls, freeloader and information costs make such transactions practically impossible. The state could protect the Marshalls and yet facilitate the sale of the land by giving the Marshalls an entitlement to prevent Taney's sale to Chase but only protecting the entitlement by a liability rule. It might, for instance, charge an excise tax on all sales of land to polluters equal to its estimate of the external cost to the Marshalls of the sale. But where there are so many injured Marshalls that the price required under the liability rule is likely to be high enough so that no one would be willing to pay it, then setting up the machinery for collective valuation will be wasteful. Barring the sale to polluters will be the most efficient result because it is clear that avoiding pollution is cheaper than paying its costs—including its costs to the Marshalls.

Another instance in which external costs may justify inalienability occurs when external costs do not lend themselves to collective measurement which is acceptably objective and nonarbitrary. This nonmonetizability is characteristic of one category of external costs which, as a practical matter, seems frequently to lead us to rules of inalienability. Such external costs are often called moralisms.

If Taney is allowed to sell himself into slavery, or to take undue risks of becoming penniless, or to sell a kidney, Marshall may be harmed, simply because Marshall is a sensitive man who is made unhappy by seeing slaves, paupers, or persons who die because they have sold a kidney. Again Marshall could pay Taney not to sell his freedom to Chase the slaveowner; but again, because Marshall is not one but many individuals, freeloader and information costs make such transactions practically impossible.

. . .

In the case of Taney selling land to Chase, the polluter, [liability rules] were inappropriate because we *knew* that the costs to Taney and the Marshalls exceeded the benefits to Chase. Here, though we are not certain of how a cost-benefit analysis would come out, liability rules are inappropriate because any monetization is, by hypothesis, out of the question. The state must, therefore, either ignore the external costs to Marshall, or if it judges them great enough, forbid the transaction that gave rise to them by making Taney's freedom inalienable.

. . .

There are two other efficiency reasons for forbidding the sale of entitlements under certain circumstances: self paternalism and true paternalism. Examples of the first are Ulysses tying himself to the mast or individuals passing a bill of rights so that they will be prevented from yielding to momentary temptations which they deem harmful to themselves. This type of limitation is not in any real sense paternalism. . . . It merely allows the individual to choose what is best in the long run rather than in the short run, even though that choice entails giving up some short run freedom of choice.

. . .

True paternalism brings us a step further toward explaining such prohibitions and those of broader kinds—for example the prohibitions on a whole range of activities by minors. Paternalism is based on the notion that at least in some situations the Marshalls know better than Taney what will make Taney better off. Here we are not talking about the offense to Marshall from Taney's choosing to read pornography, or selling himself into slavery, but rather the judgment that Taney was not in the position to choose best for himself when he made the choice for erotica or servitude. The first concept we called a moralism and is a frequent and important ground for inalienability. But it is consistent with the premises of Pareto optimality. The second, paternalism, is also an important economic efficiency reason for inalienability, but it is not consistent with the premises of Pareto optimality; the most efficient pie is no longer that which costless bargains would achieve, because a person may be better off if he is prohibited from bargaining.

Finally, just as efficiency goals sometimes dictate the use of rules of inalienability, so, of course, do distributional goals. Whether an entitlement may be sold or not often affects directly who is richer and who is poorer. Prohibiting the sale of babies makes poorer those who can cheaply produce babies and richer those who through some nonmarket device get free an "unwanted" baby.

. . .

This should suffice to put us on guard, for it suggests that direct distributional motives may lie behind asserted nondistributional grounds for inalienability, whether they be paternalism, self paternalism, or externalities. This does not mean that giving weight to distributional goals is undesirable. It clearly is desirable where on efficiency grounds society is indifferent between an alienable and an inalienable entitlement and distributional goals favor one approach or the other. It may well be desirable even when distributional goals are achieved at some efficiency costs. The danger may be, however, that what is justified on, for example, paternalism grounds is really a hidden way of accruing distributional benefits for a group whom we would not otherwise wish to benefit.

. . .

Nuisance or pollution is one of the most interesting areas where the question of who will be given an entitlement, and how it will be protected, is in frequent issue. Traditionally, . . . the nuisance-pollution problem is viewed in terms of three rules. First, Taney may not pollute unless his neighbor (his only neighbor let us assume), Marshall, allows it (Marshall may enjoin Taney's nuisance). Second, Taney may pollute but must compensate Marshall for damages caused (nuisance is found but the remedy is limited to damages). Third, Taney may pollute at will and can only be stopped by Marshall if Marshall pays him off (Taney's pollution is not held to be a nuisance to Marshall). In our terminology rules one and two (nuisance with injunction, and with damages only) are entitlements to Marshall. The first is an entitlement to be free from pollution and is protected by a property rule; the second is also an entitlement to be free from pollution but is protected only by a liability rule. Rule three (no nuisance) is instead an entitlement to Taney protected by a property rule, for only by buying Taney out at Taney's price can Marshall end the pollution.

The very statement of these rules in the context of our framework suggests that something is missing. Missing is a fourth rule representing an entitlement in Taney to pollute, but an entitlement which is protected only by a liability rule. The fourth rule . . . can be stated as follows: Marshall may stop Taney from polluting, but if he does he must compensate Taney. Unlike the first three, [the fourth rule] does not often lend itself to judicial imposition for a number of good legal process reasons. For example, even if Taney's injuries could practically be measured, apportionment of the duty of compensation among many Marshalls would present problems for which courts are not well suited. If only those Marshalls who voluntarily asserted the right to enjoin Taney's pollution were required to pay the compensation, there would be insuperable freeloader problems.

. . .

The fourth rule is thus not part of the cases legal scholars read when they study nuisance law, and is therefore easily ignored by them. But it is available, and may sometimes make more sense than any of the three competing approaches. . . . To appreciate the utility of the fourth rule and to compare it with the other three rules, we will examine why we might choose any of the given rules.

We would employ rule one (entitlement to be free from pollution protected by a property rule) from an economic efficiency point of view if we believed that the polluter, Taney, could avoid or reduce the costs of pollution more cheaply than the pollutee, Marshall. Or to put it another way, Taney would be enjoinable if he were in a better position to balance the costs of polluting against the costs of not polluting. We would employ rule three (entitlement to pollute protected by a property rule) again solely from an economic efficiency standpoint, if we made the converse judgment on who could best balance the harm of pollution against its avoidance costs. If we were wrong in our judgments and if transactions between Marshall and Taney were costless or even very cheap, the entitlement under rules one or three would be traded and an economically efficient result would occur in either case. . . . Wherever transactions between Taney and Marshall are easy, and wherever economic efficiency is our goal, we could employ entitlements protected by property rules even though we would not be sure that the entitlement chosen was the right one.

. . .

The moment we assume, however, that transactions are not cheap, the situation changes dramatically. Assume we enjoin Taney and there are 10,000 injured Marshalls. Now *even if* the right to pollute is worth more to Taney than the right to be free from pollution is to the sum of the Marshalls, the injunction will probably stand. The cost of buying out all the Marshalls, given holdout problems, is likely to be too great, and an equivalent of eminent domain in Taney would be needed to alter the initial injunction. Conversely, if we denied a nuisance remedy, the 10,000 Marshalls could only with enormous difficulty, given free-loader problems, get together to buy out even one Taney and prevent the pollution. This would be so even if the pollution harm was greater than the value to Taney of the right to pollute.

If, however, transaction costs are not symmetrical, we may still be able to use the property rule. Assume that Taney can buy the Marshall's entitlements easily because holdouts are for some reason absent, but that the Marshalls have great freeloader problems in buying out Taney. In this situation the entitlement should be granted to the Marshalls unless we are sure the Marshalls are the cheapest avoiders of pollution costs. Where we do not know the identity of the cheapest cost avoider it is better to entitle the Marshalls to be free of pollution because, even if we are wrong in our initial placement of the entitlement, that is, even if the Marshalls are the cheapest cost avoiders, Taney will buy out the Marshalls and economic efficiency will be achieved. Had we chosen the converse entitlement and been wrong, the Marshalls could not have bought out Taney.

. . .

We are likely to turn to liability rules whenever we are uncertain whether the polluter or the pollutees can most cheaply avoid the cost of pollution. We are only likely to use liability rules where we are uncertain because, if we are certain, the costs of liability rules—essentially the costs of collectively valuing the damages to all concerned plus the cost in coercion to those who would not sell at the collectively determined figure—are unnecessary. They are unnecessary because transaction costs and bargaining barriers become irrelevant when we are certain who is the cheapest cost avoider.

. . .

As a practical matter we often are uncertain who the cheapest cost avoider is. In such cases, traditional legal doctrine tends to find a nuisance but imposes only damages on Taney payable to the Marshalls. This way, if the amount of damages Taney is made to pay is close to the injury caused, economic efficiency will have had its due; if he cannot make a go of it, the nuisance was not worth its costs. The entitlement to the Marshalls to be free from pollution unless compensated, however, will have been given *not* because it was thought that polluting was probably worth less to Taney than freedom from pollution was worth to the Marshalls, nor even because on some distributional basis we preferred to charge the cost to Taney rather than to the Marshalls. It was so placed *simply because we did not know* whether Taney desired to pollute more than the Marshalls desired to be free from pollution, and the only way we thought we could test out the value of the pollution was by the only liability rule we thought we had. This was rule two, the imposition of nuisance damages on Taney. At least this would be the position of a court concerned with economic efficiency which believed itself limited to rules one, two, and three.

Rule four gives at least the possibility that the opposite entitlement may also lead to economic efficiency in a situation of uncertainty. Suppose for the moment that a mechanism exists for collectively assessing the damage resulting to Taney from being stopped from polluting by the Marshalls, and a mechanism also exists for collectively assessing the benefit to each of the Marshalls from such cessation. Then—assuming the same degree of accuracy in collective valuation as exists in rule two (the nuisance damage rule)—the

Marshalls would stop the pollution if it harmed them more than it benefited Taney. If this is possible, then even if we thought it necessary to use a liability rule, we would still be free to give the entitlement to Taney or Marshall for whatever reasons, efficiency or distributional, we desired.

Actually, the issue is still somewhat more complicated. For just as transaction costs are not necessarily symmetrical under the two converse property rule entitlements, so also the liability rule equivalents of transaction costs—the cost of valuing collectively and of coercing compliance with that valuation—may not be symmetrical under the two converse liability rules. Nuisance damages may be very hard to value, and the costs of informing all the injured of their rights and getting them into court may be prohibitive. Instead, the assessment of the objective damage to Taney from foregoing his pollution may be cheap and so might the assessment of the relative benefits to all Marshalls of such freedom from pollution. But the opposite may also be the case. As a result, just as the choice of which property entitlement may be based on the asymmetry of transaction costs and hence on the greater amenability of one property entitlement to market corrections, so might the choice between liability entitlements be based on the asymmetry of the costs of collective determination.

The introduction of distributional considerations makes the existence of the fourth possibility even more significant. One does not need to go into all the permutations of the possible tradeoffs between efficiency and distributional goals under the four rules to show this. A simple example should suffice. Assume a factory which, by using cheap coal, pollutes a very wealthy section of town and employs many low-income workers to produce a product purchased primarily by the poor; assume also a distributional goal that favors equality of wealth. Rule one—enjoin the nuisance—would possibly have desirable economic-efficiency results (if the pollution hurt the homeowners more than it saved the factory in coal costs), but it would have disastrous distribution effects. It would also have undesirable efficiency effects if the initial judgment on costs of avoidance had been wrong and transaction costs were high. Rule two—nuisance damages—would allow a testing of the economic efficiency of eliminating the pollution, even in the presence of high transaction costs, but would quite possibly put the factory out of business or diminish output and thus have the same income distribution effects as rule one. Rule three—no nuisance—would have favorable distributional effects since it might protect the income of the workers. But if the pollution harm was greater to the homeowners than the cost of avoiding it by using a better coal, and if transaction costs—holdout problems—were such that homeowners could not unite to pay the factory to use better coal, rule three would have unsatisfactory efficiency effects. Rule four—payment of damages to the factory after allowing the homeowners to compel it to use better coal, and assessment of the cost of these damages to the homeowners—would be the only one which would accomplish both the distributional and efficiency goals.

· · ·

We have said that we would say little about justice, and so we shall. But it should be clear that if rule four might enable us best to combine efficiency goals with distributional goals, it might also enable us best to combine those same efficiency goals with other goals that are often described in justice language. For example, assume that the factory in our hypothetical was using cheap coal *before* any of the wealthy houses were built. In these circumstances, rule four will not only achieve the desirable efficiency and distributional results mentioned above, but it will also accord with any "justice" significance which is attached to being there first. And this is so whether we view this justice significance as part of a distributional goal, as part of a long-run efficiency goal based on protecting expectancies, or as part of an independent concept of justice.

Thus far in this section we have ignored the possibility of employing rules of inalienability to solve pollution problems. A general policy of barring pollution does seem unrealistic. But rules of inalienability can appropriately be used to limit the levels of pollution and to control the levels of activities which cause pollution.

One argument for inalienability may be the widespread existence of moralisms against pollution. Thus it may hurt the Marshalls—gentleman farmers—to see Taney, a smoke-choked city dweller, sell his entitlement to be free of pollution. A different kind of externality or moralism may be even more important. The Marshalls may be hurt by the expectation that, while the present generation might withstand present pollution levels with no serious health dangers, future generations may well face a despoiled, hazardous environmental condition which they are powerless to reverse. And this ground for inalienability might be strengthened if a similar conclusion were reached on grounds of self-paternalism. Finally, society might restrict alienability on paternalistic grounds. The Marshalls might feel that although Taney himself does not know it, Taney will be better off if he really can see the stars at night, or if he can breathe smogless air.

Whatever the grounds for inalienability, we should reemphasize that distributional effects should be carefully evaluated in making the choice for or against inalienability. Thus the citizens of a town may be granted an entitlement to be free of water pollution caused by the waste discharges of a chemical factory; and the entitlement might be made inalienable on the grounds that the town's citizens really would be better off in the long run to have access to clean beaches. But the entitlement might also be made inalienable to assure the maintenance of a beautiful resort area for the very wealthy, at the same time putting the town's citizens out of work.

Notes and Questions

1. Calabresi and Melamed stress the ability (and sometimes the inability) of parties to bargain with each other in order to transfer property rights or even avoid liability rules. How else can one party maneuver around a legal (tort or property) rule?

Imagine a hotel owner who wants to stop a neighbor from building a tall structure because the new building will interfere with views enjoyed by the hotel guests. If the hotel owner is unable to bargain successfully with the owner of the neighboring property, where else might the hotel owner turn? What determines whether the local zoning board or politicians will favor the first or second property owner?

Imagine now that the second owner has already secured approval from the various local authorities and paid a price for the underlying property that reflects the ability to erect a new building on the site. If the first owner successfully lobbies the zoning board or the local government (which grants building permits, perhaps) to reverse its earlier decision, should the second owner sue the first owner or the government?

2. A long-standing doctrine distinguishes between public and private nuisances and allows plaintiffs to seek relief in court (as opposed to relief through more general administrative regulation or criminal prosecution) only for private nuisances. Thus no private action may be brought against the operator of an illegal casino because the harm, if it is that, is one that affects the community at large. Can you reinterpret this distinction in light of Calabresi and Melamed's framework or in political terms? Who would bargain with the owner of the casino? How would persons offended by a casino best organize to shut it down?

3. Under Calabresi and Melamed's fourth rule, Marshall may stop Taney from polluting, but if he does so he must compensate Taney. This insight has encouraged many casebooks to refer to *Spur Industries, Inc. v. Del E. Webb Development Co.*, 494 P.2d 700 (1972), in which a real-estate developer bought land near a cattle feedlot on the outskirts of Phoenix. Eventually the developer sought to market properties near the feedlot and enlisted the help of a court to shut down the feedlot (which lavished odor and flies on its neighbors) as a nuisance. The court's decision can be regarded as an example of this fourth rule because it granted a permanent injunction but required the developer to indemnify the feedlot owner.

How would you measure the liability owed by the developer to the feedlot owner? As Phoenix grew and residential development approached the feedlot, what do you suppose happened to the value of the land on which the feedlot was located? Could you make an argument that the developer's overall action has benefitted rather than injured the feedlot owner?

If the state government had come to this nuisance and taken the feedlot in order to build a school for children living in Del Webb's development, would the government be required to pay the feedlot owner the value of the land at the time of the take-over or its value before development began? Suppose the government builds a highway out to the area and takes part of the feedlot in order to build an exit ramp near the development. If property values rise further because of the new highway access, must the government pay the condemnee this higher value for the land that is taken? Returning to the feedlot-developer case, can you fashion an argument in support of a claim by the developer that the proper level of indemnification of the feedlot owner is zero?

4. The authors describe property rules as complete, initial endowments. Other commentators have extended the framework to what might be called partial property rights. Imagine that a factory produces 5,000 units of a pollutant (along with some more useful product) and that nearby residents complain. The matter goes to court, and the judge believes that the factory is negligent, although it would not have been negligent if it had been able to keep the pollution level down to about 4,000 units. But the judge is unsure, and there are many jobs and residents at stake. What are the judge's options? If the judge issues an injunction barring the factory's operations, how could the factory reopen? If the court instead ruled in favor of the factory and denied the injunction, how could the residents shut the factory down? Do you expect the court's decision to affect the likelihood that the factory will remain open?

Could the court enjoin the factory from operating at a level that generated more than, say, 4,000 units of pollution? We might regard this sort of decision as a partial property rule. Why might the parties still find room to bargain? The court could also combine property and liability rules by enjoining some level of operation and providing for damages for the injury generated by the pollution up to that level. See A. Mitchell Polinsky, Resolving Nuisance Disputes: The Simple Economics of Injunctive and Damage Remedies, 32 *Stanford Law Review* 1075 (1980). This approach might be described as a mixture of tort liability and property rules. Can you think of a means of combining tort liability and criminal rules?

A Critique of Economic
and Sociological Theories of Social Control
ROBERT C. ELLICKSON

How is it that people manage to live side by side without incessant "warre of every man against every man"? Thomas Hobbes, who won fame for posing this question, concluded that the legal system—the rules and might of Leviathan—is the wellspring of social order. Most law professors implicitly propagate this Hobbesian view, perhaps because it lends significance to what they teach. Law-and-economics scholars have been particularly prone to assert the centrality of legal doctrine. There is an opposing intellectual tradition, however, that emphasizes that social order can emerge without law. Its core theorists have been empirical sociologists and anthropologists who study stateless societies. Within the legal academy the law professors associated with the law-and-society movement have been the prime skeptics of the importance of law.

A field study that I recently conducted in Shasta County, California, provides an empirical perspective on the sources of social order. The study investigated how rural residents resolve disputes arising over damage done by stray livestock. Much of the Shasta County evidence casts doubt on the Hobbesian view. In areas of Shasta County that are legally designated as "closed range," a victim of cattle trespass is legally entitled to recover damages from the livestock owner on a strict-liability basis. In "open-range" areas, by contrast, a victim of trespass across an unfenced boundary has no rights to legal redress. I found that the rural residents of Shasta County (and even the insurance adjusters who settle minor trespass claims) pay almost no heed to these legal rules. The operative rule in both open- and closed-range areas is an informal

Abridged and reprinted without footnotes by permission from 16 *Journal of Legal Studies* 67 (1986). Copyright © 1986 by The University of Chicago Press. Reprinted in slightly different form in Robert C. Ellickson, *Order Without Law: How Neighbors Settle Disputes,* chs. 10–11 (Harvard Univ. Press, 1991).

norm that the owner of livestock is responsible for the conduct of his animals. Shasta County residents rely mainly on self-help to enforce this norm.

In other domains of Shasta County life, however, the legal system provides the supreme rules. Claimants who have suffered property damage arising out of highway collisions between vehicles and livestock routinely file claims with liability insurers; when they have suffered personal injuries, they hire attorneys who file complaints in court. Ranchers who would never sue neighbors whose cattle trespassed would not hesitate to sue to protect water rights.

When I sought to place my findings from Shasta County in theoretical perspective, I found current theories of social control inadequate. . . .

The System of Social Control

A system of social control consists of *rules* of normatively appropriate human behavior. These rules are enforced through *sanctions,* whose administration is governed by other rules. To help describe alternative systems of social control, I distinguish between two types of sanctions, five controllers that administer sanctions and make rules, and five types of rules. My taxonomy will honor, when possible, the current scholarly vocabulary of social control.

Rewards and Punishments: The Sanctions Suitable to a Tripartite Normative Classification of Human Actions

Systems of social control typically employ both *rewards* and *punishments*— both carrots and sticks—to influence behavior. In administering these positive and negative sanctions, enforcers usually apply rules that divide the universe of human behavior into three categories: (1) good behavior that is to be rewarded, (2) bad behavior that is to be punished, and (3) ordinary behavior that warrants no response. Figure 1, which illustrates a triparite classification system of this sort, employs the standard sociological adjectives "prosocial" and "antisocial" to describe out-of-the-ordinary behavior. Economists would use "goods" and, when pressed, "bads" to describe these two extremes. People who act antisocially are "deviants"; people who act prosocially are "surpassers."

If the members of a social group were to wish to move behavior in a prosocial direction (upward in Figure 1), they conceivably could employ fewer than three normative classifications. For example, they could use a bifurcated

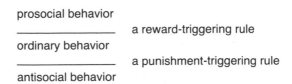

Figure 1. A tripartite classification of human behavior

prosocial behavior

_____   reward/punishment switch-point rule

antisocial behavior

Figure 2. A bifurcated classification of human behavior

system that dropped the ordinary-behavior category and that thus looked like the one in Figure 2. Or they could employ only punishments (rewards) as sanctions and have a bifurcated system consisting of two categories, that is, ordinary and punished (rewarded) behavior.

Finally, one could imagine a social control system that placed all behavior in the same normative category and that thus eliminated the need for the establishment of substantive rules whose role was to trigger changes in sanctions. For example, a society could establish an unachievable standard of perfect behavior for human conduct and levy penalties on all behavior, with the penalties presumably growing in magnitude as the deviation from perfection increased.

These unitary and bifurcated systems seem alien because in most social contexts people employ tripartite normative systems that make use of rewards, punishments, and no sanctions at all. Baseball fans, for example, cheer a shortstop's fielding gems, boo his errors, and sit on their hands when he handles a routine ground ball. Or suppose an automobile were to stall, block traffic in one of the two northbound lanes of a congested limited-access highway, and create a mile of backed-up vehicles. Most drivers would probably perceive that another motorist who stopped to direct traffic would be acting prosocially, that motorists who quietly waited out the jam would be acting ordinarily, and that motorists who leaned on their horns while they waited would be acting antisocially.

As both examples indicate, the rules used to evaluate human behavior tend to be set so that the "ordinary" category encompasses most conduct that occurs. This approach has the advantage of reducing the costs of administering sanctions: that which is most common requires no response.

Because behavior that ordinarily occurs typically warrants no punishment, the word "norm" is ordinarily used in English in a potentially ambiguous way. "Norm" denotes both behavior that *is* normal and behavior that people *should* mimic to avoid being punished. These two usages are potentially conflicting because almost everyone laments some features of the status quo. That the word "norm" has been able to maintain these two usages is a linguistic clue that ordinary behavior is only rarely regarded as antisocial behavior.

Sources of Rules and Sanctions: The Controllers Involved in Creating Social Order

It is useful to distinguish between five *controllers* that may be sources of both rules of behavior and sanctions that back up those rules. There are first-party, second-party, and third-party controllers. An *actor* who imposes rules and sanctions on himself is exercising first-party control. A promisee-enforced executory contract is a system of second-party control over the contingencies that the contract explicitly covers; the *person acted on* administers rewards and punishments depending on whether the promisor adheres to the promised course of behavior. Third-party control differs from second-party control in that the rules are ones to which the actor never explicitly agreed, and the sanctions are administered by persons not involved in the initial interaction. Third-party controllers can be either nonhierarchically organized *social forces; organizations* (nongovernmental hierarchies that make and enforce rules); or *governments* (state hierarchies).

. . .

I refer to the rules that emanate from first-party controllers as *personal ethics,* those from second-party controllers as *contracts,* those from social forces as *norms,* those from organizations as *organization rules,* and those from governments as *law.*

. . .

Max Weber, surely one of the most impressive theorists of social control, applied a somewhat different taxonomy. Weber defined law as the rules enforced by bureaucrats who specialize in social control activity. Weber's approach strains ordinary language because it is insensitive to the identity of the controller who has made or who is enforcing the rules. For example, employees of debt-collection agencies are specialized bureaucratic enforcers, but one ordinarily thinks of them as enforcers of contracts, not of laws. Similarly, if the Catholic church were to use specialized bureaucrats to enforce announced church policy, one would ordinarily view this not as the legal system in action but as something else, that is, what I call organization control. Following Donald Black, I therefore use "law" only to describe governmental social control. Drawing on a definition of Frank Michelman's, I define a government as a hierarchical organization that is widely regarded as having the legitimate authority to inflict detriments on persons (within its geographically defined jurisdiction) who have not necessarily voluntarily submitted themselves to its authority.

Ordinary human conduct will be referred to here as "primary behavior." Social control activity (such as the administration of sanctions), carried out in response to (or in anticipation of) primary behavior, will be termed "secondary behavior." Rules govern secondary behavior as well as primary behavior. An enforcer who improperly responds to another's primary behavior may himself suffer punishments. Tertiary behavior is social control activity carried out in response to secondary behavior. This classification system could be extended tier by tier, in principle to an infinite number of levels of social control.

These distinctions among different levels of behavior can contribute to a better understanding of how to prove the existence of a rule. A guideline for human conduct is a rule only if the existence of the guideline in fact influences the behavior either of those to whom it is addressed or of those who detect others breaching the guideline. The best and always sufficient evidence that a rule is operative is the routine (although not necessarily inevitable) administration of sanctions, whether rewards or punishments, on people detected breaking the rule. For example, the best evidence of a primary rule against dishonestly is a pattern of secondary behavior: the regular punishment of people discovered to be dishonest. Conversely, the total absence of enforcement actions against detected violators of a guideline is conclusive evidence that the guideline is not a rule.

An operative punishment-triggering rule may be so effective that it is never violated. There might then be no enforcement activity to observe to prove the rule's existence. In these situations other, less reliable evidence may prove the existence of a rule. For example, an observer may sometimes be justified in inferring primary rules from patterns of primary behavior. Thus an alien who visited England could infer, without observing any enforcement activity, that there are rules that people should shake hands with their right hands but drive on the left side of the road. Observing primary behavior is, however, a risky way of determining the rules that govern primary behavior. False negatives and false positives are both possible. A false negative is most likely when detection of acts of deviancy is extremely difficult, but there is nevertheless an operative punishment-triggering rule that is regularly enforced against the few discovered deviants. For example, if the IRS were regularly to treat proven income from tips as taxable income, then that would be an operative rule even though few tips are in fact reported as income because IRS agents are rarely able to prove who received tips. False positives are possible because not all behavior is normatively constrained. That people regularly sleep does not indicate that there is a rule that they should sleep. Only the regular punishment of detected nonsleepers or the regular rewarding of sleepers would provide ironclad evidence that rules govern the primary activity of sleeping.

People often make aspirational statements about appropriate human conduct. These statements appear in statutes, in rule books for games, in association bylaws, in the adages of everyday speech, and so on. An aspirational statement is evidence of an operative rule but evidence that is rebutted when patterns of primary or secondary behavior flout the aspirational statement. What people do should be taken as more significant than what they say. For example, a criminal statute that prohibits unmarried adults from fornicating is not a rule as that term is used here if detected violators would not regularly be punished. Similarly, Polonius's adage, "Neither a borrower, nor a lender be," suggests a normatively appropriate course of primary behavior, but patterns of both primary and secondary behavior show that it is not a rule in the United States today. Aspirational statements are likely to provide the best evidence of rules only when patterns of primary and secondary behavior are unknown. For example, because little is known of ancient times, the Ten Commandments and equivalent aspirational statements in the sacred texts of other cultures are (weak) evidence of rules that prevailed in antiquity.

A rule can exist even though the people influenced by the rule are unable to articulate the rule in an aspirational statement. Children can learn to speak a language correctly without being able to recite any rules of grammar. Adults who daily honor a complex set of norms that govern dress would be startled if asked to lay out the main principles that constrain their choice of apparel. Rural residents of Shasta County had trouble articulating the norms that governed how they shared the costs of boundary fences. An observer of regular patterns of secondary (and perhaps primary) behavior may nevertheless be able to identify the content of unarticulated rules.

The existence of legal rules is usually easier to prove than is the existence of norms. Court dockets, police reports, and the like reveal law-enforcement efforts, and a law library contains most of the relevant (if often ambiguous) aspirational statements.

Norms are harder to verify because no particular individuals have special authority to proclaim norms and because the enforcement of norms is highly decentralized. After studying Shasta County I nevertheless became willing to assert that rural residents there honor a norm that an owner of livestock is responsible for the conduct of his animals. The fact that many of the people I interviewed said that a good neighbor would supervise his livestock was only weak evidence of the norm. That most of them did mind their animals said only a little more. The best evidence that this norm existed was that Shasta County residents regularly punished, with gossip and ultimately with violent self-help, ranchers who failed to control cattle. . . .

The Scope of a General Theory of Social Control

On the basis of independent variables describing a society, a general theory of social control would predict the content of the society's rules.

. . .

Because a society's operative rules are best revealed by the characteristics of the events that regularly trigger enforcement activity, the general theory would thus predict which events would trigger sanctions, what the sanctions would be, how controllers would gather information, and which controller would administer sanctions in a given instance. To put forth even a rudimentary theory an analyst would have to incorporate theories of the behavior of the five controllers. In other words, a general theory of social control requires subtheories of human nature, of market transactions, of social interactions, of organizations, and of governments. For starters a theorist thus needs a command of psychology, economics, sociology, organization theory, and political science.

No small challenge, this. My object in the balance of this paper is more modest. To suggest the utility of the taxonomy just developed, I will employ parts of it to highlight major shortcomings in mainstream law-and-economics and law-and-society theories of social control.

Beyond Legal Centralism: The Shortcomings of Law-and-Economics Theory

Most law-and-economics scholars and other legal instrumentalists have underappreciated the role that nonlegal systems play in achieving social order.

Their articles are full of law-centered discussions of disputes, such as cattle trepass disputes between farmers and ranchers, whose resolution is in fact largely beyond the influence of governmental rules. There are, of course, notable exceptions. Law-and-economics stalwarts such as Harold Demsetz and Richard Posner have understood that property rights may evolve in primitive societies without the involvement of a visible sovereign. Several economists have emphasized that promisees can enforce express contracts without the help of the state. Yet many scholars who work in law and economics still seem to regard the state as the dominant, perhaps even the exclusive, controller.

The Legal-Centralist Tradition

Economist Oliver Williamson has used the phrase "legal centralism" to describe the belief that governments are the chief sources of rules and enforcement efforts. The quintessential legal centralist was of course Thomas Hobbes. Hobbes thought that in a society without a sovereign all would be chaos. To quote some of Hobbes's best-known lines, without Leviathan one would observe

> continual feare, and danger of violent death; And the life of man, solitary, poore, nasty, brutish, and short. . . . To this warre of every man against every man, this is also consequent; that nothing can be Unjust. The notions of Right and Wrong, Justice and Injustice have no place. Where there is no common Power, there is no Law; where no Law, no Injustice. . . . It is consequent also to the same condition, that there be no Propriety, no Dominion, no *Mine* and *Thine* distinct; but only that to be every mans that he can get; and for so long, as he can keep it.

Hobbes apparently saw no possibility that some nonlegal system of social control, such as the decentralized enforcement of norms, might bring about at least a modicum of order even under conditions of anarchy. (The term "anarchy" is used here in its root sense of a lack of government rather than in its vulgar sense of a state of disorder. Only a legal centralist would equate the two.)

The seminal works in law and economics hew to the Hobbesian tradition of legal centralism. Economist Ronald Coase's work is an interesting example. Throughout his scholarly career, Coase has emphasized the capacity of individuals to work out mutually advantageous arrangements without the aid of a central coordinator. Yet in his famous article *The Problem of Social Cost,* Coase fell into a line of analysis that was wholly in the Hobbesian tradition. In analyzing the effect that changes in law might have on human interactions, Coase implicitly assumed that governments have a monopoly on rule-making functions. In a representative passage, Coase wrote, "It is always possible to modify by transactions on the market the initial *legal* delimitation of rights. And, of course, if such market transactions are costless, such a rearrangement of rights will always take place if it would lead to an increase in the value of production" (emphasis added). Even in the parts of his article where he took transaction costs into account, Coase failed to note that in some contexts rights themselves might initially be delimited not by visible sovereigns but rather through decentralized norm-making processes.

In another of the classic works in law and economics, Calabresi and Melamed similarly regard "the state" as the sole source of social order:

> The first issue which must be faced by an legal system is one we call the problem of "entitlement." Whenever a state is presented with the conflicting interests of two or more people, or two or more groups of people, it must decide which side to favor. Absent such a decision, access to goods, services, and life itself will be decided on the basis of "might makes right"—whoever is stronger or shrewder will win. Hence the fundamental thing that the law does is to decide which of the conflicting parties will be entitled to prevail.
>
> Having made its initial choice, society must enforce that choice. Simply setting the entitlement does not avoid the problem of "might makes right"; a minimum of state intervention is always necessary. Our conventional notions make this easy to comprehend with respect to private property. If Taney owns a cabbage patch and Marshall, who is bigger, wants a cabbage, he will get it unless the state intervenes.

Although they doubtless know better, in these passages Calabresi and Melamed lapse into an extreme legal centralism that denies the possibility that controllers other than "the state" can generate and enforce entitlements.

. . .

Perhaps because legal centralists overrate the role of law, they seem unduly prone to assume that actors know and honor legal rules. Economists know that information is costly, and a growing number emphasize that humans have cognitive limitations. Yet in making assessments of the instrumental value of alternative legal approaches, respected law-and-economics scholars have assumed that drivers and pedestrians are fully aware of the substance of personal-injury law; that, when purchasing a home appliance whose use may injure bystanders, consumers know enough products-liability law to be able to assess the significance of a manufacturer's warranty provision that disclaims liability to bystanders; and that people who set fires fully understand the rules of causation that courts apply when two fires, one natural and the other man-made, conjoin and do damage.

Some Evidence That Refutes Legal Centralism

Much of the evidence I gathered in Shasta County refutes legal centralism. When adjoining landowners there decided how to split the costs of boundary fences, they reached their solutions in almost total ignorance of their substantive legal rights. Moreover, as already noted, when resolving cattle trespass disputes, virtually all rural residents applied a norm that an animal owner is responsible for the behavior of his livestock, even in situations where they knew that a cattleman would not be legally liable for trespass damages. Although I found that governmental rules and processes were often important in the resolution of disputes arising out of highway collisions between vehicles and livestock, even in those situations most rural residents badly misperceived the applicable substantive law. As I will now show, other empiricists have come up with analogous findings.

Substantive Norms Often Supplant Substantive Laws.   Law-and-society scholars have long known that in many contexts people look primarily to norms, not to law, to determine substantive entitlements. In a path-breaking study published in 1963, Stewart Macaulay found that norms of fair dealing constrained the behavior of Wisconsin business firms at least as much as did substantive legal rules. H. Laurence Ross's study of how insurance adjusters settled claims arising from traffic accidents similarly found that the law in action differed substantially from the law on the books. For example, Ross discovered that adjusters applied rules of comparative negligence even in jurisdictions where the formal law made contributory negligence a complete defense.

Vilhelm Aubert investigated the effect of the Norwegian Housemaid Law of 1948. That statute limited a maid's working hours to a maximum of 10 hours per day, gave maids entitlements to holidays and overtime pay, and imposed other labor standards on employers of housemaids. Although the ceiling of 10 hours per working day was violated in about half the households studied and the overtime pay provisions in almost 90 percent, Aubert found that no lawsuits had been brought under this statute within the first two years of its enactment. Aubert concluded that a housemaid's basic mechanism for controlling employer abuse was a nonlegal one, namely, her power to exit the relationship by obtaining employment in another household. He concluded that the statute's effect had been modest at most.

· · ·

Laboratory evidence also casts doubt on legal centralism. Hoffman and Spitzer fortuitously discovered the importance of substantive norms during their laboratory experiments on the dynamics of Coasean bargaining. In an early experiment Hoffman and Spitzer endowed their laboratory-game players with unequal initial monetary entitlements. The game rules allowed the players to negotiate contracts that would increase their joint monetary proceeds from the game. The contracts could include provisions for side payments. Hoffman and Spitzer expected to observe only Pareto-superior contracts, that is, ones under which no party to the contract would come out monetarily worse off. In the two-person games, most players (especially those who knew that they would play against each other at least twice) were instead inclined to split equally the gross proceeds from a game, even when an equal split was Pareto inferior for one of them. Intrigued by this result, Hoffman and Spitzer conducted another experiment that they concluded revealed that a set of informal norms, what they called "Lockean ethics," helped govern when players were prone to equalize the gross proceeds. In short, Hoffman and Spitzer tried to be sovereigns but found that norms (or conceivably personal ethics) often trumped their initial distributions of property rights.

The Pervasiveness of Self-Help Enforcement. Legal centralists regard governments as the chief enforcers of entitlements. . . . As mentioned, I found that
self-help was rife in Shasta County. Ranchers who refused to mind their cattle,
or to bear a proper share of boundary-fence costs, or to continue labor to a
controlled burn did so at the risk of suffering the sting of negative gossip or
some other gentle form of neighbor retaliation. Rural residents were also
eventually willing to resort to violent self-help against the trepassing livestock
of ranchers who were repeatedly unmindful.

. . .

Sociologists have long been aware of the power of gossip and ostracism.
Donald Black has recently gathered cross-cultural evidence that violent self-
help is also common. Black asserts that much of what is ordinarily classified as
crime is in fact retaliatory action aimed at achieving social control. The law
itself explicitly authorizes self-help in many situations. Both tort and criminal
law, for example, authorize a threatened person to use reasonable force to
repel an assailant. The legal-centralist assertion that the state monopolizes the
use of force is, to put it bluntly, absurd.

The Scantiness of Legal Knowledge. Ordinary people know little of the
private substantive law that applies to personal interactions. Motorists may
possibly learn that the failure to wear a seatbelt is a misdemeanor, but only
personal-injury lawyers are likely to know whether the tort law of their state
makes an injured motorist's failure to wear a seatbelt a defense in a civil
action. First-year law students may complain that what they are encountering
is boring but never that it is old hat.

Surveys of popular knowledge of law relevant to ordinary household transactions, such as the leasing of housing or the purchase of consumer goods,
invariably show that respondents have scant working knowledge of private
law. For example, when interviewers asked some 300 Austin households 30
yes-or-no questions about Texas civil law, "high-income Anglos" answered
correctly an average of 19 out of 30 and "low-income Mexicans" an average of
13 out of 30 (a performance worse than chance). Another survey revealed that
a solid majority of Texas therapy patients did not know that they were protected by a legal privilege of nondisclosure, perhaps because "[f]or 96 percent
of the patients the therapist's ethics, not the state of the law, provided assur-
ances of confidentiality." In Vilhelm Aubert's aforementioned study of the
Norwegian Housemaid Law of 1948, housemaids and housewives were asked
if they were aware of nine specific clauses in the statute, two of which it did
not in fact contain. The respondents "recognized" the two fictitious clauses
somewhat more frequently than the seven real ones.

Highly educated specialists could be expected to have a somewhat better
grasp of the private-law rules that impinge on their professional practices.
Givelber, Bowers, and Blitch conducted a national survey of nearly 3,000
therapists to measure knowledge of the California Supreme Court's 1975
Tarasoff decision, which dealt with the tort duties of therapists when patients
utter threats against third parties. They found that although 96 percent of

California therapists and 87 percent of therapists in other states knew of the *Tarasoff* decision by name, the great majority wrongly construed it as imposing an absolute duty to warn rather than a duty to warn only when a warning would be the reasonable response under the circumstances. . . .

Beyond Exogenous Norms: The Shortcomings of Law-and-Society Theory

In contrast to the law-and-economics scholars, law-and-society scholars have long been aware that norms and self-help play important roles in coordinating human affairs. Perhaps because their vision of reality is so rich, however, sociologists and their allies have been handicapped because they do not agree on, and often do not show much interest in developing, basic theoretical building blocks. Anyone who widely reads in both law-and-economics and law-and-society literature is bound to come away feeling that economists, although often dismayingly blind to realities, are clearer, more scientific, and more successful in building on prior work. The late Arthur Leff, who read extensively in both, called law-and-economics a desert and law-and-society a swamp. Having probed the causes of the aridity, I now inquire into the causes of the swampiness.

Sociological Theories of the Interaction of Law and Norms

Legal Peripheralism and Evidence That Refutes It.   Some sociologists are extreme legal peripheralists who dismiss the legal system as utterly uninfluential. Legal peripheralism dates back at least to the Roman historian Tacitus, who is still quoted for having asked, "Quid leges sine moribus?" ("What are laws without morals?") This view was particularly popular a century ago, when the social Darwinist William Graham Sumner was emphasizing the role that "folkways" played in achieving social order.

Extreme legal peripheralism is as untenable as is extreme legal centralism. Although law may often be overrated as an instrument of social engineering, it is not invariably toothless. For example, after the Russian Revolution the Communists who took over the state apparatus were eventually able to use law to alter (although hardly totally to transform) life in Moslem central Asia. Similarly, most observers would agree that changes in federal civil rights law during the 1950s and the 1960s helped to undercut social traditions of racial segregation in the South.

Focused field studies of the effect of changes in private substantive law also refute extreme legal peripheralism. I found that the adoption of closed-range ordinances in Shasta County deterred some ranchers from running their herds at large because they thought that the ordinances would increase their liabilities to motorists who collided with their cattle. Prior empirical studies have found, among other things, that the allocation of legal property rights in the intertidal zone affects labor productivity in the oyster industry, that the structure of workers' compensation systems influences the frequency of workplace fatalities, and that the content of medical malpractice law has an effect on how claims are settled. The mainstream law-and-society position today seems to be the sensible one that both law and norms can influence behavior.

. . .

Most law-and-society scholars shy away from all theories of the content of norms. For example, in his justly famous study of contractural relations among Wisconsin business firms, Stewart Macaulay identified two principal norms that governed interfirm behavior: (1) "one ought to produce a good product and stand behind it"; and (2) "commitments are to be honored in almost all situations." Essentially viewing people as rational actors who try to maximize their net gains, Macaulay seemed to regard most of the behavior he observed as somehow adaptive. Yet Macaulay offered no explanation for the emergence of the particular norms he observed. Why had the business culture not generated norms of "caveat emptor" and "there is no such thing as a binding commitment"? Macaulay did not venture to say. He passed up his opportunity to offer a theory of the content of norms and communicated only the (important) message that controller-selecting norms can discourage actors from using the legal system.

Similarly, in his enormously readable book *The Human Group,* sociologist George Homans identified "a norm that is one of the world's commonest: if a man does a favor for you, you must do a roughly equivalent favor for him in return." Homans drew on William Foote Whyte's *Street Corner Society* to illustrate this norm of reciprocity. Whyte had studied the Norton Street Gang, a group of young men of Italian descent living in a Boston slum. The Nortons believed in mutual aid but also in keeping accounts square: "In bad times as in good, if you have a few extra dimes you are expected to give them to your friend when he asks for them. You give them to him because he is your friend; at the same time the gift creates an obligation in him. He must help you when you need it, and the balance of favors must be roughly equal. The felt obligation is always present, and you will be rudely reminded of it if you fail to return a favor." Although Homans stands out among sociologists for his clearheadedness, in the *The Human Group* and his other writings he treats particular norms, even one of the world's most common, as exogenous facts of life.

As these examples show, just as microeconomists tend to take consumers' tastes as given and limit themselves to the study of market processes, so sociologists tend to work not on what norms are but on how norms are transmitted.

Notes and Questions

1. How does the author explain the normal pattern of punishing antisocial behavior, rather than rewarding behavior that is not antisocial? Do you find this explanation

convincing? Can you think of counterexamples where the law regulates behavior by rewarding someone who is not antisocial?

2. What conclusions do you draw from the fact that landowners were ignorant of the substantive legal rules applicable to trespasses by cattle? How do you compare this to contrary indications in other areas, such as evidence that most people are well aware of the legal rules governing rescue? Do you have a theory that helps predict when there is knowledge or ignorance of the law? If people were ignorant of the law of rescue, would observers who favored the imposition of a duty to rescue change their views?

3. What are the penalties imposed by social groups for serious violations of community norms? Are these penalties utilized in formal lawmaking? Is there more similarity between formal (legal) and social (customary) remedies or between formal and social norms?

The Social Consequences of Common-Law Rules

RICHARD EPSTEIN

The Theme

The past generation of legal scholars has been preoccupied with the social and economic consequences of common-law rules. One manifestation of this concern is found in the work of the law-and-economics movement, which has repeatedly sought to demonstrate that certain common-law rules facilitate "efficient" resource use. In a parallel development, some legal historians have insisted that courts have shaped the common law to promote or subsidize industrial growth and development, and hence to advance the interests of certain classes at the expense of others.

Both of these arguments assume that a choice between competing common-law rules can have a significant effect on the allocation of resources and the distribution of wealth in society. This assumption often rests on the unstated premise that the issues subject to frequent appellate litigation have an institutional importance that equals their intellectual difficulty. But this premise is deeply flawed. Even if these choices do in theory redistribute wealth between social classes or encourage efficient behavior, their actual social impact is minimal.

The central theme of this selection is that the intellectual and institutional constraints on common-law adjudication require one to be very cautious in

Abridged and reprinted without footnotes by permission from 95 *Harvard Law Review* 1717 (1982).

attributing major social and economic consequences to common-law rules. . . . In my view, the prolonged debates over the first principles of tort and contract law remain the subjects of common-law adjudication precisely because the stakes are too small to provoke efforts to achieve legislative reversal of the common law outcomes. . . .

My argument is that structural features limit what the manipulation of common-law rules can achieve. The more focused and sustained methods of legislation and regulation are apt to have more dramatic effects than does alteration of common-law rules and thus will attract the primary efforts of those trying to use the law to promote their own interests.

This view is called into question by Morton Horwitz's assertion that business interests preferred to arrange subsidies through the common-law system because adjudication is less politically conspicuous than the legislative process. However, the contention that nineteenth-century commercial interests and legislatures shrank from public debates on subsidy issues is questionable simply as a matter of historical fact. Even on theoretical grounds the arguments seem weak. Common-law decisions are public, not covert. And even if it were thought that judicial rhetoric could conceal a common-law subsidy, commercial interests often have no choice but to do battle in the legislative and administrative arenas, whether their reception is friendly or hostile.

In order to understand better the dominance of the legislative and administrative systems, consider the institutional barriers to effective wealth redistribution through the manipulation of common-law rules. Control over the judicial process is divided between courts and litigants; no one group can dictate the case agenda. In addition, most common-law rules are not cast in class form; there is no easy one-to-one correspondence between a given rule and the advancement of a particular social class. The ability to work substantial transfers of wealth between social classes is also severely hampered by the demand for public justification by written opinion that lies at the heart of the common-law process. The protection that the resulting generality affords is far from absolute, for the degree of generality can vary widely with context: a rule that refers to the rights of A and B without qualification has greater generality than one referring to promisor and promisee, which in turn is more general than one referring to creditor and debtor. The systematic demand for formal generality and neutrality in the common law may not guarantee that the proper substantive result will be reached in any individual case, but it does provide one bulwark against the invidious application of legal rules. The easiest way to oppress the poor, or to confiscate from the rich, is by laws directed at the rich and poor as such; the clandestine use of formally neutral principles is a poor second choice. . . .

Similar difficulties plague efforts to attribute substantial allocative effects to particular common law rules. It is often said that the common law aims to minimize (perhaps within a constraint of justice) the total costs of accidents, accident avoidance, and administration of the legal system. But it is far easier to state this condition than it is to discern a given rule that satisfies it. It is, for example, rarely easy to identify the cheapest cost-avoider or to trace the incentive effects of a given rule. The effort to make the defendant bear the social costs of his own activity will complicate efforts to regulate the plaintiff's conduct. And whatever trade-off is made on the question of incentives will be in systematic tension with the demand to economize on administrative costs, which rise with the sophistication and complexity of the common-law liability rule. Without a precise way to measure both the size and direction of the countervailing costs and benefits, one cannot determine with confidence which rule is preferable, let alone show that the rule finally chosen substantially influences efficiency.

The allocative effects of choices between common-law rules are, in any event, often small in comparison to what is accomplished by direct government action. Statutory controls can utilize a range of sanctions that are unavailable at common law: taxes, fines, inspections, filing requirements, and specific bans and orders with wide and dramatic effects. Even so simple a matter as placing limitation periods on private actions requires a statute; no common-law principle explains why a cause of action valid on one day should be barred the next. Private-law remedies are a limited arsenal in comparison: the private law of nuisance and the Clean Air Act are very different modes of social control.

Having thus raised doubts about the assumptions underlying the claims of both schools, how can we measure the social consequences of various common law rules? . . .

One such test is whether a legal rule places large numbers of cases within or without the common-law system. Often these rules involve issues akin to standing or immunity. . . . These "gatekeeper" rules are presumptively of enormous importance even if their status is but infrequently litigated. Unlike general tort-liability standards, these rules gain their effect because they can be directed to specific social classes or groups.

A second measure of the importance of a given decision lies in the willingness of those disadvantaged by the rule to expend resources to persuade a legislative or administrative agency to change it. The lobbying and coalition formation essential to secure legislative or regulatory intervention usually require a major effort, often in the face of determined opposition from those who benefit from the common law status quo. Of course, the absence of such an effort does not necessarily establish the insignificance of a choice of rules, for those who are disadvantaged may lack the economic or political resources to make themselves heard. But when there *is* such an effort to undo common-law outcomes, we can have some confidence that the stakes in question are commensurate with the expense incurred. . . .

To assess the validity of the incautious claims made about the allocative and distributional effects of certain common-law doctrines . . . this selection examine[s] several doctrinal disputes with an eye not to their proper substantive outcomes, but to their social importance. . . .

Tort: Negligence and Strict Liability

It has often been said that the crucial choice for the tort system is that between the rules of negligence and strict liability in cases of accidental harm. . . . The negligence rule—which has dominated the scene for the past 100 to 125 years— has been credited on one hand with subsidizing industrial development and on the other with maximizing the efficient use of scarce resources. The first of these assertions—especially in the work of Lawrence Friedman and Morton Horwitz—has been associated with a progressive, not to say radical, critique of the common law. The second, especially in the work of my colleague Richard Posner, has been associated with its conservative defense. My purpose here is not to take sides in this debate, but to show that both sides share the error of attributing to the issue greater significance than it warrants. . . .

Distributional Consequences

The claim that the common law of negligence provided a "subsidy" for industrial development rests upon two separate premises. The first is that a rule of strict liability—the conceptual rival of negligence—constitutes the neutral baseline against which the subsidy is measured. The second is that the negligence rule effected a truly substantial redistribution of wealth into industrial channels. . . . There is, however, reason to doubt that the effects of any general tort rule are either substantial or systematic.

Proportion of Cases Affected. In both property-damage and personal-injury actions, the doctrinal differences between negligence and strict liability are likely to alter the outcomes only of marginal cases. Under either standard, the plaintiff must establish causation in fact and offer a calculation of damages. The difference between the two regimes therefore goes to the single and doctrinally important issue whether the defendant breached a duty to exercise "reasonable" care. But in practice, if negligence were held legally necessary, it could usually be proved; as a brute fact of nature, most accidents occur because someone has failed to take the elementary precautions needed to prevent them. What gaps remain between the two rules can be and have been bridged by res ipsa loquitur and other evidentiary presumptions.

In addition to this convergence of the two rules, other considerations cast doubt on the significance of the choice of liability standards. At least in property-damage cases, the economic importance of the choice is further reduced by the apparent infrequency of substantial damage claims. Major private nuisance or blasting cases must be rather uncommon events for any given tract of land, even in a dense urban environment. One useful way to test the importance of liability rules is to ask whether they become embedded in the price of land. The utter indifference of assessors and purchasers to general liability rules confirms that the choice has almost no effect.

It is therefore difficult to find between the rules the large variation in outcomes, either for individual cases or in the aggregate, that constitutes a minimum condition for the redistributive thesis. Any definitive statement about the economic significance of these liability rules would require more detailed knowledge than is currently available of the number of suits filed, the amounts in controversy, and the settlements and dispositions under each system. Nonetheless, the convergence of negligence and strict liability suggests that the choice of one general rule of liability over the other will not result in major transfers of wealth.

Class-Directed Benefits. Whatever wealth transfers do occur provide evidence for the redistributive thesis only if they systematically favor one class over another. It is one thing, for example, to demonstrate that a rule alters the balance between insurance companies and their commercial insureds; it is quite another to show that a rule favors the rich at the expense of the poor. . . .

Property Damage [and] the Calculus of Potential Gains. The very sweep of the intellectual debate over negligence and strict liability cuts against the redistributive thesis in real-property damage cases. It is very difficult for anyone to turn a general liability rule to his private ends. Even if the choice between negligence and strict liability regimes would make a difference, which rule should a rational firm choose? An individual firm engaged in one lawsuit, and thus immediately faced with a choice of rules, does not know what its posture will be in future actions. It must therefore perform the nearly impossible task of discounting the expected value of each (unknown) future suit to decide which rule best suits its interest. If it opts for the negligence rule in the expectation that it will usually be a defendant, it runs the risk of compromising any actions it may later bring for damage to its plant. If it chooses strict liability, its own future construction and operating costs will be increased by the added risks of suit.

This dilemma raises important problems for the redistributive thesis. If a firm lacks meaningful statistical data on how often it will be a plaintiff and how often a defendant, it cannot determine which liability rule works in its favor, and hence is unable to press for a subsidy even if one were theoretically possible. And even if a firm is prepared to expend the resources to acquire accurate data, the potential for redistribution is quite narrowly confined. Thus, if a firm does not find itself in one role more frequently than in the other, a significant subsidy cannot be achieved by manipulating general tort-liability rules. . . .

These barriers to the redistributive thesis are further strengthened by the fact that many, if not most, property damage suits are likely to set one firm against another, so that any wealth transfer that does occur will be confined within a single class. Industries tend to concentrate in certain regions, if only because of the enormous advantages of being close to customers, suppliers, and even competitors. It may have been coincidental that the plaintiff in *Rylands v. Fletcher* was a miner like the defendant, but there was never any reason to assume that most plaintiffs in the flooding cases would be farmers. All the parties in the two flooding cases that preceded *Rylands v. Fletcher* were miners. Because businesses are concentrated in industrial zones or on land suited to the intended enterprise, property-damage actions will commonly be fought by parties in *some,* if not the same, business: flooding cases arose between miners, just as cattle-trespass cases arose between neighboring cattle owners.

Finally, the difficulties of attaining subsidies through common-law rules are further complicated by the "public good" problem. Let it be supposed that a general rule, whatever its content, provides benefits to a large number of parties who are not before the court. Yet a single firm must bear all the costs of developing the rule in the individual case. What reason is there to expect that any firm will shoulder all the costs when it captures only a tiny fraction of the benefit? . . .

To avoid all of these difficulties, commentators sometimes say that judges can be counted on to turn the common law to the task of subsidizing industrial development. Even if judges were so inclined, however, they would have to overcome the same problem that stymies individual firms: how to know which rule will advance the aggregate interests of the class they wish to help.

Moreover, there is little historical evidence of a consistent judicial bias in favor of industry. In both strict liability and negligence cases, courts showed a marked hostility to the idea that even statutorily authorized activities, such as railroads and canals, could be immunized from damage actions. And disputes over the more general choice of liability rule reveal that the judiciary has never been of a single mind on the proper foundations of tort liability, possibly because of disagreement on more fundamental issues of social policy. Nineteenth century opinions are notable primarily for their marked divergence on the issue of negligence versus strict liability. . . .

Institutional Competence. In addition to the primarily economic objections developed above, the structural limitations of the common-law system weaken the subsidy thesis by providing an alternative—and cogent—explanation for judicial behavior, especially in the land-use cases. The courts have recognized for centuries that only legislative bodies possess the means to devise effective and comprehensive solutions to many of the most serious issues in the field of land use; in most instances, the judiciary has accordingly refrained from attempts to regulate matters that require systematic treatment. . . .

In *Lexington & Ohio Railroad v. Applegate,* the plaintiffs sought to enjoin the operation of a railroad that ran through the center of Louisville. Horwitz quotes in isolation the passage from the opinion that speaks of the importance of railroads for economic growth; if read alone, the excerpt might be taken to indicate that the courts strove to subsidize various forms of industrial development. But the passage itself is obiter dicta, as four separate reasons taken together led the court to deny the injunction. The first reason, immediately relevant here, was that the legislature had authorized the operation of the route and thus that the grant of an injunction would not simply have been a decision on the merits, but a direct challenge to a coordinate branch of government. Judicial deference to legislative action, one of the most powerful institutional constraints on a court, therefore explains the case better than does any desire to subsidize industrial growth. The decision in *Applegate* should be understood to embody a sensible division of responsibility, similar to the balance that obtains today, between the courts—and the legislature in land-use matters. Damage actions remain available for actual nuisances, and injunctions may by obtained to prevent imminent or ongoing harm. But the relative stakes should not be forgotten. A single legislative decision to permit development is in the end far more important than frequent litigation on the permissible scope of industrial operations. . . .

Personal Injuries. In principle, personal-injury cases offer a greater opportunity for systematic redistribution than do property damage suits. In most personal injury actions, when a firm is involved, it is as the defendant against an individual plaintiff; thus, many of the information costs and countervailing interests that operated to undermine the redistributive thesis in the property area are problems in the personal-injury field. Of course, the convergence of negligence and strict liability makes it probable that any aggregate effect, whether or not redistributive, will not be great. Several other factors raise additional doubts about the relative magnitude of the transfers involved.

Injuries to Strangers. In the nineteenth century, as in the twentieth, the highway accident cases—in which Horwitz himself locates the origins of the negligence standard—typically involved private parties on both sides. Whatever the precise numbers, any efforts to arrange a subsidy for industrial defendants would have been constrained by the need to develop a set of rules that would also do adequate service between private citizens, some of whom would themselves be well-to-do. The only way to create a powerful subsidy for industry in a common-law setting would have been to develop a lower standard of care for firms than for private individuals. Here the analysis of judicial behavior must ultimately rest on empirical evidence. To the extent that variations in the standard of reasonable care can be detected, Gary Schwartz's masterful and exhaustive study of New Hampshire and California cases demonstrates that, in fact, very *high* standards were often expected of institutions—manufacturers, railroads, and electric companies, among others—whose expertise was taken for granted by the courts. In addition, it is not clear that industrialists would have enthusiastically supported such a double standard; no corporate executive could be certain that he would not himself be injured by another firm's defective elevator or brakeless streetcar. Would he have been willing to give up his own right to sue for the resulting loss of his earning power?

Other doctrinal choices support this same view. The redistributive thesis should predict an aggressive expansion of the contributory negligence defense. Of perhaps even greater importance, the thesis should also predict a refusal to adopt strict-liability principles in vicarious-liability cases, in order to help to insulate the corporation from suits by strangers, for a stringent strict-liability standard would let many suits through the gate. Yet these predictions also fail empirically; in neither of these areas do we see a steady, confident progression of the cases to a single end. The nineteenth-century judiciary viewed contributory negligence with some ambivalence and applied vicarious liability expansively. . . . The uncertain course of judicial decisions over the generations accurately reflects both the difficulty of the subject matter and its marginal economic impact.

Industrial Accidents. At first blush, the redistributive thesis does find support in the field of industrial accidents. But the reason that common-law doctrines mattered in this setting is that they were narrowly drawn and functioned as "gatekeeper" rules. In the extensive nineteenth-century discussions of employers' liability, the choice between negligence and strict liability was overwhelmed by the enormous disagreement over whether the assumption of risk defense, especially when "implied" from conduct, was sufficient to bar an employee's negligence cause of action. . . .

But it comes as no surprise that major structural changes in the treatment of industrial accidents, changes that did not occur until the early twentieth century, resulted from legislation. The revolution was, of course, the adoption of comprehensive workers' compensation laws, which—because of their complex administrative provisions and detailed benefit rules—could not have been introduced by judicial decision. Although one must not discount the importance of ideological causes of the change, the economic stakes involved are indicated by the willingness of all interested groups to bear the substantial expenses of organizing coalitions, testifying, lobbying, and drafting and passing legislation. Throughout this debate, legislative attention was never focused on the choice between negligence and strict liability. The legislative solution had discarded the entire tort-liability system in the field of industrial accidents and started anew. . . .

Efficiency

A recognition of the major difficulties of the subsidy hypothesis clears the way for an examination of its rival social explanation—the view that the adoption of the negligence system maximizes the wealth of society at large. Many of the arguments developed in the previous section also apply to the efficiency thesis. For example, if a shift from negligence to strict liability—or vice versa—does not have a significant impact on the outcome of cases, it can hardly be expected to cause major allocative effects.

Incentives. The law-and-economics approach lays great stress on the incentives for cost-minimizing behavior that legal rules are said to create. The supposed beauty of the negligence system is that it requires defendants only to take cost-justified precautions, because greater precautions produce net social losses. Yet the strict-liability regime demands no more; whatever liability rule governs, some accidents that are too costly to prevent will occur. Strict liability simply requires the defendant to compensate those harmed by such accidents.

It is sometimes urged that negligence achieves different—and better—results than strict liability, because the defenses of contributory negligence and assumption of risk spur prospective plaintiffs to avoid accidents that they could prevent more cheaply than could prospective defendants. The effect of these financial incentives is difficult to determine, particularly in personal-injury cases in which the plaintiffs' inherent instincts of self-preservation play a central role. But even if it is established that such incentives make a difference in plaintiffs' conduct at the margin, there is nothing in a strict-liability system to preclude the adoption of robust defenses, be they cast in the form of assumption of risk, contributory negligence, or, as I have suggested elsewhere, contributory causation. . . .

Administrative Costs. Incentives, moreover, are not the only issue. To be efficient, the liability rules must take administrative costs into account. But the negligence rule burdens every case with additional issues of categorization and proof, which naturally add expense to the process. In the end, as Landes and Posner have suggested, this burden might be worth bearing if the negligence standard takes considerable numbers of accident cases out of the legal system. In the absence of useful data about the extent to which the negligence rule reduces the frequency of suit and increases the administrative costs of actions that are brought, however, all one can say a priori is that the two effects tend to cancel each other out. Moreover, the frequency and expense of litigation may be determined less by substantive standards than by procedural rules. For example, frequency of suit must surely be influenced strongly by the rule of costs adopted, although it is unclear whether the American rule (each party pays his own) or the English rule (winner is reimbursed) stirs up more litigation. To make any confident predictions about the efficiency of substantive tort rules in isolation from procedural rules is to push the evidence beyond its decent limits. . . .

Conclusion

In the course of this selection, I have examined a number of common-law rules solely to determine whether their social and economic consequences are a fruitful subject for further investigation. Based on this examination, I have concluded that they frequently are not, and that legal scholars of whatever persuasion must exercise considerable care in choosing suitable objects for their inquiries. The point here does not go to the merits of any issue. One may regard large nineteenth-century industrialists as the architects of a golden age, as simple robber barons, or perhaps as a bit of both. But whatever view ultimately prevails, the general common-law rules of tort and contract constituted only a tiny portion of the legal turf over which their battles, or those of any generation, were fought.

My argument is not that the common law does not have enormous intrinsic worth both as a system of thought and as a minimum condition for social order. The true lesson is that the legal points so troublesome to both judges and lawyers in a common-law setting need not mirror the pressing issues of society at large.

Notes and Questions

1. In our society, who seems more powerful in terms of generating real social and economic consequences: an individual judge or an individual legislator; the judiciary or the legislature as a whole?

2. The unabridged version of this selection offers examples from other areas of law in support of the central thesis—the relative inconsequence of common-law rules. Can you think of doctrinal disputes in contract or other law that consume judges but are unlikely to have great consequences?

3. What if judges abolished tort law (or at least as much of this law as is in their control), so that there were no (common law) liability? If initial entitlements were allocated in this matter, would there be considerable social consequences? What would be the likely private and political reactions to the withdrawal of common-law tort liability?

VIII

Insurance, No-Fault, Products Liability, and Tort Reform

In each area of law the question arises whether the law and the society it governs would be better improved by small, scattered changes or by radical reform. Proposals that lawyers regard as radical, requiring the most cautious study and treatment, often strike nonlawyers as hair-splitting ideas for change that fail to address the real need for a better legal system. Particularly in the area of tort law, many observers, lawyers, and nonlawyers alike stand prepared to discuss fairly significant reforms. Few lawyers would assert that the world we know is the best of all possible tort worlds. Uncompensated victims, bankrupt defendants, rising insurance bills, and expensive legal (and court) costs all appear as pieces of evidence in favor of change. If defenders of the tort system could demonstrate that it compensates for these problems by reducing mortality, improving the environment, averting mass disasters, or accomplishing some other heroic aim in a way that is unmatched by other legal systems, calls for reform would be less frequent. Instead, this may be the only area of law where experiments in other countries are followed with great interest and where problems of transition are often labeled crises that should spur reform.

The first reading examines the cause of the insurance "crisis" in the mid-1980s, when there were notable disruptions in the provision of insurance to commercial enterprises and local governments. It provides a summary of tort-law developments and a sophisticated explanation of the functioning of insurance markets.

The next selection accepts the popular notion that the fault-based tort system is an "appallingly inefficient" means of compensating victims. It stresses the effects of various other compensation schemes, in particular "no-fault" systems, on incentives to reduce the frequency and severity of accidents. Along the way the reader is treated to a description of a variety of schemes (proposed, or in some cases in place, both in the United States and abroad) and to several empirical studies of the effects of these schemes.

The third reading turns to the important subject of products liability and its reform, arguing that the issue be viewed from a normative perspective based on consumer sovereignty. One suggestion is that the move toward stricter liability for products has been misguided.

The fourth reading returns to the theme of incremental versus radical reform and offers a chilling assessment of the likelihood of successful, radical change.

The final reading introduces the language and analytic style of "positive political theory," also known as public choice analysis. This new branch of legal theory explores voting mechanisms, grapples with the question of why political coalitions form in favor of particular kinds of legislation (or other decision-making), compares the strengths and weaknesses of interest-group politics, and studies the strategic relationships among the branches of government. It is an approach that has had increasing impact in such diverse areas as constitutional law and corporate law, and it has been the focus of new law school courses in legislation. The reading introduces some of the tools of this approach and uses them to suggest that certain kinds of incremental tort reform are more likely than radical reform. The authors also refer to contemporary constitutional law literature to introduce a different, perhaps opposite, school of thought—"republicanism"—which sees an important role for government and the courts in encouraging citizens and public servants to engage in virtuous decision-making that is less selfish and more concerned with the common good.

The Current Insurance Crisis and Modern Tort Law

GEORGE L. PRIEST

This selection is an effort to understand the source of the crisis in insurance that has recently disrupted product and service markets in the United States. From press accounts, the crisis seemed to peak in the early months of 1986, when reports became common of extraordinary changes in commercial casualty insurance markets. Insurers had increased premiums drastically for an unusual set of products, such as vaccines, general aircraft, and sports equipment, and for an equally diverse set of services, such as obstetrics, ski lifts, and commercial trucking. In still other cases—intrauterine devices, wine tasting, and day care,—insurers had refused to offer coverage at any premium, forcing these products and services to be withdrawn from the market.

The crisis extended beyond commercial enterprises. Municipalities and other governmental entities faced similarly extreme premium increases or the unavailability of market insurance coverage altogether. Some cities closed jails and suspended police patrols until insurance coverage was obtained. Parks and forest preserves were closed. Fourth of July celebrations were cancelled because of concerns over uninsured liability.

Over time, product manufacturers and service providers adjusted to these insurance difficulties. Recently, some commentators have interpreted the decline in the volume of complaints as evidence that the insurance crisis has subsided. But this is an incomplete and unrealistic view of the problem. Although there has allegedly been some increase in insurance availability, the huge premium increases of early 1986 are largely intact, necessitating sustained price increases and reducing demand for affected products and services. Moreover, many of the effects of the crisis have become permanent. A recent Conference Board study of liability problems of the country's largest corporations—those whose size and self-insurance capability make them least vulnerable to changes in commercial insurance markets—reports that 25 percent had removed products or withdrawn services from the market. The effect on national income of increased prices and the withdrawal of products and services is obvious.

Similarly, there have been long-term effects of the insurance crisis on municipalities and government entities. Many municipalities have responded to the crisis by altering the array of public services, sometimes radically. Some cities have removed all playground equipment from their public parks. Diving boards have been removed from school swimming pools. Many municipalities denied market insurance (as well as some commercial entities) have continued operations after joining mutual-insurance groups. As we shall see, the actual insur-

Abridged and reprinted without footnotes by permission of The Yale Law Journal Company and Fred B. Rothman & Company from *The Yale Law Journal,* vol. 96, pp. 1521–1590 (1987).

ance protection provided by industry-wide or municipal mutual-insurance pools may largely be illusory, only postponing the effects of expected liability for some uncertain period. At the least, the sudden growth in the frequency of mutual insurance is reflective of the extremely fragile character of today's commercial casualty insurance industry.

Three sets of theories have been put forth to explain the insurance crisis and to propose its cure. The first set views the crisis as collusively engineered by insurance companies, either by explicit price-fixing or by financial manipulation of insurance reserve accounts. A second set of theories explains the crisis with reference to the cyclical effect of changes in interest rates on investment returns and, thus, on insurance premiums. According to this theory, the crisis is the direct result of recent declines in interest rates which have forced insurance premiums to increase. In a slightly different version, the crisis is attributed to an earlier period of high-interest rates during which insurers, shortsightedly, underpriced risks to gain premiums for investment, risks which must be recalculated at higher premiums today. The collusion and the insurance cycle theories both imply that these insurance problems will disappear in time, either through the breakdown of the cartel (or by antitrust enforcement or regulation), through the cyclical return of higher interest rates, or through the more accurate evaluation of risks by chastened insurers.

A third theory, most prominently espoused by the Justice Department, attributes the crisis to modern tort law's expansion of corporate liability exposure, which has necessitated insurance premium increases. According to the Justice Department, only wide-ranging tort law reform can cure the crisis. Within the past 18 months, on the basis of the department's and other attributions of the crisis to tort law, 42 states have enacted tort reform or insurance legislation. Although there are differences among the states, the general similarity in approach and the sudden spontaneity of the response represent the most extraordinary state-law development having national impact since the states' unanimous adoption of the Uniform Commercial Code. Indeed, the Justice Department now claims that this legislation—though in effect only for scant months—has largely cured the insurance problem.

None of these three theories, however, fully or adequately explains the phenomena of the recent insurance crisis. Virtually all commentators, in characterizing the crisis, have focused solely on the sudden increases in insurance premiums and the occasional withdrawal of insurance coverage during early 1986. The crisis, however, has been attended by a much broader group of related phenomena, which must be addressed by any comprehensive explanation. For example, long before 1986, there began to occur increasing shifts toward corporate self-insurance for expected tort liability. In addition, in recent months, while simultaneously increasing premiums, insurers have fundamentally redrafted the basic commercial insurance policy—raising deductibles, lowering levels of aggregate coverage, and revising specific policy exclusions. A comprehensive theory of the crisis must account for all of these phenomena, and must explain their peculiar concentration within commercial casualty lines only, rather than across the broad range of insurance offerings. None of the three existing theories of the insurance crisis can do so.

This selection provides a different theory of the crisis. It attributes these changes in insurance coverage to modern tort law, but by a mechanism much different from that suggested by the Justice Department. In my view, it is simplistic to no more than assert a connection between expanding tort liability and the disruption of insurance coverage. The relationship between legal liability and the insurance function obviously is more complicated: The demand for commercial insurance coverage derives from tort liability. The expansion of tort liability might well create opportunity, rather than crisis, for the insurance industry.

This selection argues that the characteristic of contemporary tort law most crucial to understanding the current crisis is the judicial compulsion of greater and greater levels of provider third-party insurance for victims. The progressive shift to third-party corporate insurance coverage, since its beginnings in the mid-1960s, has systematically undermined insurance markets. The decline in interest rates within the past two years has led the most fragile of these markets—those for which third-party coverage is least supportable—to collapse. The collapse is signalled by the accelerating conversion to self-insurance. This conversion, in turn, forces insurers to exact drastic premium increases, as well as to restructure the terms of the basic insurance policy, in order to salvage a market among remaining insureds. Where these salvage efforts have proven unsuccessful, insurers have refused to offer coverage altogether.

This explanation of the crisis uncovers what I believe to be a tragic paradox of our modern civil liability regime. The expansion of liability since the mid-1960s has been chiefly motivated by the concern of our courts to provide insurance to victims who have suffered personal injury. The most fundamental of the conceptual foundations of our modern law is that the expansion of tort liability will lead to the provision of insurance along with the sale of the product or service itself, with a portion of the insurance premium passed along in the product or service price. Expanded tort liability, thus, is a method of providing insurance to individuals, especially the poor, who have not purchased or cannot purchase insurance themselves. This insurance rationale suffuses our modern civil law, and must be acknowledged as one of the great humanitarian expressions of our time.

The paradox exposed by my theory is that the expansion of tort liability has had exactly the opposite effect. The insurance crisis demonstrates graphically that continued expansion of tort liability on insurance grounds leads to a reduction in total insurance coverage available to the society, rather than to an increase. The theory also shows that the parties most drastically affected by expanded liability and by the current insurance crisis are the low-income and poor, exactly the parties that courts had hoped most to aid. . . .

Modern Tort Law and Its Economic Effects

Since the early 1960s, courts have steadily expanded tort liability for injuries suffered in the context of product and service use. These changes in the law result from the acceptance of a coherent and powerful theory that justifies the use of a tort law to compensate injured parties, a theory its founders called "enterprise liability." According to the theory, expanded provider liability serves three important functions: to establish incentives for injury prevention; to provide insurance for injuries that cannot be prevented; and to modulate levels of activity by internalizing costs, including injury costs.

The second feature of the theory—the importance of providing insurance for unpreventable losses—is most crucial for understanding our current insurance crisis. According to enterprise liability theory, expanded legal liability does more than achieve optimal control of accident and activity rates. Expanded tort liability improves social welfare, in addition, because it provides a form of compensation insurance to consumers. A provider, especially a corporate provider, is in a substantially better position than a consumer to obtain insurance for product- or service-related losses, because a provider can either self-insure or can enter one insurance contract covering all consumers—in comparison to the thousands of insurance contracts the set of consumers would need—and can easily pass the proportionate insurance premium along in the product or service price. Most importantly, to tie insurance to the sale of the product or service will provide insurance coverage to consumers who might not otherwise obtain first-party coverage, in particular the poor or low-income among the consuming population.

The insurance rationale was central to the first judicial adoption of enterprise liability theory, by the California Supreme Court in *Greenman v. Yuba Power Products,* in 1963. The approach was rapidly extended across the various state jurisdictions, first in the products liability field and, later, in other areas of tort law. Briefly, however, enterprise liability theory has justified both restrictions in available legal defenses and expansion of substantive liability standards.

As examples, since the mid-1960s, courts have invoked the insurance rationale to limit the defenses of contributory negligence, assumption of risk, and consumer product misuse; to eliminate defenses related to the status of the victim (as in actions brought by licensees and trespassers against property owners); and to restrict the effectiveness of statutes of limitation. At the same time, the theory has supported the affirmative extension of liability through the adoption of standards of strict liability, retrospective liability, and unreasonably dangerous per se, relaxation of causation requirements, and, more generally, through the near-universal acceptance of comparative negligence, which permits a jury to render judgments against defendants even if they are responsible only in some small proportion for a plaintiff's injury.

These legal changes have particularly affected insurers because courts have simultaneously expanded the scope of the insurance contract. Again, under the influence of enterprise liability theory, courts have adopted the maximization of insurance coverage as the principal interpretive objective in insurance contract disputes. Thus, courts have interpreted policy coverage provisions broadly and policy exclusions narrowly to achieve the compensation goal.

As a result of these doctrinal developments, both the range of defendants against whom a plaintiff might collect a judgment and the range of plaintiffs to whom a corporate defendant might ultimately be liable have expanded. Courts have also expanded the range of losses for which compensation might be sought, chiefly by allowing increased recovery of emotional and other noneconomic losses.

Understanding the insurance crisis requires that the economic effects of expanded liability be analyzed precisely. Putting aside administrative costs, a legal rule or agency regulation can have two principal economic effects: (1) it can influence investments in loss prevention to affect the accident rate; and (2) it can influence the provision of insurance for losses not prevented. These two effects, prevention and insurance, are the only important economic consequences of any legal or regulatory policy. It is irrelevant which goals a court, a regulatory agency, or society is trying to achieve. It is unimportant if the motivation for a legal policy is some moral imperative, some vague or specific sense of justice, or personal or pecuniary advantage. The only important economic effects than any legal policy can have are effects on levels of investments in prevention of losses and of insurance for losses not prevented.

It is well accepted that the optimal level of accident prevention will be attained if incentives are created for both the provider and consumer to make additional safety investments up to the point at which the marginal costs of such investments equal their marginal benefits. Regardless of context, the accident-prevention question is whether it was cost-effective for the provider or for the consumer to have made greater investments to prevent the loss. If the finder of fact determines that neither party could have prevented the loss in a cost-effective manner, then the resolution of the dispute in favor of the plaintiff or the defendant serves only to determine which party is to be the insurer for the loss, that is, where the loss should lie, given that the loss could not have been efficiently prevented.

Some courts and agencies appear to believe that increasing the obligation of providers to compensate victims will always increase these providers' investments in safety. Where damage measures are compensatory, however, there is a very definite ceiling to the accident-prevention investments that any provider will make. Providers will make investments to reduce the accident rate only to the extent that such investments are cost-effective—that is, that the marginal costs of preventive investments are less than the marginal gain in expected accident cost-reduction. Beyond that point, it is less costly to pay damage claims or judgments than to prevent accidents from occurring. Thus, extending liability beyond that point will not affect the level of safety investments.

There are insurance implications, however, of every liability decision. If a liability standard were set only to require cost-effective preventive investments by a provider, some losses would still occur. Victims would bear these losses, and thus would bear the obligation to obtain insurance for them through some first-party insurance mechanism. In contrast, if a provider were held liable for losses which the provider could not cost-effectively have prevented, the provider would bear the obligation to provide insurance. Either through self-insurance or some form of third-party insurance, the provider insures victims for losses that cannot efficiently be prevented.

The economic effects of steadily increasing provider liability thus are quite simple in structure. A liability rule can compel providers of products and services to make investments that reduce the accident rate up to the level of optimal (cost-effective) investments. After providers have invested optimally in prevention, however, any further assignment of liability affects only the provision of insurance. More extensive provider liability will generate more extensive provider insurance and nothing more.

The expansion of liability under modern tort law has obviously increased the provision of provider insurance. Any standard beyond a bare cost-benefit test (often identified with negligence) will provide an insurance effect. Courts, of course, have extended liability far beyond the simple cost-benefit standard. Thus, modern tort law compels a very substantial level of provider insurance.

More precisely, modern tort law has broadly shifted the insurance obligation from first-party insurance to third-party or self-insurance by providers. Even a bare-bones cost-benefit standard has insurance consequences: Such a standard creates an obligation of potential victims to obtain market insurance or to self-insure for unpreventable losses. Modern tort law has shifted that obligation to providers, requiring providers either to obtain third-party market insurance or to self-insure for the losses suffered by consumers of their products or services. The expansion of tort liability since the mid-1960s has expanded the range of contexts in which provider insurance must be offered. Courts understand this point perfectly. Much of the modern extension of tort liability has been expressly justified by the salutary insurance consequences that are supposed to result. Thus it is a paradox that the modern regime somehow has led to the reduction of insurance availability. The next section examines more closely the operation of insurance markets in an attempt to resolve the paradox. . . .

The Insurance Crisis Observed

This section demonstrates how the shift to third-party tort insurance coverage and the consequent increase in risk-pool variance has weakened existing insurance markets and led to the unraveling of existing insurance risk pools. The section examines first how consumer pools and second how provider pools have been affected by the expansion of corporate tort liability and the shift from first- to third-party insurance sources. It shows why providers have changed the nature of their product and service mix, including ceasing production altogether; why insurers have changed the terms of insurance policies to reduce coverage levels; why insurers have refused coverage on any terms in some commercial lines; and why still other providers have been forced to form industry-wide mutuals.

The Effect of Increased Variance on Consumer Risk Pools

The previous section showed that the shift to third-party insurance sources increases the variance of consumer risk pools. An increase in risk-pool variance obviously leads to an increase in real insurance costs. A firm subject to competition must add these costs to the product or service price.

Adverse selection in consumer risk pools occurs when some group of consumers chooses not to pay increased prices that include the higher level of third-party insurance. These consumers decline to purchase the product or service and, thus, drop out of the existing risk pool. The consumers most likely to drop out are the low-risk within the pool. In general, within any pool there will be two low-risk sets: those systematically less likely to suffer injuries at all and those whose injuries systematically generate lower real costs.

Consumers who systematically face a lower injury probability are likely to find the insurance provided with the product or service not worth its added premium. Many commentators have tended to view product- or service-related injuries as occurring randomly, generating an equal injury probability to each consumer. Many product- and service-related injuries, however, are systematically associated with particular product uses. Most modern products can be employed in a wide range of diverse activities. Those consumers who use products in typically less, rather than more, risky ways are likely to drop out of the consumer pool if tort law requires the manufacturer to insure all consumer uses. These consumers will shift to alternative products or services that cannot be used in equally risky ways—products which, as a consequence, will be cheaper because of the lower attendant insurance premiums.

A familiar modern example is consumers of four-wheel-drive vehicles. In recent years, the liability of manufacturers of such vehicles has been expanded under design defect and warning law and, more generally, as courts have limited the defenses of contributory negligence, misuse, and assumption of risk. Manufacturers have been held liable, for example, for injuries suffered when these vehicles have rolled or flipped in contexts of extreme mountain driving on grounds that the manufacturer could either design the product to better protect the consumer or insure the consumer for the loss.

Manufacturers must respond to this increased liability either by changing product design to protect drivers in extreme conditions or by increasing insurance coverage for the consumer set as a whole. Whether the manufacturer changes the design or merely increases insurance coverage, product costs will increase and the product price will increase. The price increase, of course, may seem desirable for consumers who drive in extreme backroad conditions. But consumers who purchase four-wheel-drive vehicles for other purposes—say, easier driving on snowy or muddy roads—may not find the increased price worthwhile. These consumers could be lured away if they were offered a four-wheel-drive vehicle suitable for snow and mud, but not for extreme grades which, if only because of the lower attendant insurance premium, could be offered at a lower price. It is not surprising that many manufacturers have begun offering van and station wagon models with a four-wheel-drive option.

This process is adverse selection in the product market. Prior to the expansion of manufacturer liability, the vehicle was sold without insurance for losses resulting from flips or rolls. Consumers who enjoyed backcountry travel insured themselves for such losses in first-party insurance markets. The expansion of manufacturer liability shifted the insurance source to the third-party mechanism. Because the manufacturer was prohibited by product-liability law from making this additional insurance optional, low-risk consumers within the pool—those not intending to expose themselves to backcountry risks—dropped out of the pool, either by shifting to domestic four-wheels, or by declining to buy the product at all. When low-risk consumers drop out, the insurance premium added to the price of backcountry four-wheels must be increased by an ever greater amount.

The second set of low-risk consumers affected by the expansion of provider liability are the low-income or poor, who bring low risks to a liability insurance pool because of the lower damages they will receive because of their lower income and poorer future employment prospects. As the insurance premiums tied to products and services increase, these consumers also drop out because the price they must pay is increasingly greater than the value received. Such consumers will shift to substitute products for which the consumer set is more homogeneous in terms of income or for which the underlying risk level is lower. Of course, an income effect is also likely to be important here: Such consumers will discontinue buying products with large insurance premiums because, given their available resources, they cannot continue to afford them. Again, as this set of low-risk consumers drops out, the attendant insurance premiums must be commensurately increased.

Adverse selection in consumer risk pools explains why the increase in insurance premiums has been extreme for products and services in recent years. It also provides the only explanation of why increases in corporate tort liability compel providers to withdraw products and services from markets altogether. Again, if there were no adverse selection, increases in insurance premiums or self-insurance costs could largely be passed on to consumers. Sales may decline, as must be expected from any price increase, but there would be no reason to withdraw products from the market. There is a different effect, however, where a price increase derives from increasing risk-pool variance. Increasing variance generates adverse selection by low-risk consumers who successively drop out of the pool. The pool, as a consequence, unravels. At some point, demand for the product sold with the necessary insurance premium simply disappears. There remains no set of identifiable consumers to whom the product or service is worth its price.

There is substantial evidence of these phenomena in the recent insurance crisis. Increases in insurance premiums, sometimes extraordinary increases, have been reported. In many cases, providers have been able to pass on these premium increases to consumers. Some ski areas have increased lift tickets by $2 to $3 per day. Insurance costs now add $80,000 to each Beech aircraft and $75,000 to each Piper aircraft.

Many firms, however, have had difficulty passing on increases in premiums or self-insurance costs. Some firms have made greater investments in product or model differentiation in order to attempt to segregate low-risk from high-risk consumers, such as in the differentiation of domestic from backcountry four-wheel-drive vehicles, described above. It is difficult for others than industry experts to distinguish insurance from consumer preference grounds for product differentiation, but there is some strong evidence of this trend, at least with respect to more extreme risks. Some bus companies, for example, are reported to have terminated charter transit to ski areas because of the relatively higher risks of icy roads. In addition, as noted earlier, one New York vineyard has terminated or severely curtailed wine-tasting from fear of host liability for alcohol-related injuries. Public parks that have removed playground equipment or school districts that have removed diving boards from swimming pools are other examples of the effect. By excluding such services, these entities reduce the variance of risks related to the services they continue to provide.

. . .

It is important to distinguish, at least analytically, between products and services withdrawn for insurance reasons and for deterrence or incentive reasons. Some products or services may simply generate so many injuries that continued production is infeasible. A legal standard that attached liability where the marginal expected injury costs exceeded the marginal costs of preventing the injury (here, the bare cost of production) would be sufficient to drive such products and services from the market and, thus, increase social welfare.

In contrast, the product and service withdrawal that harms consumers and reduces social welfare derives from the expansion of liability on insurance grounds, where the shift from first- to third-party insurance increases the variance or reduces the independence of risks enough to cause the risk pool to unravel. For example, many vaccines are known to be beneficial to large majorities who receive them, but to cause adverse reactions in some small number of individuals. If the adverse reactions truly cannot be prevented, and if the benefits of the vaccine exceed production and injury costs, then liability for adverse reactions is solely an insurance question. The important issue, however, is whether it is cheaper to insure for these reactions through either a first-party or third-party mechanism. As shown above, first-party mechanisms are typically superior in defining the level of insurance coverage and in segregating consumers according to levels of coverage appropriate to their income. The shift from first- to third-party insurance recovery can lead to the unraveling of the pool and the withdrawal of the vaccine from the market.

The point is that, for some products and services, insurance pools could be maintained on a first-party basis that could not be maintained on a third-party basis because of adverse selection. It is for these products and services that the expansion of tort liability on insurance grounds—beyond that level set to create optimal safety incentives—is most harmful. In contrast, products too risky or injurious to produce at all could not support either first- or third-party insurance markets. Asbestos may be such a product. Obviously, asbestos production in the United States has stopped entirely because of the substantial third-party manufacturer liability. Quite possibly, asbestos-related injuries could not be insured in first-party markets either. Diseases without probabilistic features—inevitable diseases—are uninsurable because risks are not uncorrelated. It is difficult to believe that park slides, diving boards, vaccines, and other products and services recently withdrawn share this feature.

The Effect of Increased Variance on Provider Risk Pools

The increase in the variance of consumer risk pools from the shift toward greater third-party insurance delivery, of course, increases the risk of each individual provider. The previous section showed how this effect enhanced adverse selection and how the concomitant increase in correlation of corporate risks aggravated the problem. Adverse selection in provider risk pools consists of the relatively low-risk providers dropping out of the pool to self-insure.

What are the forms of self-insurance that low-risk providers have selected? There has been a massive shift in recent years toward self-insurance through the formation of captive insurance subsidiaries offshore in the Caribbean. Apparently, the subsidiary form assures deductibility of reserves; off-shore jurisdictions are attractive because of liberalized reserve requirements. A slightly different form of self-insurance in response to the crisis has been the formation of industry-wide mutuals, such as by doctors and surgeons, who believe they are better able to judge their exposure (or control their exposure) than insurers believe. Although the extent to which low-risk providers have recently shifted toward self-insurance is poorly documented, it is known that self-insurance rather than market-insurance constitutes a very substantial share of many specific commercial insurance lines.

The shift of the low-risk toward self-insurance, of course, places substantial pressure on the market insurance function: at the extreme, the pool unravels and the affected commercial risks become uninsurable. Insurers have strong interests in preventing this course of development. Since the early 1980s, insurers have been progressively changing the terms of the basic commercial insurance policy in order to make market insurance more attractive to low-risk providers in order to keep them in market insurance pools. These changes in basic levels of coverage have been accelerated since 1986.

Insurers have changed coverage terms in commercial policies in three separate ways: increasing deductible and coinsurance levels, lowering aggregate policy limits, and expanding coverage exclusions. . . . [E]ach of these changes reflects an effort to make the commercial policy more attractive to the relatively low-risk within provider pools. In effect, these changes reduce the level of commercial coverage offered to providers. Yet, despite the effective reduction—in many cases, the sharp reduction—in the level of commercial insurance coverage, insurers have simultaneously been forced to raise commercial premiums substantially and in some markets ultimately to refuse to offer coverage altogether. Most commentators have focused on the increase in premiums and the refusal to offer coverage as characteristic of the crisis. The premium increases and coverage withdrawals, however, are only more prominent than the other fundamental changes in commercial casualty policies; each is a symptom of the progressive unraveling of provider pools. Each reflects an effort by insurers to fight off modern tort law's stimulus of adverse selection in order to try to maintain a commercial insurance market. . . . Specifically, these contractual devices reduce variance in risk for the purpose of making insurance available to low-risk parties in the population of insureds. They are efforts to reduce adverse selection, and thus to prevent the unraveling of insurance markets that would occur if low-risk members of insurance pools were to exit.

The unraveling process would consist of the lowest-risk members of the pool dropping out, which, in turn, would necessitate premium increases. The premium increases would be followed by a new set of lowest-risk members dropping out; then, further increases in premiums; and so on, in successive episodes of withdrawals and premium rises. Of course, as low-risk members withdraw, the constituency of the pool becomes further concentrated among high-risk members. . . .

Who has suffered most from these developments? It is clear in my mind that the greatest harm from the expansion of tort liability and the consequent shift from first- to third-party tort insurance coverage has been suffered by the poor and low-income within the consuming population. The increase in market insurance and self-insurance costs leads to increases in the price level of virtually all commodities. In some cases, these increases will effectively price low-income consumers out of the market for the product altogether. Increases in product prices shrink the purchasing dollar and, in proportionate terms, shrink it more severely for the poor.

The increase in insurance costs reduces the availability of products or services for the poor in still more ways. The unwillingness of obstetricians to perform normal deliveries for less than $1,200, noted earlier, more severely affects those individuals for whose deliveries Medicare will pay $507 than those consumers able to pay the full cost by private negotiation with the doctor. Similarly, the termination of midwife services because of the absence of insurance is relatively less troublesome to those who can afford to substitute licensed physicians or full-priced obstetricians.

More generally, even if the low-income are able to continue to pay the insurance-enhanced prices of products, they are harmed. The poor and low-income are much more likely to be paying a premium that exceeds the risk they bring to the pool under third-party than under first-party insurance coverage, because of the difficulty providers face in segregating risk pools. No one today would seriously propose that all citizens pay an identical life-insurance premium where some of the beneficiaries—the high-income—would receive greater total benefits. No one would urge that all individuals pay the same disability premium regardless of expected income. No one would seriously propose that all homeowners pay the same fire premium regardless of home value, or that all drivers pay the same auto premium regardless of car value.

Yet this is exactly the regime of life and disability insurance provided by modern tort law. Modern tort law forces all consumers to pay the same third-party insurance premium for product- and service-related coverage for losses leading to injury, disability and death.

Wealthy and high-income members of the consuming population are also harmed by the shift to third-party coverage. Admittedly, in individual instances, relatively high-income members may gain from paying a premium averaged for a pool containing some relatively low-income members. More generally, however, the broadening of risk pools characteristic of the shift to third-party coverage harms the high-risk of the insured population as well, because it reduces the availability of insurance in all contexts. The high-income do not benefit when jails and parks are closed or, for that matter, when relatively low-cost products and services, such as vaccines, are withdrawn from the product market.

Nevertheless, it is the poor who are most severely harmed. It is not surprising that the crisis has heavily disrupted the provision of municipal and other governmental services. Because such services are generally provided without charge, it is impossible directly to pass on to users the costs of increased insurance premiums. Yet it is also because municipal services are provided without charge that they are of particular importance to the lives of the poor.

The motivation for the judicial expansion of third-party liability during the early 1960s was to protect the poor within the consuming population by providing them with a form of insurance that they might not otherwise obtain. At the time of the first adoption of the strict-liability standard, however, our society had not introduced the array of social insurance programs available today, and the courts may not have perceived the wide range of privately available first-party health and disability coverage. Today, the number failing to possess basic health coverage is very small.

Our society covers the poorest of the poor through Medicaid, Medicare for the aged, and General Assistance. Although it is estimated that 18 million individuals totally lack health-care insurance, it is unknown how many fail to qualify for Medicaid while healthy and able to maintain employment, but would qualify if, through injury or illness, their employment were to cease. This group is uninsured only formally: They are uninsured until they need health insurance, at which time they become insured. A more precise 1982 telephone survey of families found only 1.5 percent in which any family member had been denied health care for financial reasons. Disability coverage in the United States appears more spotty, at least for non-job-related injuries. But there are no studies indicating the number of individuals failing to possess private disability coverage who would not be covered by government disability programs. It is a crucially important empirical question exactly how many in our society remain unprotected for basic health and disability losses, although the number appears to be quite small and the definition of the set quite peculiar. It is at the least unusual to believe that an effective and comprehensive way to provide care to such individuals is through tort-system recoveries for injuries from products or services.

Notes and Questions

1. According to the analysis presented in this selection, would you expect insurance policies to be written for long periods or short periods of time? Some insurance policies cover risks arising in a six-month period and others cover three years. Which would you expect to be true of policies covering burglary; automobile liability?

2. It is understandable why municipalities and other organizations would hesitate to operate facilities when insurance is simply unavailable. Could insurance companies offer a policy during these crisis periods with an unknown premium? Could the parties agree to supply and pay for coverage at a rate to be determined by an arbitrator after the fact or by some after-the-fact price index—and in light of national experience or premiums that are set when the market settles? See Kenneth S. Abraham, "Environmental Liability and the Limits of Insurance," 88 *Columbia Law Review* 942 (1988).

3. The reading emphasizes the shift from first-party to third-party insurance. Is it possible that an increased pool of buyers of third-party liability insurance would strengthen rather than weaken insurance markets? What if a kind of liability insurance is mandatory; would you expect premiums to spiral upward or to remain stable?

4. There may be some adverse selection in products markets, but it is an open question whether there are many examples of such reactions in other markets. Do you think many physicians, swimming-pool operators, or midwives (examples of providers of services who found themselves unable to secure insurance in the crisis period discussed in the reading) identify themselves as high risks? If not, what would explain the temporary collapse of these insurance markets?

Incentive Issues in the Design of "No-Fault" Compensation Systems

MICHAEL J. TREBILCOCK

Introduction

The efficacy of tort law as a central policy instrument for compensating or deterring personal injuries has come under increasing criticism in recent years. It is argued that many personal injuries go uncompensated, and that even those victims who receive compensation face enormous costs, delays, and psychological traumas that are inconsistent with a humane and efficient method of compensation. It is also argued that as a system of deterrence tort law has not been proved to influence significantly behavioural incentives, so that its deficiencies as a compensation system are not offset by its efficacy as a deterrence system. The critics suggest that the compensation and deterrence

Abridged and reprinted without footnotes by permission from 39 *University of Toronto Law Journal* 19 (1989) and Michael Trebilcock.

goals of the tort system should be assigned to separate legal instruments—the compensation goal to some form of no-fault or social insurance system, and the deterrence goal to criminal and regulatory instruments.

It is the central thesis of this selection that no compensation scheme can responsibly disregard incentive effects, particularly incentives to reduce the frequency and severity of injury claims. I accept that the tort system as a system of compensation is appallingly inefficient, and that a strong prima facie case exists for the adoption of some form of nontort method of compensation; but when we begin to turn our minds to ways in which alternative compensation schemes might be designed, we are quickly driven to the realization that the compensation and deterrence goals ascribed to the tort system cannot be separated and will require reconciliation in any compensation scheme. If we are to effect this reconciliation, many of the problems that currently plague the tort system will have to be confronted in new legal or institutional contexts. It may well be possible for us to achieve a better set of trade-offs than the tort system has achieved or is even likely to achieve. But to pretend that the trade-offs do not exist, or to assume that safety issues can be confidently remitted elsewhere in the legal system when the evidence does not warrant such confidence, is to espouse the untenable proposition that compensating for accidents is better than preventing them.

The Central Predicates of the Case for No-Fault Compensation Schemes

Proponents of no-fault insurance schemes have often relied on one or more of the following three premises to advance their case. Each of these premises seriously risks obscuring the role of conduct variables in the design of non-tort insurance schemes.

Horizontal Equity

A pervasive theme running through many critiques of the tort system is that it fundamentally violates ethical notions of horizontal equity. "Horizontal equity" in this context is taken to require that people with equivalent needs should receive equivalent benefits. The tort system is said to create a normatively indefensible "accident preference" whereby those injured by the negligence of others may receive full compensation, but those sustaining injuries in other circumstances receive nothing from the system and must rely on their own resources or on meagre social-welfare benefits, despite the fact that their financial needs are equivalent to those injured as a result of the negligence of others. Similarly, it is argued that workers' compensation systems create an "industrial preference" whereby workers injured on the job typically receive more generous benefits than workers injured off the job. These concerns then extrapolate to an argument that sickness-created financial need should be treated similarly to accident-induced financial need, and indeed that other sources or causes of financial need such as congenital disabilities, unemployment, retirement, or a breadwinner's desertion or death should all be treated similarly.

This emphasis on horizontal equity has a long and distinguished intellectual pedigree. It centrally informs the UK Beveridge Report on social insurance and allied services, which provided the intellectual underpinnings for much of the post war British social-welfare policy (and welfare policies in many other countries), and it is central to the "comprehensive entitlement" concept espoused in the New Zealand and Australian Woodhouse reports. In a much-quoted passage, Sir William Beveridge argued as follows:

> If the matter were now being considered in a clear field, it might well be argued that the general principle of a flat rate of compensation for interruption of earnings adopted for all other forms of interruption should be applied also without reserve or qualification to the results of industrial accident and disease, leaving those who felt the need for greater security, by voluntary insurance, to provide an addition to the flat subsistence guaranteed by the State. If a workman loses his leg in an accident, his needs are the same whether the accident occurred in a factory or in the street; if he is killed, the needs of his widow and other dependants are the same, however the death occurred. Acceptance of this argument and adoption of a flat rate of compensation for disability, however caused, would avoid the anomaly of treating equal needs differently and the administrative and legal difficulties of defining just what injuries were to be treated as arising out of and in the course of employment. Interpretation of these words has been a fruitful cause of disputes in the past; whatever words are chosen, difficulties and anomalies are bound to arise. A complete solution is to be found only in a completely unified scheme for disability without demarcation by the cause of disability.

The ethical appeal of this argument for an "integrated," "comprehensive," or "universal" social insurance policy is powerful, and is rendered more so by the following examples cited by Weiler:

> The question of which category a disability fits into is crucial in determining the level and adequacy of the income replacement which an injured victim will receive. As an example, suppose that a single person earning $30,000 a year is permanently disabled in an automobile accident on his way to work. If he can establish that someone else is entirely at fault, he will collect tort damages calculated at $2,500 a month, nontaxable. If the accident occurred while he was at work and he cannot establish that the other driver was at fault, he will collect $600 a month in 'no fault' auto benefits. If he was injured at home as a result of a crime, e.g., a burglary, he will collect $500 a month. But if he was injured at home due to nobody's fault, and must rely solely on the CPP or GAINS programs, he will get only $300 a month, a sum which is taxable (at least in principle).

Indeed, by the same line of reasoning, there seems to be no justification for treating physical disablement, whether from accident or illness, differently from any other misfortune. As Liebman argues, "Our society may revere medicine and sympathize with the sick, but it holds no view that could explain distinctions between persons totally unable to work according to whether their condition results from an illness or, on the other hand, from limited natural abilities, decades of racism or sexism, homosexuality, family burdens, technological change, a broken home or national fiscal policy."

What distinguishes this view of horizontal equity is that it emphasizes equality of *outcomes* (for example, either maintaining similar financial floors for everyone in need or maintaining everyone's financial status quo ante) by focusing exclusively on the financial condition of claimants (*status variables*), and largely discounts the relevance of *conduct variables* (relating to either injurer or victim) that may have led to the financial status of the claimant, at least as a determinant of entitlement to assistance; this view seems also largely to discount conduct variables that may influence the capacity of a claimant in the post-misfortune state to adopt responsive strategies. I use the term "conduct variables" to embrace both care-level and activity-level considerations as analysed in standard law and economics treatments of accident law.

Community Responsibility

Closely linked to the arguments from horizontal equity for no-fault insurance schemes is the argument that providing for, or redressing, many of the major negative contingencies of life should be viewed not solely or even primarily as a matter of individual responsibility, but as a collective responsibility. This theme suffuses the Beveridge Report, at least at the level of the need for providing basic social security nets against physical want, and dominates the thinking of the Woodhouse reports in proposing earnings-related income guarantees in the event of injury or disability. In the New Zealand report, Sir Owen Woodhouse (then a prominent New Zealand judge) stated the first guiding principle of a compensation system:

> First, in the national interest, and as a matter of national obligation, the community must protect all citizens (including the self-employed) and the housewives who sustain them from the burden of sudden individual losses when their ability to contribute to the general welfare by their work has been interrupted by physical incapacity. . . .
>
> The first principle is fundamental. It rests on a double argument. Just as a modern society benefits from the productive work of its citizens, so should society accept responsibility for those willing to work but prevented from doing so by physical incapacity. And, since we all persist in following community activities, which year by year exact a predictable and inevitable price in bodily injury, so should we all share in sustaining those who become the random but statistically necessary victims. The inherent cost of these community purposes should be borne on a basis of equity by the community.

The distinguishing characteristics of this view of collective rather than individual responsibility for accidents are at least twofold. First, accidents are viewed as relatively random, unavoidable consequences of life in an industrialized, interdependent society; just as society is the collective beneficiary of the economic activities carried on in it, so should it bear collective responsibility for the costs. Second, to relate this view to the view of horizontal equity described above, the collective underwriting of the costs of accidents (and presumably other kinds of misfortune, such as sickness) on a horizontally equitable, integrated basis will almost certainly lead to the public administration of the compensation scheme and the relegation of private insurance and other private means of accommodating these contingencies to a minor or secondary role. Although I accept . . . that redistributive goals should preclude treating an inability to pay contributions to a benefit scheme as a disqualification per se from receipt of benefits, and that adverse selection problems in many contexts may dictate mandatory participation, major questions are still left open. How is any compensation scheme to be financed? What scale of benefits is to be provided to claimants? In the context of tort reform, should there be any role for private rights of action to redress personal injuries?

That these issues remain open is amply attested to by major differences of philosophy between Beveridge and Woodhouse, despite their common espousal of the views of horizontal equity and community responsibility sketched above. Beveridge favoured flat-rate benefits sufficient to provide subsistence means of support and flat-rate contributions by employers and employees to a general social insurance scheme, earnings-related benefits for industrial accidents only, and some experience-rating of employers in fixing their contributions to reflect differentials by industry in the costs of work-related accidents. Beveridge made no recommendation for abridging the role of the tort system (other than the subtraction of social benefits from tort awards). Woodhouse, in contrast, favoured much higher levels of (earnings-related) benefits to provide protection against income interruption, rather than mere subsistence-level benefits, and recommended the imposition of non-experience-rated, flat-rate levies on motorists and employers as major sources of financing for the scheme. Woodhouse also recommended the abolition of tort claims for most classes of personal injury. The Woodhouse recommendations were implemented with few changes in the New Zealand Accident Compensation Act, 1972, which covers work-related accidents and disease, accidents on the highway, at home, or elsewhere, and medical misadventure. These differences of viewpoint on financing (pricing), benefit levels, and the residual role of the tort system are a central focus of this paper.

Different Instruments for Different Objectives

As noted above, a pervasive theme of much of the literature that is critical of the tort system is that we should not ask a single legal or policy instrument to serve multiple and often divergent objectives. To the extent that we wish to reduce the incidence of accidents, we should rely on (and, if necessary, reform) criminal and regulatory sanctions. To the extent that we wish to compensate victims of accidents more completely and at a lower administrative cost, we should set up compensation schemes directed exclusively to this end. This criticism of the tort system resonates more broadly with the injunction of economists that separate policy objectives require separate policy instruments, and indeed with my own criticism of modern North American tort law as attempting to assign increasingly prominent weight to social insurance objectives along with the more traditional deterrence and corrective justice objectives, which often lead in opposite directions. If the tort system cannot promote social insurance considerations as a major goal coherently with deterrence and corrective justice considerations, can a no-fault or social insurance system ignore deterrence considerations? . . .

Financing-Pricing Issues

A contributory scheme, whether private or public, of no-fault or social insurance—for example, workers' compensation, or automobile no-fault—raises questions of how to determine contributions to the scheme. In other words, to the extent that such a scheme explicitly or implicitly purports to be a form of insurance, how is the insurance to be priced?

An economic perspective on this issue would argue that internalizing the accident costs to the source of the activity that causes them serves an important social function by creating incentives for those carrying on the activities to act more carefully, where this is feasible, or, where it is not, to reduce the level of activities that are inherently risky. The civil liability system may promote these objectives in its assignments of liability; however, even in the absence of a civil liability regime, the determination of contribution levels to a compensation fund may serve similar ends. The economic analysis of liability rules has revealed the problem of determining which activity or actor should be "taxed" with accident costs when an accident is typically the product of a multi-causal chain of necessary conditions. Even in simple two-party interactions both parties may be able to control their care and activity levels in ways that will reduce the probability or severity of accidents. Problems of attribution and causation will similarly afflict efforts to allocate contributions to a compensation scheme in proportion to the expected accident costs that each contributor is likely to generate for the scheme.

Notwithstanding these difficulties, in exploring the potential for utilizing differential pricing/contribution schedules to reduce accident costs, it is useful to distinguish contributions made to cover prospective losses to third parties (third-party contributions), and contributions made to cover prospective losses to oneself (first-party contributions). Workers' compensation schemes exemplify the former; automobile no-fault insurance schemes exemplify the latter.

Workers' Compensation Schemes

With respect to the financing of workers' compensation schemes, Beveridge recommended the retention of a measure of experience-rating of employers through industry classification differentials, notwithstanding his general commitment to flat-rate levies for social insurance benefits. Woodhouse recommended the latter, although the New Zealand Accident Compensation Act, 1972, has retained a system of industry differentials. This is the most common form of experience-rating in jurisdictions that have adopted workers' compensation schemes. Studies of the effects of this form of experience-rating on accident rates have been somewhat inconclusive. In a recent Ontario study, Weiler recommended that the contributions of larger employers, whose past accident experience is a statistically meaningful predictor of future accident experience, should be rated on an individualized basis. In 1982 this would have resulted in one automobile manufacturer's receiving a rebate of $1.6 million, while another firm in the same industry would have paid an additional $2.6 million surcharge. Similarly, full experience-rating in nickel and smelting would produce a net advantage for one employer of nearly $3 million over its principal competitor; in meat-packing, the discrepancy between one company and another would have been $2 million. On the strength of differentials of this magnitude, Weiler concluded that individualized experience-rating was likely to have a significant impact on workplace safety.

This is not to suggest that the process of experience-rating is free of difficulty. The problem of small employers whose past accident record is not a statistically reliable indicator of risk potential has already been noted. Concerns have also been voiced as to whether full experience-rating will create socially perverse incentives for employers to suppress claims by means of improper coercion of workers. In the case of industrial health hazards with long latency periods, experience-rating may not be feasible. Alternative or complementary approaches, such as penalty assessments by reference to observed safety conditions, penalties levied against specific employers for acts of gross negligence, or hazardous-emissions fees or taxes, may mitigate some of these difficulties, although residual problems will remain. For example, penalty assessments for observed safety conditions raise questions about the criteria and processes by which these assessments are to be determined and the appropriate quantification procedures that are able to capture the potential health hazard costs that the conditions entail. If penalty levies for gross negligence are to be imposed, one must be able to define with some predictability the relevant standard of liability, and to quantify the levies to reflect the expected social costs of acts of gross negligence if the deficiencies of the tort system are not to be replicated. Hazaradous-emissions fees or taxes create difficult measurement and monitoring problems, and setting the level of the fee or tax will be extremely speculative, if not impossible, if a damage schedule reflecting potential health hazards associated with different levels of emissions cannot be determined with confidence. The adverse health effects of many substances may not become known until some distant point in the future.

However, the central question remains whether a no-fault insurance scheme such as a workers' compensation scheme can afford to eschew serious efforts at experience-rating individual employers' contributions to the scheme, wherever feasible, in order to sharpen incentives to increase care levels and to reduce inherently risky activity levels. A negative response predicated on the incompleteness and imperfections of a full experience-rating system and on the availability of alternative criminal and regulatory strategies for reducing workplace accidents ignores the fact that the effectiveness of the alternative strategies is also far from complete; indeed, the impact of occupational health and safety regulation has in many respects proved disappointing. It has been insufficiently recognized by critics of the tort system that the causation problems that afflict the tort system cannot be avoided in the design of any experience-rating scheme associated with a compensating system, or in the design of any system of regulatory sanctions for reducing injury rates.

No-Fault Automobile Insurance Schemes

With respect to no-fault automobile insurance, it is important to appreciate the different incentive structures of third- and first-party insurance. In a pure third-party tort-insurance regime, motorists' incentives to exercise care towards others are influenced by third-party liability insurance premiums and uninsured excess liability, and they are motivated to take self-protective measures by the fact that their own losses are uninsured. In a pure first-party regime that does not recognize third-party claims (that is, first-party no-fault), motorists have no direct incentives to exercise care towards third parties, but do face incentives to take self-protective measures—in part because of the implications for first-party insurance premiums and in part because, under most typical first-party systems, all pecuniary and non-pecuniary losses are not fully compensated. What is the empirical evidence of the safety implications of moving from a fault-based third-party insurance regime to various forms of no-fault first-party insurance regimes?

Landes examined data from fifteen American states that had adopted partial no-fault schemes, and found that fatalities in each state had increased by between 376 and 1,009 a year. All U.S. no-fault schemes preserve the right to sue in some contexts. Add-on no-fault regimes (like that in Ontario) permit suit in all cases, but require the subtraction of no-fault benefits from awards against third parties. Threshold no-fault regimes prohibit suit under the stipulated threshold, but do permit suit above the thresholds. These thresholds vary widely. They are typically either monetary or verbal. In the case of monetary thresholds, medical expenses incurred by an accident victim beyond a specified dollar threshold trigger the right to sue. In the case of verbal thresholds, injuries that entail, for example, "serious and permanent physical impairment" trigger such a right. Landes claims that a monetary threshold of as low as $1,500 is likely to result in a 10 percent increase in fatalities. Most commentators find this conclusion implausible, given the relatively modest no-fault elements of the insurance mix in many states, particularly the preservation of the right to sue for most serious injuries, and have been critical of aspects of Landes's empirical methodology. In addition, private first-party insurance schemes retain significant risk-rating characteristics that are likely to discourage many of the forms of driving behaviour that are discouraged by risk-rating under third-party insurance. Three subsequent studies of the U.S. data found no significant effects on fatality rates from the adoption of no-fault regimes, although a fourth study found that no-fault states have experienced increases in their loss ratios of between 3.9 percent and 7.4 percent, depending on the extent of the no-fault features of the schemes.

Brown examined auto fatality rates in New Zealand following the adoption of a universal accident compensation scheme in 1972 (which effectively abolished all tort claims for personal injuries) and found that the rate continued to decrease after the adoption of the scheme. He acknowledged, however, that other safety factors may also have been influential during this period—the mandatory use of seat-belts, crack-downs on drunken driving, speed limit reductions, and gasoline rationing, so that in the absence of a multivariate analysis (which he did not undertake), his findings are of limited value. In contrast, McEwin, in his analysis of the effects of no-fault regimes in Australia and New Zealand, found that in the jurisdictions that completely abolished tort claims and adopted non-risk-rated first-party premium structures (such as New Zealand), road fatalities increased by 16 percent per capita.

Gaudry conducted a detailed multivariate analysis of accident and fatality rates in Quebec before and after the adoption of a pure no-fault automobile insurance scheme in 1978. The Quebec scheme, which is administered by an agency of the provincial government, bans all third-party personal injury claims arising out of auto accidents, provides for a high level of earnings-related benefits up to an income ceiling, and finances the scheme by imposing flat-rate levies on all motorists. Gaudry found that bodily injury accidents increased by 26.3 percent a year after the adoption of the scheme, and fatalities by 6.8 percent (equivalent to 100 additional deaths a year). While the increase in accident rates may partly reflect a reporting bias, the fatality rate does not, and Gaudry's findings in this respect have attracted widespread attention. Gaudry attributes the increase in accident and fatality rates in part to the more stringent enforcement of compulsory insurance requirements, which causes previously uninsured motorists to drive less carefully, and in part to the adoption of a flat premium-pricing regime that drastically reduces the cost of automobile insurance for high-risk drivers (for example, young males), which encourages them to drive when previously they were "priced off" the Quebec highways. The assumption here appears to be that young male drivers are risk-preferrers and will be more influenced in their behaviour by a certainty of present costs—substantial insurance premiums—than by the mere probability of a substantial penalty for misconduct, even if the penalty entails the same expected costs as the insurance premium. Alternatively, young drivers, because they have fewer resources than other drivers, have more price-elastic demands for driving (including compulsory insurance) and are disproportionately influenced by changes in prices.

Devlin, in a recent detailed analysis of the Quebec experience, reached even more striking conclusions than Gaudry. She found that the adoption of the no-fault regime in Quebec led to more high-risk and fewer lower-risk drivers on Quebec highways, accounting for a small increase (1 or 2 percent) in fatalities; more important, she found that average care levels fell substantially after the introduction of no-fault, resulting in a 9.6 percent increase in fatal accidents (or 154 more deaths a year). She concluded that "a liability system for automobile accidents which operates in the presence of liability insurance still provides incentives for more prudent driving than does a no-fault system with insurance. Furthermore, the tying of insurance premiums to driving behaviour is essential if one wants individuals to exercise more care when driving."

In the light of the studies cited . . . what assessment is possible of the effects of tort–insurance regimes on incentives for accident reduction? One view is that tort liability, even standing alone, is unlikely to have major incentive effects, and once third-party insurance is introduced it will have no significant effects, so that the movement to a first-party no-fault system is unlikely to entail any reduction in safety. This view turns largely on perceptions of the nature of errors involved in most automobile accidents. A 1970 study by the U.S. Department of Transportation concluded that "the vast majority of accidents involve that large group of drivers with low accident likelihoods—drivers who, over their lifetime, will be involved in no more than a handful of state reportable accidents." The study found that it is a fallacious view that the "drivers who 'cause' accidents are . . . a small identifiable group that is guilty of hazardous driving, vastly overrepresented in accident statistics, and responsible for the 'accident problem'. . . . Most drivers are often 'guilty' of driver error. A certain magnitude of driver error is representative of the behaviour of the general average of drivers, and must be considered as normal, even though such behaviour departs from 'standard,' 'correct,' or 'ideal' behaviour." One study estimates that a driver makes 200 observations per mile, 20 decisions per mile, and one error every 2 miles. Those errors result in a near collision once every 500 miles, a collision once every 61,000 miles, a personal injury to some individual once every 430,000 miles, and a fatal accident once every 16 million miles.

However, this view proves too much. In its extreme form, it asserts that not only civil sanctions but criminal and administrative sanctions are likely to prove futile. Driver error is simply a manifestation of inherent human fallibility. A similar argument is sometimes made that conduct on the highway that is dangerous to others is also typically dangerous to oneself, and no one wants to injure himself or herself. Thus, civil and presumably other sanctions add nothing to this desire for self-preservation. Again, this argument, if true, makes all sanction regimes for driving conduct futile. But some drivers are likely to be less diligent in assessing risks than others; some drivers may have a stronger taste for risks than others. In these cases, sanctions, including civil sanctions, that raise the cost of risk-taking may have an influence on behaviour. The appropriate question is whether, at the margin, *some* driver errors will be reduced by civil and other sanctions. While the presence of third- or first-party insurance will mute the incentive effects of civil liability in various ways, attention must be redirected to what risks are insured, what risks are uninsured, and, in particular, how the insurance is priced.

To the extent that accident rates are a function of activity or exposure levels (that is, miles driven), as several empirical studies have found, then, to the extent that insurance-premium levels geared to exposure reduce activity levels, accident rates are likely to fall. In this respect, premiums ideally would be geared to the number of miles driven and to the reasons for and conditions in which the driving is done (business or leisure, urban or rural). Territorial classifications and business-leisure driving distinctions are able to capture (albeit crudely) some of the qualitative aspects of exposure levels, but the number of miles driven is difficult and expensive to preduct and monitor and is typically not employed as a primary rating variable. Rating variables of age and sex also function as crude proxies for both quantitative and qualitative aspects of exposure levels; for example, young men as a class drive more miles and engage in more seriously risky behaviour than young women, and both are overrepresented in accident statistics relative to the adult population at large. The Gaudry study of the Quebec no-fault system suggests that reducing the driving activity levels of high-risk driver classes is an important function of insurance pricing (in practice, by in effect forcing a binary choice between driving unlimited amounts or not driving at all, given that the number of miles driven by members of high-risk driving classes cannot be closely monitored and priced). Thus, proposals to abolish age and sex as automobile insurance rating variables for either third- or first-party insurance seem seriously misconceived from a safety perspective. From an ethical perspective, however, it will be argued that these categories are over-inclusive, and penalize or discriminate against low-risk individuals within these categories because of non-controllable ascriptive charactcristics.

To the extent that accident rates are a function not only of activity levels but also of care levels, then, as Devlin's study of the Quebec no-fault system found, the absence of significant deductibles and the failure to price insurance so as to reflect the differential risks that individual drivers present to others (third-party insurance) or themselves (first-party insurance) are likely to reduce care levels significantly. Thus, under a first-party no-fault system, a failure to differentiate premiums by referrence to previous accident experience, traffic violations, or demerit points is likely to increase accident rates significantly, even though such differentiation entails the de facto importation of notions of fault or accident propensity into the setting of no-fault premiums, albeit not the right to claim compensation. This will inevitably lead to due process concerns—demands for arbitration and appeal procedures—over rate classification or increase decisions. By definition, those decisions will be cruder under any risk-classification system than highly individuated tortious determinations of fault. Even with extensive risk-rating of premiums under a no-fault ("no blame") first-party insurance scheme, it is possible that safety incentives would be attenuated relative to tort–third-party insurance regimes, where tort law still operates to stigmatize certain kinds of conduct and arguably helps to reinforce socially responsible attitudes to risk-taking.

However, the central question that must be faced is whether a no-fault automobile compensation scheme can afford entirely to ignore conduct variables in the pricing of its coverage (as Woodhouse recommended and as is done in New Zealand and Quebec) if traffic safety issues are to be taken seriously. Again, it is not enough to say that other policy instruments are available and widely invoked to control improper behaviour on the highway. As with workplace accidents, the empirical evidence on the efficacy of alternative traffic safety measures is often indeterminate and in many cases disappointing. In both cases, criminal and regulatory sanctions typically entail financial penalties (often quite modest), and it is not clear why these economic disincentives should be expected to work better than or indeed as well as those entailed in serious risk-rating policies under a compensation scheme. . . .

A Residual Role for the Tort System

On the assumption that the compensatory implications of most accidents (and perhaps other kinds of misfortune, such as illness) should be dealt with under no-fault or social insurance schemes, the question remains whether there is a useful residual role for the tort system, particularly in mitigating some of the incentive losses that may be associated with the complete abrogation of the tort system for personal injuries. It will be recalled that the views of Beveridge and Woodhouse diverged sharply on that question. A number of proposals have been advanced, and in some cases implemented, for mixed fault–no-fault systems. The principal classes of such proposals are briefly reviewed below, with particular emphasis on their incentive effects.

Damage Thresholds

In 16 of the 24 American states with some form of automobile no-fault system, actions are permitted for full tort measures of damage, including pain and suffering, where damages have been suffered that exceed a threshold. Some of the thresholds are monetary (medical expenses incurred above a financial threshold); some are verbal ("serious and permanent impairment" or some variant thereof).

Below these thresholds, accident victims typically receive scheduled benefits for medical costs and income losses, but not for non-pecuniary losses. These savings, coupled with savings in transaction costs by virtue of an agency's not having to determine issues of fault, are expected to finance entitlements of accident victims who hitherto had been denied recovery under the traditional tort system.

The rationale for preserving full tort entitlement above these thresholds is somewhat murky. The pragmatic or political rationale is that large damage claims need to be kept in the tort system in order to mute political resistance from the personal-injury bar to the adoption of a no-fault system. A more principled rationale is that traumatically injured victims suffer such enormous diminution in the enjoyment of life that money may provide alternative sources of pleasure or satisfaction. This argument implies that the pain and suffering of less traumatically injured victims is less "real" or more transitory and does not warrant financial recognition. Whether this distinction is defensible in principle may be open to debate. Alternatively, it might be argued that traumatic injuries more cogently attract the deterrent or corrective rationales for the tort system, but this argument seems suspect given that such injuries may be caused equally by egregious misconduct or minor inadvertence on the part of the injurer. In either case, if less serious injuries result, threshold no-fault systems ignore the insurer's conduct.

Conduct Thresholds

The failure of damage thresholds to correlate an injurer's liability with his or her conduct suggests an alternative view of no-fault–fault thresholds—that is, a focus on the egregious or delinquent conduct of the injurer instead of, or in addition to, a focus on the seriousness of the victim's injuries. Such conduct thresholds have been recognized, even in no-fault legal regimes that ostensibly ban all lawsuits by victims against injurers.

For example, even under the New Zealand Accident Compensation Act, which purports to abolish tort claims for all personal injuries, the New Zealand Court of Appeal has held that an action for exemplary damages could be maintained by a victim against an injurer who had committed assault and battery (an attack with a hammer). The court held that the purpose of exemplary damages is "to punish and deter" egregious misconduct; the purpose of awards in personal-injury cases entailing merely error or inadvertence is primarily to compensate. Again, U.S. courts have held that workers' compensation legislation that ostensibly bars tort claims by employees against employers for workplace injuries does not preclude such claims when an employer has fraudulently concealed from an employee the fact that the employee has contracted a serious disease as a result of workplace exposure.

What seems to emerge from these cases is a recognition that some combination of egregiously delinquent conduct and traumatic consequences for the victim will provoke powerful retributive, deterrent, and corrective justice rationales for permitting unconstrained tort recoveries by such victims against such injurers. In short, conduct variables will not easily be denied.

Add-On No-Fault

Eight American states with some form of no-fault automobile insurance system provide for add-on first-party no-fault benefits that in no way constrain the ability of a recipient to sue a negligent third party for full tort damages (subject to subtracting the no-fault benefits already received from any award or settlement). A number of Canadian provinces, including Ontario, have adopted similar systems. The Osborne Commission has recently recommended the substantial enrichment of Ontario's first-party no-fault benefits.

The Pearson Royal Commission recommended a similar system for the United Kingdom.

The U.S. experience with add-on no-fault benefits, at least where those benefits are reasonably generous and cover most forms of economic loss and medical costs, has been that insurance premiums have escalated dramatically—far faster than they have under most threshold no-fault systems and traditional tort–third-party insurance systems. From 1976 to 1983 the amount of the average automobile insurance premium increased almost twice as much (91 percent versus 50 percent) in the average no-fault state as in the average traditional state. Increases have been particularly pronounced in generous add-on no-fault jurisdictions or jurisdictions with very low lawsuit thresholds.

O'Connell and Joost deduce empirically that most of the premium increases have related to third-party bodily injury liability claims, not to first-party no-fault benefits. They attribute this in part to the fact that once a claimant's economic and medical costs have been covered on a no-fault basis, he or she is likely to become a more aggressive litigator of residual claims, and will have less incentive to settle the claim or to compromise his or her demands for noneconomic damages than a victim under a traditional tort law–third party insurance regime. As noted earlier, savings in non-pecuniary damages are looked to in all no-fault automobile schemes as a primary source for the financing of expanded entitlements to no-fault benefits. To the extent that these savings are forgone by an expansive right of suit, the only alternative source of financing in a contributory system is premium increases.

Given greater consumer sensitivity to price (a matter that confronts all drivers annually) rather than coverage gaps (which confront a small subset of drivers occasionally), any contributory no-fault insurance system or mixed fault and no-fault system that entails substantial premium increases to finance broader coverage seems unlikely to be politically acceptable. Therefore, a generous add-on no-fault system seems particularly likely to violate pricing constraints.

Elective (Neo–) No-Fault

Jeffrey O'Connell has developed a series of ingenious proposals for permitting either injurers or victims (in product, automobile, medical, and sports contexts) to choose between offering or securing expeditious compensation for economic and medical costs on a no-fault basis and waiving rights to tort claims, or waiving the no-fault benefits and pursuing tort claims for full tort measures of damages. In the case of third-party elective no-fault, his proposals have been fairly criticized as raising serious adverse selection problems. For example, product manufacturers would offer only pre-accident no-fault insurance on their most dangerous products, for which, in the event of personal injury, the probability of tortious liability is high. It is not clear why consumers would find it to their advantage to accept such offers, or, if they did, whether courts would view the agreements as the products of informed choices. Similarly, in the case of proposals to encourage postaccident offers of compensation for pecuniary losses and plaintiffs' legal costs (but not non-pecuniary losses), where the incentive is the loss of the contributory negligence defence and the reversal of the onus of proof of liability in the absence of such an offer, defendants would be likely to make such offers only when the probability of their being found liable in tort is high. Again, it is not clear why plaintiffs would find it in their interests to accept such offers. O'Connell has recently proposed various elective first-party no-fault schemes as an alternative to elective third-party schemes.

For example, in the auto accident context, he envisages a scheme whereby drivers can choose to purchase either first-party no-fault insurance or third-party liability insurance. If two people with first-party no-fault insurance collide with each other, each looks to his or her own insurer to cover his or her losses. If two people with third-party coverage collide, entitlements to recovery would be resolved as they are at present under traditional tort–third-party insurance systems. If a person with first-party no-fault insurance collides with someone with third-party insurance, the first person looks to his or her own insurer to cover his or her own losses, and is not liable for the other person's losses, even when the first person was negligent. The second person would look to his or her own insurer for full tort measures of compensation if negligence on the part of the first person could be proved. This coverage would be treated as analogous to the conventional coverage of claims against uninsured motorists.

These proposals have been advanced by O'Connell partly because of a recognition of the political resistance in the United States to further adoption of mandatory no-fault compensation schemes. This resistance no doubt reflects the influence of special interest groups, especially the personal-injury bar, to further curtailments of tort claims. O'Connell now seems prepared to gamble that if no-fault cannot prevail in the political market-place, it may well prevail as a matter of consumer choice in the economic market-place. Although he loads the characterizations somewhat, he also acknowledges that there may be genuine divergences in consumer preferences as to tort-insurance regimes, which preclude a judgment that all accident victims are clearly better off under one system or the other:

> The liability option might be a "better deal" than the no-fault option for individuals who have very high limits of both health insurance and income continuation insurance. It might also be preferred by individuals who view auto accidents as opportunities to "win" a large sum of money, rather than as misfortunes that can result in permanent injury, and who believe that there is no real danger of losing, because Medicare, Medicaid, and welfare will provide for them if they do not win a liability award. The liability option would seem to be more attractive to people who like to sue, who dream of making a "big killing," or who distrust insurance companies. The liability option would pay relatively more money than the no-fault option if the accident victim involved is one who has suffered only moderate injuries, such as back strain, whiplash, or fracture. Moderate injuries

cause only limited medical expense and work loss, but they can cause a great deal of pain and suffering, compensable under liability insurance but not under no-fault insurance.

The no-fault option would be better for individuals who want to be sure that catastrophic losses will not outstrip their own coverages no matter how high. It would certainly be better for individuals who do not have very high limits of both health insurance and income continuation insurance. It would also be the preferred choice of motorists who do not want to be paid for their pain and suffering but who do want swift, sure, and complete payment of their accident-caused personal economic losses. This option would be more attractive to people who dislike gambling, who eschew litigation, or who distrust lawyers. In general, the no-fault option would assure relatively more money than the liability option to an accident victim who suffers very severe injuries such as a severing of the spinal cord or a serious brain concussion. The economic loss of such victims can easily outstrip the limits of the average liability insurance policy.

This is not an entirely fair or complete characterization of the factors that would influence individual choice between the fault and no-fault options. In particular, it ignores the fact that low-income and elderly individuals would prefer first-party no-fault because they are insuring lower than average prospective income losses. It also ignores the fact that high-risk drivers and drivers of heavy vehicles (like trucks) who cause disproportionate injuries to third parties (drivers of lighter vehicles, motorcyclists, bicyclists, pedestrians) relative to themselves would prefer the no-fault option because it insulates them from liability (and hence coverage costs) to negligently injured third parties. In this respect, like his third-party elective no-fault proposals, O'Connell's first-party elective no-fault proposals may also create significant adverse selection problems. Outside the auto accident context, these problems with elective first-party no-fault insurance schemes are likely to become more severe. For example, in the case of elective disability insurance for sickness, adverse selection problems caused by severe information asymmetries between insurers and prospective insured parties are likely to lead to a suboptimal provision of insurance. Wilson has shown that in some circumstances mandatory group coverage may make both high-risk and low-risk individuals better off. Similar, and indeed more serious, adverse selection problems might be anticipated in the case of elective first-party spousal desertion insurance, unemployment insurance, or retirement insurance, and may well cause private, voluntary insurance markets to unravel.

Implications

In reviewing the various residual roles that might be assigned to the tort system, at least with respect to personal injuries, in contexts where no-fault or social insurance schemes have been adopted, the strongest case can be made for some combination of conduct and damage thresholds—for example, intentionally, recklessly, or grossly negligently inflicted injuries leading to serious and permanent physical impairment. Retributive, corrective justice, and deterrence rationales for liability seem to present themselves in their most powerful form by focusing on conduct that entails a high probability of causing serious harm and therefore high expected costs. Where liability exists, victims would recover full tort measures of damages, including damages for pain and suffering, and in some contexts punitive damages. From an economic perspective, preserving tort claims in these circumstances would avoid the total forfeiture of whatever socially desirable injury-reduction incentives the tort system creates, particularly for parties who are especially sensitive to the costs entailed—employers, product manufacturers, physicians, and hospitals; individuals such as the drunken driver whose conduct, for example, renders a child paraplegic might also face full tortious liability. By preserving these residual tort claims, the role of the tort system as ombudsman, whereby the system acts as a check on the performance and assiduousness of regulators and other public enforcement agencies (monitoring the monitors), would also be maintained to some degree.

Notes and Questions

1. Does the prospect of high premiums affect your own driving behavior? Do high premiums result in fewer automobiles or drivers on the road? How might an empirical study differentiate between these effects?

2. An important feature of tort liability is its attempt to compensate victims of tortious behavior for their lost earnings (among other things). In contrast, no-fault systems with scheduled damages severely limit recovery for lost earnings.

Whose lost earnings are reflected in the insurance premiums charged to manufacturers of consumer products? Do you expect lower-income consumers to prefer lower prices or the prospect of more generous recoveries under products-liability law? Might these consumers be described as subsidizing higher-income consumers? How might manufacturers offer consumers a choice between lower prices and higher tort recoveries (for high-earning consumers)?

Proposals for Products Liability Reform:
A Theoretical Synthesis

ALAN SCHWARTZ

. . .

The Consumer Sovereignty Norm in Products Liability Law

The consumer sovereignty norm may govern cases of both actual and hypothetical consent. In the former case, well-informed, uncoerced consumers
actually consent to particular contract clauses. The consumer sovereignty
norm uncontroversially supports enforcement of these clauses; the various
moral theories to which Americans adhere respect truly consensual arrangements. In the latter case, affected consumers do not actually consent to the
contract clauses for which enforcement is sought. Consent could be lacking
because consumers are unaware of the clauses or their effects, or because the
clauses are required by the state. In this second case, consumer sovereignty
holds that courts should enforce only contract clauses to which well-informed,
uncoerced consumers would have assented: when actual assent is lacking,
courts should enforce clauses to which hypothetical consent is given. These
clauses can be of two types: default rules, which apply when the contract at
bar is silent, and required clauses, which courts impose on all contracting
parties.

Consumer sovereignty also is an attractive norm when consent is hypothetical. Initially, giving hypothetical consent the force of real consent is justifiable
on utilitarian grounds when the contract clauses that courts adopt as default
rules or rules of law would maximize the utility of affected persons. Clauses
achieving this result are identified first by supposing that persons prefer contracts that maximize their expected utility, and then by deriving the terms that
this preference best implies.

. . .

There are two candidates for the meta rule that should determine products
liability law: maximin and utility maximization. The former directs persons to
secure the best possible outcome in the worst state that could occur. People in
the present version of an "original position" would realize that the worst state
is to be injured and poor. To choose maximin would imply that the law should
provide complete compensation in all possible future states, so as to include
this worst state. Maximin is attractive to the very risk averse because its
criterion makes only the worst conceivable social states relevant to the choice
of rules. Less risk-averse persons would reject maximin because it can lead to

Abridged and reprinted without footnotes by permission of The Yale Law Journal Company and
Fred B. Rothman & Company from *The Yale Law Journal*, vol. 97, pp. 353–419 (1988).

quite costly laws. For example, . . . a full compensation rule would be likely to force even normally risk-averse persons to purchase excessive insurance.

People who are being asked to choose a meta rule for products liability law, rather than settling on a basic social structure, would probably reject maximin in favor of utility maximization. They would know that the worst outcome—incurring serious, completely uncompensated injury—is substantially mitigated by existing social safety nets. Hence, the strongest motivation for choosing maximin is absent. The costs of maximin, such as having to make excessive insurance payments in the form of high prices, also bear most harshly on the poor, who by definition have the least disposable income. Thus, maximin is not especially desirable even though one may turn out to be poor. Further, the utility maximization criterion does take ordinary risk aversion into account and implies the least costly rules. Consequently, people choosing in the circumstances presented here would prefer the utility maximization meta rule, the rule that the consumer sovereignty norm suggests. The rules derived by use of the consumer sovereignty method thus seem justifiable from an impartial point of view. It again appears that courts could use this method to establish default rules or the content of a strict liability doctrine.

A competing conception of tort law holds that the state should pursue corrective justice. According to this conception, victims should be permitted to sue those who have caused harm, unless defendants' actions were, in some morally relevant sense, privileged or otherwise justifiable. Like the consumer sovereignty norm, the corrective justice theory favors enforcement when uncoerced, well-informed persons actually consent, before any injury occurs, to a contract that imposes substantial risks on them. The central premise of corrective justice is that it is unjust to deprive someone of property or personal integrity against that person's will. This premise supports permitting persons to waive their right to compensation—to agree to bear risks—in appropriate circumstances.

In the absence of actual consent, the question is whether corrective justice theory implies different legal rules than the hypothetical consent rules supported by the consumer sovereignty norm. Corrective justice theorists seem to have devoted little attention to this issue. The common corrective justice paradigms involve cases in which the parties do not bargain—where no markets exist. Corrective justice and consumer sovereignty seem consistent in bargaining situations because the consumer sovereignty norm seeks to replicate the choices of the same persons whose full integrity and autonomy corrective justice theorists want to protect. Hence, the conclusions this selection reaches with respect to products liability law, which governs relations between parties interacting in markets, should be taken as consistent with corrective justice view of tort law, at least until those views are given further expression.

. . .

A firm that compensates consumers for the harms its product causes will reflect the expected compensation cost in the purchase price. An element of the price thus is an insurance premium, whose size ideally varies with the amount of "coverage" against loss that consumers demand. The provisions of the optimal contract respecting defect risks therefore will reflect the amount of insurance against these risks that consumers prefer. It is customary to identify this amount on the basis of three principal assumptions. First, a consumer will choose an insurance contract that maximizes his expected utility. Second, consumers' utility functions are "state dependent": They depend on the state of affairs arising after purchase of the product. The consumer's utility is lower in the state of the world in which the product is defective than in the state of the world in which it works perfectly. Third, firms offer insurance at actuarially fair prices; the amount of their premium equals the expected value of the risk against which the person insures.

Insurance . . . shifts wealth from the state of the world in which the marginal utility for money is relatively low—the state in which no injury occurs—to the state in which it is relatively high—the state in which an injury happens. This analysis predicts, therefore, that consumers will insure against those risks whose materialization would increase their marginal utility for money. Identifying these risks can be a difficult empirical inquiry, but some aspects of the question seem obvious.

Certain forms of loss have two significant properties: They increase the marginal utility of money, and they are replaceable, dollar for dollar, by insurance. An accident that causes a consumer to lose wages creates such a loss. The consumer's marginal utility for wealth is higher in the state in which such an accident occurs than in the state in which it does not because the consumer has less wealth in the former state than in the latter and so will use marginal dollars to satisfy more urgent needs, such as for shelter or medicine. Further, since the accident causes only a monetary loss, it is fully replaceable by insurance. In consequence, the consumer could equalize his marginal utility for wealth across states of the world by purchasing full insurance against losses—for example, of wages—that both increase the consumer's marginal utility of money and can be completely erased by monetary payments. Losses with these two properties constitute what lawyers call "pecuniary" loss or harm; insurance theory thus predicts the existence of substantial private insurance against pecuniary loss. The wide use of major medical and disability insurance is consistent with this prediction. Therefore, the default rule should require firms to compensate consumers fully for pecuniary loss.

A more difficult question is whether, and to what extent, consumers would insure against other forms of loss. It sometimes is difficult to know whether accidents will increase or reduce a person's marginal utility for money. Consider a business executive who runs recreationally and who loses a foot in an accident. Suppose that she insured fully against her "replaceable" losses, such as medical expenses and temporary lost wages. Apart from these losses, the injury could increase the marginal utility of money for this consumer if it caused her to substitute travel or the symphony for running because these activities are more expensive. Her marginal utility could fall, however, if she substitutes reading for running. In the latter case, the consumer not only would want no insurance, but would prefer to shift dollars from the injury to the noninjury state—the reverse of the cases above—by betting against the accident happening. It is therefore difficult to say, as a general proposition, that people will insure against events that would only induce them to substitute other activities for those activities that accidents preclude.

. . .

Some evidence suggests that people may want coverage against pure suffering. Consumers, for example, sometimes purchase accidental death and dismemberment insurance that protects against particular dramatic events whose occurrence is easy to verify, such as the loss of a leg. A partial motive for this insurance probably is to receive dollars that in some sense will ease the mental pain of these traumatic losses. On the other hand, the premium volume for this insurance is so small that the insurance cannot reflect a large pain and suffering component. And people do not routinely insure against the loss of children, even though such losses cause great emotional pain. It thus is difficult to infer whether people want insurance against pain and suffering losses from observing what they actually buy.

In addition to these empirical uncertainties, other factors also argue against including insurance for pain and suffering in a compensation-based rule. Consumers would prefer less than full insurance against accidents that cause only mental pain, even when these accidents would increase the marginal utility of money, because of "income effects." Recall that such accidents increase the marginal utility of money only when victims will purchase expensive substitute activities to assuage the utility losses from suffering. Such substitutes are sought in the states of the world in which accidents happen. The demand for most goods and services has positive income elasticity; people increase their consumption as their incomes rise. Because accidents make people poorer in a utility sense, people will purchase lesser amounts of substitute activities in "accident states" than they would have purchased if they had not been injured but instead had to give up goods that they then valued as much as they valued not suffering. Informed consumers will anticipate wanting lesser amounts of substitute activities in accident states than they would otherwise want, and so will make provision to buy less. In other words, consumers will not purchase full insurance. Therefore, the ideal legal rule regulating accidents causing mental losses that increase people's marginal utility for money would award victims partial damages. These damages would reflect the partial insurance consumers would want ex ante. The level of partial insurance consumers would want varies among people, however, so the law's manageable choices are full insurance—overcompensation—or no insurance—undercompensation. The issue is on which side to err. The remaining factors that should influence courts in deciding what consumers want imply erring on the side of undercompensation.

Intuition suggests that people would want to buy slight or no coverage against purely mental harms. As the runner illustration above showed, there is no good reason to suppose that, apart from causing pecuniary harm, accidents commonly increase persons' marginal utility for money. In addition, to buy mental loss coverage is, in effect, to sacrifice considerable present wealth in the form of insurance premiums to consume expensive vacations that will assuage whatever emotional distress accidents may cause. In the absence of evidence that spending money is a typical, or even common, response to grief and suffering, this motive for insurance seems unlikely. Finally, pain and suffering losses are difficult for firms to anticipate and verify. The likely response of firms to these problems is to charge high prices for the coverage. These prices make pain and suffering insurance a bad buy for most people, whether it is sold by manufacturers or by insurance companies. These three factors together imply that the more purely mental the loss, the less likely a consumer will want to insure against it.

To summarize, the optimal contract concerning product-related risks would pay firms to provide insurance against the core pecuniary losses that defective products could cause, such as lost wages, medical expenses, or property damage. It is unlikely, though not certain, that this contract would require any insurance against what now are sizable and common elements of the standard products liability judgment: pain and suffering and emotional distress. The appropriate default rule, then, probably should allocate the risk of incurring pecuniary harm to firms, and the risk of incurring nonpecuniary harm to consumers. The appropriate prescriptive rule under strict liability must be consistent with the default rule if the animating norm is consumer sovereignty. Hence, given current evidence, the aspect of strict liability that prohibits firms from shifting the risk of incurring nonpecuniary harm to consumers cannot be justified by reference to the goal of compensating consumers for harm.

The Product Safety Decision

The insurance analysis above asked how the optimal contract should allocate the risk that harm will result from a firm's breach of its obligation to produce a safe product. The analysis here asks how an optimal contract would specify this safety obligation. It is helpful first to ask what an efficient safety obligation is. For convenience, suppose that consumers cannot affect safety but that firms can. Assume that if a certain accident were to occur, it would reduce consumer A's utility by an amount equal to the loss of $5,000. Consumer A would pay the firm up to $50 to reduce this accident's probability by one percent. An efficient contractual safety obligation, then, would induce the firm to make its product this much safer if the cost of doing so would be below $50, but not if it would cost more than this. An explicit contract clause therefore could require the firm to make "cost-justified reductions in the risk of harm." If these words were absent, the court should impose this obligation as the default rule.

Informed parties would seldom choose such a rule explicitly and would often contract out of the rule if it were to apply in default. This is because . . . full application of this rule requires the court or jury to ask how much particular consumers would pay ex ante to reduce risks of harm by specific amounts, information which is extremely difficult to reconstruct ex post. Parties commonly respond to this difficulty by specifying the quality obligation. Contracts exclude implied warranties and then make express warranties or explicit prom- ises. Such clauses are the equivalent of the generic risk-reduction clause de- rived above, because informed consumers will only pay for safety levels or features that cost less than the risk reductions that consumers regard as worth- while, and firms will make only those safety investments whose costs can be recovered from the consumers.

This analysis has several implications. First, the judicial role should be limited to enforcing the parties' contracts respecting product quality; there is little need to create default rules because contracts seldom are silent about quality. Second, courts should be reluctant to enforce contracts when consum- ers are uninformed; it is difficult to regard particular express warranties or quality promises as reflecting optimal tradeoffs between cost and risk reduc- tion when consumers lack the data to assess risks. Thirs, significant difficulties arise if consumers actually are uninformed. In this event, the consumer sover- eignty norm directs courts to create and enforce the contracts that informed parties would make. As shown above, there are two possibilities. First, courts could devise a version of the generic risk-reduction rule, assessing a firm's safety performance against the criterion of consumers' willingness to pay for safety. This could be done by choosing rules for damages that would induce firms, when making products, to equate the marginal cost of investments in safety with the marginal willingness of consumers to pay for safety. Second, courts could specify the quality levels below which products cannot fall or the safety features products must contain. Courts in products liability cases often do the latter: Design defect litigation seeks to specify minimum quality levels and required safety features. Both this effort and the generic risk-reduction method seem undesirable policy responses, for they presuppose that courts and juries can assess consumer tradeoffs between safety and risks, and there is no reason to believe that courts possess this ability to any great degree.

Strict Liability and Market Failure

The Strict Liability Doctrine

Strict liability is a judicially created doctrine that regulates the risk allocation aspects of transactions between manufacturers and consumers and between manufacturers and employees. Its rationale can be summarized by three propositions. First, the law should serve two functions: to create incentives for manufacturers to produce safe products and to ensure the provision of a peculiarly private form of social insurance, supplied by firms. Second, these two functions are complementary, not conflicting; both imply strict manufac- turer liability. Third, private markets cannot provide appropriate safety incen- tives or sufficient insurance, largely because the consumers and employees who operate in them are imperfectly informed. Hence, courts should impose the risks of product accidents so as best to serve the law's safety and insurance functions.

. . .

The contract-proscribing aspects of strict liability drive a wedge between the law's compensation and safety goals. The compensation goal dictates that firms should be held strictly liable only for pecuniary harm, because this is all the insurance consumers probably want. The safety goal suggests that firms should be liable for more harm than this in order to create appropriate incentives for firms to invest in safety. If firms are liable for more harm, however, they will reflect this liability in their prices, and consumers will be forced to purchase excessive insurance. Second, the product regulation facet of strict liability—in particular, the design defect aspect—reflects judicial attempts to set quality standards, a task, it was suggested above, for which courts may be ill-suited. . . . Strict liability is justifiable if firms routinely use suboptimal contracts. [Contracts will be optimal if consumers can shop, or search, for desirable terms, and if they can price these terms and contractual risks precisely.]

. . .

Though the data show consumer search to be fairly extensive, especially for expensive items, shopping is costly. Also, shopping costs vary directly with the number of product "attributes," including contract clauses, that consumers inspect. Since complex products such as cars often have several attributes that consumers find salient, it seems sensibly cautious to suppose that consumers search too little to ensure that contract clauses perfectly reflect their preferences. Firms may exploit the power that a lack of shopping confers on them to raise prices above competitive levels.

. . .

There seems little reason to believe that firms also exploit insufficient consumer shopping by "degrading" contract "quality"—by not selling insurance at all. The question facing a firm is whether it will do better—that is, maximize profits—by selling the contract clauses consumers want at excessive prices, or by selling clauses that consumers do not want at excessive prices. If consumers have a noticeable preference for a particular clause—in economic terms, have a significant willingness to pay—firms will do better, other things being equal, by satisfying this preference: the greater each consumer's willingness to pay, the fewer consumers each firm needs to recover costs and make profits. On the other hand, if the cost to a firm of supplying a particular clause is high in relation to consumers' willingness to pay for it, the clause may not be supplied. There is little reason to believe that cost considerations outweigh consumer preferences for standard insurance coverage against economic losses—after all, a great deal of this insurance is now sold. The same analysis also applies to clauses specifying quality obligations. Therefore, the likely response of firms to a lack of consumer shopping is to offer consumers the contract clauses they prefer, though at excessive prices.

To summarize, the rejection of freedom of contract that strict liability reflects is difficult to justify on the ground that imperfect information exists, if imperfect information is taken to mean that consumers do not know what their contracts say or do not search sufficiently for the contract clauses they prefer. The contract language problem is best remedied by legislative or administrative actions to improve contract readability, not by episodic judicial bans of particularly hard-to-read language. The consumer search problem is irrelevant to products liability altogether, since firms commonly exploit the existence of high search costs through price, not through inefficient warranty clauses. Further, the appropriate state response to the existence of high search costs is to reduce them, which courts cannot do. Thus, strict liability does not appear necessary on the grounds that the first two information assumptions made above may not be satisfied. Strict liability may be justified, however, if the third assumption—that consumers know risks of harm—is false.

. . .

[As for pricing risks,] the data are best approached by considering five null hypotheses: (1) Consumers misperceive the true level of risk they face; (2) consumers' risk perceptions change inappropriately when true risks change; (3) consumers accurately perceive risk levels (the reverse of (1)); (4) consumers accurately perceive changes in risk levels (the reverse of (2)); (5) consumer risk perceptions can be made acceptably accurate by disclosure. The discussion below focuses on hypotheses (1), (3), and (5), because apparently no evidence at all has been collected on hypotheses (2) and (4). This is troubling, for without such evidence conclusions as to the desirability of strict liability are inevitably suspect.

The evidence fails to show that consumers misperceive risk levels to the extent that undesirable equilibria exist. Though one well-known study showed that farmers underinsured against flood damage, later studies drew opposite conclusions about whether people correctly perceive risks. In one study, housing prices correctly reflected earthquake risks, being appropriately lower when these risks rose. In another study, housing prices appropriately decreased with increases in the probability of defects. Recent survey data showed that consumers have a substantial willingness to pay to reduce the injury rate from common household products such as drain openers; this willingness implied that survey respondents valued a hand burn from a chemical drain opener at $120,000, for example. This figure seems excessive, and thus suggests either that persons overestimate the costs of harm or are very risk averse when the issue is personal injury. These explanations in turn imply that people will not be underinsured and will put strong pressure on firms to produce safe products. Further, studies relating wages to risks of particular jobs show that workers appreciate risks to life and health and exact substantial wage premiums for bearing them; the greater the threat of injury and death from a job, the higher the wage demanded by workers. This evidence implies, perhaps faintly, that markets in accident risks work adequately. Finally, consumers routinely purchase extended warranty coverage when buying expensive items such as cars and computers. This practice suggests an awareness of risk and a proper response by markets to this awareness. In sum, evidence drawn from surveys and actual market behavior more strongly supports the view that consumers are informed than the view that they are ignorant.

Commentators sometimes claim that the cognitive psychology literature, which shows that people in laboratory experiments make systematic errors when processing information, supports the claim that consumers behave optimistically in markets. The references to this literature seem overdrawn. If the psychologists had a general theory about how people make decisions, and the theory generated predictions about what people will do in various circumstances, their experiments could be regarded as testing these predictions. Then, if the predictions actually were confirmed in the laboratory, it would be plausible to claim that the theory's view of how people decide matters in real life is correct, at least until the facts prove it false. Psychologists lack such a theory, however. They have instead a large set of observations about how experimental subjects behave. The external validity of this data is now in controversy for two related reasons. First, decision strategies that are inappropriate to laboratory settings often could be appropriate in real life. Second, the tasks people are assigned in laboratories sometimes seem too artificial to support a strong inference that persons routinely misperform important tasks in their actual lives. Consequently, it seems premature to make this experimental data the factual premise of important legal rules.

Further, the most relevant psychological claim and the data supporting it suggest that consumers are likely to be pessimists to the extent that they err in a particular direction. The claim holds that people process information in the order of its salience to them. This strategy may be misleading because salience supposedly is determined by application of the "availability heuristic," which directs people's attention to information that is vivid and easily summoned to mind rather than to more reliable statistical data. Because negative information about products is often alarming, it has high salience. This psychological claim implies, and supporting evidence predicts, that consumers will attach disproportionate weight to negative data, and thus overreact to product-related risks. Pessimism as to risk levels, if it exists, is less troublesome than optimism because it is less stable; firms have an incentive to dissipate it, since pessimistic consumers tend to purchase less. Therefore, the psychological studies suggest, if anything, that consumers are likely to err in the direction that would cause them the least harm.

The data also suggest that people can be taught to forego bad decision strategies. Further, there is considerable evidence that people respond rationally to the provision of information about risk. Nor are people likely to "overload" when given a substantial amount of such information. Disclosure solutions thus seem feasible.

Recall now the null hypotheses set out [previously]. The evidence supports rejection of hypothesis (1); it has not been proven that consumers misperceive risk levels. The evidence fails to support rejection of hypothesis (5); disclosure solutions and education seem promising ways to cure consumer errors. The evidence as to hypothesis (3), that consumers perceive risk levels accurately, is less clear. It apparently is more difficult to justify rejecting hypothesis (3) than to justify accepting it, but the issue is close.

. . .

Since this selection's purpose is as much clarification as recommendation of reforms, it will conclude this discussion with the observation that the answer one gives to the question whether courts should impose strict liability or not depends on . . . the values one is least willing to sacrifice—supposing the facts could leave a reasonable person in doubt. . . . If strict liability is chosen, the consumer sovereignty norm should determine the rules that constitute the strict liability doctrine. . . .

Proposals for Reform

A Substitute for Defect Tests: True Strict Liability with Contributory Negligence

The Substitute Described. . . . Strict liability with a contributory negligence defense is attractive because it yields efficient results. Assume that consumers know what safe use entails and that total accident costs would be optimally reduced if both consumers and firms exercised care. A consumer who takes care incurs no accident costs; the proposed legal rule holds firms liable unless consumers are careless. On the other hand, a consumer who fails to take care bears all accident costs; contributory negligence is a complete defense. Since a consumer's failure to take care will constitute contributory negligence only if the consumer's expected costs of care are lower than the expected costs of the accidents that care would avoid, these informed consumers will by and large adopt the cost-minimizing strategy of taking care. Therefore, firms will incur all accident costs their designs cause. The cost-minimizing strategy for firms in this circumstance is to make optimal investments in safety—to invest in safety until the cost of further investments equals the gain in the reduction of expected accident costs. Because it induces both consumers and firms to reduce accident costs optimally, strict liability with contributory negligence is efficient.

This analysis makes strong assumptions about what consumers know. It is an open question whether its conclusion still holds when these assumptions are relaxed. To pursue this question, one should realize that what constitutes due care for a consumer is often a function of the firm's safety efforts. For example, if a car has good brakes, driving 60 miles an hour on a highway is not careless, although driving 100 miles an hour might be; if the car has bad brakes, driving 40 miles an hour might be careless. If contributory negligence is the failure to take the care that the circumstances warrant, a consumer who fails to drive slowly when the brakes are bad would be contributorily negligent. The analysis above argued that consumers will behave optimally in light of the actions taken by firms. It thus assumed that consumers know what steps firms have taken—that is, consumers know just how safe or dangerous products are. The assumption that consumers have perfect information is illegitimate, however; if consumers know how safe products are, there is no need for any form of strict liability, yet we suppose that some form of strict liability is justified because information is imperfect.

An appropriate definition of contributory negligence will cure this difficulty: Contributory negligence should be defined as a consumer's failure to take due care when using a product that is optimally safe. In the illustration above, this rule would permit consumers to drive 60 miles per hour, but not 100 miles per hour, regardless of the actual state of the brakes. A consumer who drove 60 miles per hour and was injured because the brakes were bad would not be contributorily negligent, and the firm would bear all costs. If consumers do not know how safe products are but do know how to use safe products, they will take the cost-justified care that safe products require for the reasons given. Consequently, firms will face all accident costs and will respond by taking due care. Hence, the appropriate contributory negligence rule solves the problem of consumers misspecifying risks.

Current law deviates in two ways from the strict liability with contributory negligence solution. First, a sizable minority of states allow only a contributory-negligence defense that resembles assumption of risk. A consumer is required only to take the care that her actual knowledge of product safety requires. She is negligent if she drives 60 miles per hour knowing the brakes are bad, but is not negligent if she is ignorant of the true state of the brakes, yet speeds. This rule is inefficient. It creates an incentive for consumers to be careless and thus leads to excessive accidents and excessive investments in safety by firms to make up for consumer carelessness.

Second, jurisdictions that do require consumers to behave non-negligently commonly use comparative negligence. Under such a regime, consumer misbehavior reduces, but does not bar, recovery. In the products liability context, a consumer is negligent when she fails to take the care that is required given an optimally safe product, and the firm is "negligent" if the product fails the risk/benefit test. When both parties are found negligent, the jury apportions liability between them. Comparative negligence, however, is an undesirable rule in products liability law.

To see why, realize first that comparative negligence is efficient if firms bear no accident costs when they are not negligent. In this event, firms can and will avoid all losses by complying with the due care standard, supposing it to be correctly set at the level where the marginal cost of safety equals its marginal benefit. Consumers will then bear all losses and also take due care because they cannot do better than avoiding only those losses it is worth it to them to avoid. Interestingly, the result that comparative negligence induces optimal behavior has been shown to hold in the case where both parties are negligent and the accident cost is apportioned between them according to some sharing rule. A comparative negligence regime thus is unobjectionable, and perhaps on equity grounds desirable, when juries can assess the negligence of firms. The risk/benefit test is supposed to enable juries to do this in products liability cases, but this test is unworkable. If it is abandoned, comparative negligence must be abandoned as well because, when a consumer is negligent, there will remain no way to decide whether the firm should bear any portion of the loss; inquiries into the behavior of the firm are foreclosed ex hypothesis. In a regime of true strict liability for firms, therefore, an optimal amount of safety will be produced only if contributory negligence is a complete defense to liability.

This legal regime will be more administratively efficient than current law because true strict liability greatly reduces the information demands on juries. The substantive issue in litigation will be: What constitutes optimal consumer care given that the product is safe? This question is much easier for juries to answer than the questions posed by the risk/benefit test.

Some may object that this question is as difficult for juries as those they now face. The objection is this: A consumer is contributorily negligent when the costs to her of behaving carefully are less than the expected accident costs that care would avoid. Yet consumer care costs in considerable part are opportunity costs: the detriment to a consumer of not exceeding the speed limit is the pleasure foregone from driving fast or the unhappiness of arriving late. Juries, the argument goes, are no more able to assess these opportunity costs than they are to measure consumer benefits from various designs. Hence, the solution of strict liability with a contributory negligence defense simply substitutes one insoluble measurement problem—determining consumer care—for another—applying the risk/benefit test.

This objection is unpersuasive because mental states have never been relevant in determining human actors' negligence. Considering mental states not only introduces a severe measurement difficulty, but also raises the utility monster problem. Thus, a tort defendant cannot claim that driving 100 miles an hour in a school zone was not negligence because she derived enough pleasure from frightening children to outweigh the utility losses to her victims and their parents. The negligence of real persons—here consumers—is assessed by asking whether they failed to follow directions, ignored warnings, or took actions that most people would regard as being more risky to others than the actions were worth. Whether a consumer has acted in this way in a particular case poses an easier question than those posed by the risk/benefit test. Accordingly, the suggested reform will have the administrative advantages claimed for it.

There is, however, a problem in putting such stress on defenses. The analysis above supposed that consumers always know what constitutes due care in connection with optimally safe products. Consumers sometimes may be less informed. For example, the appropriate dosage for a particular drug is seldom self-evident. This difficulty is met today by requiring firms to provide warnings.

· · ·

The . . . [arguments in this selection] suggest that courts should reestablish a contributory-negligence defense and be more reluctant than they now are to let juries speculate about the adequacy of warnings. . . . To test the plausibility of the proposed reform, consider how it would resolve four typical cases. In case (a), a consumer purchases a Volkswagen Golf and is injured when, with no negligence on her part, the Golf crashes into a tree. She sues, claiming that if the car had been composed of much heavier metal, she would not have been injured. Under the proposed rule—and current law—the consumer loses; she knew or should have known that Volkswagens are less crashworthy than Rolls Royces, and therefore assumed the risk.

In case (b), a magic metal which existed when the Volkswagen was sold would have made the Volkswagen as safe in crashes as today's Rolls Royce, though the Volkswagen would have weighed no more than normal Volkswagens. The manufacturer did not use the magic metal because it would have added $500 to the car's costs and cars made of magic metal are much more expensive to repaint and hammer out in the event of scrapes and dents. Under current law, the manufacturer would be liable only if Golfs with magic metal did better on the risk/benefit test than Golfs without it. Under the proposed test, the manufacturer would be liable unless it gave an adequate warning that informed consumers of the additional risk of not using the magic metal. The consumer did not assume the risk in this second case because, without such a warning, she was unaware of the full set of options provided by the market.

In case (a), consumers are as well informed as the firm about the design choice. They know that Rolls Royces are safer but more expensive than Volkswagens. Hence, the VW maker's design choice would be optimal, for it had to satisfy informed persons. In case (b), the consumer is not similarly informed. Ordinary people lack the information and expertise to make the technological choices involved in metal selection. Hence, the firm's design choice cannot be presumed to be optimal.

True strict liability is justified in case (b) for three reasons: (i) If use of the magic metal is optimal, given that consumers behave nonnegligently, the firm will be induced to use it; (ii) If use of the metal is not cost justified, the price of Golfs nevertheless will more accurately reflect their accident costs, thereby better informing consumers of the risk of driving (just as a warning would); (iii) The administrative costs of trials would be reduced and their results made more predictable because the only issue would be the adequacy of the warning (if one were given). The law would not demand a complex risk/benefit test.

In case (c), magic metal is invented after the Golf is sold. A majority of jurisdictions today, applying the state-of-the-art defense, would exculpate the manufacturer on the ground that consumers could not expect products to contain features whose use was infeasible when the products were made. Scholars criticize this rule as creating insufficient incentives for firms to develop safety improvements. Two alternative rules may be considered. One rule would hold that liability depends on the effectiveness of a firm's research program. If a firm had an optimal safety research program and did not discover a safety improvement, then it would not be held liable. Such a program would require a firm to equate the marginal costs of safety research with its marginal gains. These gains would be the increased profits resulting from lowering expected accident costs through improved safety information. The other alternative would be to hold the firm liable absolutely, that is, liable regardless of when knowledge of the safety improvement could have been learned.

If courts can accurately set the due-care research standard required by the first rule, the two alternative rules create identical and optimal incentives for firms to do safety research. Should firms be held liable regardless of how much research they do, as under the second rule, they will conduct research until the marginal costs of further research equal the marginal gains from increased safety information. The first rule ideally sets the due-care standard at the point at which the marginal costs and gains of further research are equal. Hence, if firms believe that the due-care standard will be accurately established and applied, they will do as much, and no more, research under it as under a rule of absolute liability. Equating the marginal costs and benefits of research programs, however, seems more difficult than an ordinary negligence question. If some inaccuracy in courts' decisions on this matter is assumed, choosing between absolute liability and negligence as a replacement for the current state-of-the-art rule is difficult. On the one hand, absolute liability is easy to administer and creates the correct private incentives for research. On the other hand, absolute liability creates considerable uncertainty; firms cannot predict their future liability. The incentives created under this standard also may not be socially optimal. This is because research is a public good that firms doing the privately optimal amount of research are likely to underprovide. Thus, inaccuracy in the judicial application of a negligence rule would be desirable if it induced firms to research somewhat excessively. There is no simple way to resolve these competing considerations, but the probable likelihood of overcompliance with a negligence rule argues for that rule's desirability.

. . .

Case (d) is case (b) with multiple actors: The Golf lacked the magic metal, though the firm knew of it, but the consumer was injured when her car shot off an insufficiently banked curve. If the consumer was nonnegligent, she should not bear the loss because she is presumably uninformed and both injurers, the firm and the relevant county, could reduce the likelihood of such losses by warning or redesign. Under current law, the plaintiff could sue either injurer and collect her full loss from one of them, or could request that the loss be apportioned between the two injurers. Whether a particular injurer who is sued for the full loss can collect any part of the judgment from the other is a difficult legal question. There was no contribution among joint tortfeasors at common law, but this rule has been altered by statute in most states. Courts today vary in allowing an injurer to obtain contribution (or indemnity) under these statutes.

The appropriate rule for apportioning the loss would be to assess the injurers on a comparative basis. One should realize that the county responsible for roads in this illustration is subject to a negligence rule. Under the strict liability contributory-negligence solution proposed above, the manufacturer has an incentive to invest in safety until the marginal cost of doing so equals the marginal gain—which the firm will likely calculate not as benefits to consumers but as the reduction in its expected accident costs resulting from additional expenditures on safety. A court can treat the failure of a firm to optimize safety according to this rule as negligence, in the spirit of the Learned Hand test, because it can retrieve the variables the firm used in its calculations. Since both actors in this illustration can be subject to a negligence rule and comparative negligence is efficient, a satisfactory rule would allow the consumer to sue either the county or Volkswagen, with comparative negligence governing any claim between the nonconsumer parties.

The proposed solution of strict liability for all design harms, with contributory negligence and assumption of risk defenses, would replace current efforts to specify the firm's quality obligation through adjudication with a rule that holds a firm liable whenever its design causes harm. This new rule seems faithful to the law's safety and compensation goals, resolves most cases relatively easily, and should make the law less arbitrary in its application. . . .

Conclusion: Reform and the Products Liability Crisis

Modern products liability law subsitutes tort regulation of defective product problems for the contract regulation of the past on the ground that consumers are imperfectly informed. This shift in the legal treatment of defective products is difficult to justify because the evidence that consumers misspecify risks in ways that disadvantage them is weak. Consequently, those who believe that private choice should be restrained only if good reasons have been established for doing so should strive to reverse the abandonment of contract for tort. Decisionmakers whose concern is to compensate victims unless there are good reasons for not doing so could justifiably reject this conclusion, since the evidence that consumers are well-informed is persuasive but not compelling. Even so, tort regulation of defective products could be much improved as measured against the contractual norm that the state should provide consumers with the sales contracts that they would choose if they were better informed. According to this norm, courts have made three major errors. First, they sometimes have responded to the presumed existence of imperfect information by directly regulating product quality in design defect cases. This regulation is unworkable, and should be replaced by a rule that holds firms liable whenever their products cause harm. Second, courts have relaxed consumers' obligation to take care in the mistaken belief that meeting this obligation requires information that consumers lack. The law should restore consumers' obligation to be careful because consumers are competent to fulfill this obligation, and because holding firms liable unless consumers are careless will produce the degree of safety that well-informed consumers would prefer. Third, courts misconceive the relation between tort law's compensation and deterrence goals. Compensating consumers fully for pecuniary harm satisfies both of these goals, but compensating consumers for nonpecuniary harm does not; informed consumers choosing ex ante would refuse both the extra insurance and extra safety that strict liability for nonpecuniary harm now requires them to purchase. Hence, firms should be strictly liable only for pecuniary harm.

Notes and Questions

1. What is meant by strict liability for defective products? What if, while climbing a ladder after taking normal steps to secure its footing, A falls and injures himself?

Does a rule of strict products liability mean that the manufacturer of the ladder is liable? What if lightning causes A to fall from the ladder?

2. If two manufacturers are involved in a single accident, does negligence play any role in a strict liability regime? If, for instance, two cars skid on an icy road through no fault of the drivers, how should liability be allocated? Is there really such a thing as strict liability for products?

3. Another way to go about arguing for less insurance (associated with the purchase of products) is to show that consumers rarely buy such insurance on their own. Can you think of any situations in which consumers voluntarily choose and pay for a regime in which they are fully reimbursed for injuries or other losses? Can you generalize as to when such coverage is available for purchase?

4. Perhaps markets respond to marginal consumers rather than to average consumers. Can you construct an argument that the law should require insurance—that is, provide for generous product liability—because there is reason to satisfy the average consumer and not the marginal consumer?

Some Reflections on the Process of Tort Reform
ROBERT L. RABIN

. . .

The Lessons of History

The literature of tort reform is notably ahistorical. Perusing the multitude of commission reports, magazine articles and newspaper accounts devoted to the "crisis" in tort law, one might well conclude that there was once a Golden Age of Tort, still clearly identifiable not so long ago, known as the Negligence Era. It was a period in which the principle of liability based on fault provided a sound moral and economic basis for deciding unintentional accident cases. For nearly two centuries, the fault principle reigned supreme. The horse and buggy may have been replaced by the ubiquitous automobile, cottage industry may have lost out in the competitive struggle to the mass producers of a vast array of consumer products, and the family doctor may have given way to the high-powered specialist, capable of identifying and treating a far wider array of ailments. But until relatively recently, the literature would suggest, accident law—through the due care formula—provided a serviceable system for compensating unintentional harm. Whatever changes modernity had wrought, the foundations of the system remained intact: the two-party structure of tort litigation, the relative ease of determining causation, and the intrinsic fairness of the fault principle.

Copyright 1988 San Diego Law Review Association. Abridged and reprinted without footnotes from 25 *San Diego Law Review* 13 (1988) with permission of the *San Diego Law Review.*

As any serious student of tort law knows, this image of a universally acclaimed fault principle, done in by runaway juries and the siren song of enterprise liability, is a considerable distortion of the historical picture. Deep-seated dissatisfaction with accident law dates back at least to the rise of the railroad as an engine of destruction, and peaked again as a consequence of the second revolution in mass transport—the centrality of the ever-present automobile.

But what can be learned from these earlier experiences? Do the chronicles of past tort reform movements help to delineate the character of more recent developments? Are the prospects for achieving change in the system clarified by examining earlier efforts to overhaul tort law? The would-be tort reformer is hardly in a position to set his or her sights ambitiously (comprehensive reform) or modestly (incremental reform) without reflection on these questions.

The evolution of the workers' compensation movement is, in a sense, a familiar story. The period following the Civil War marks the beginning of rapid demographic change and industrial growth in this country. The railroads spread throughout the land and new industries were founded at an unprecedented pace. A labor force harnessed to the machinery of industrial growth was soon found to be seriously at risk. In the last quarter of the nineteenth century, workplace injuries became a common feature of the landscape of manufacture and transport. As consequence, a persistent and rising number of claims for redress were brought before the courts.

Characteristically, the initial judicial response was quite conservative. The courts denied many claims, relying heavily upon a trinity of defenses to the negligence action—contributory negligence, assumed risk, and the fellow servant rule—which created a veritable obstacle-course to recovery for workplace injuries. But the toll of industrial accidents grew ever greater and the claims for relief became more insistent, and the courts, in classic common-law fashion, began to soften their unyielding position. The wall of hostility became riddled with qualifications: some courts adopted a vice-principal exception to the fellow servant rule, some fashioned a safe place restriction on the assumed risk defense, and so on. Qualifications built on exceptions, and soon the law of industrial injury was characterized as much by ambiguity and uncertainty as by harshness. In this milieu, not only did the tide of claims continue to rise, but the attendant administrative costs crested as well.

Shortly after the turn of the century, in reaction to this unsatisfactory state of affairs, the workers' compensation movement swept the country. Between 1910 and 1921, 42 states passed industrial injury legislation, replacing tort law with an administrative system affording compensation for accidental injuries arising on the job.

Such is the story of workers' compensation narrowly told: A pioneering effort to replace tort law with a no-fault compensation scheme. In fact, however, there is a more embracing version of the workplace injury story that identifies workers' compensation as one strand in a broader pattern of legislative activity associated with the Progressive Era. Without attending to this story, the present day significance of the historical record is obscured.

To begin with, a wide-ranging commitment to workplace reform was a central tenet of Progressivism. At the height of the movement, there was a flood of child labor and women's rights legislation aimed at regulating the maximum hours and workplace conditions afforded to groups perceived to be particularly vulnerable. Similarly, detailed regulatory limitations were enacted in a wide variety of occupations in which special health and safety concerns existed—from miners to cigar makers. Indeed, health and safety concerns occupied a central position on the Progressive agenda entirely apart from the employment nexus; witness the enactment, in quick succession, of landmark congressional regulation in the area of food and drugs, as well as meat inspection legislation.

Workers' compensation represented a convergence of these two core Progressive concerns: it addressed a perceived inequity in the treatment of laborers, and, at the same time, responded to the concern about health and safety conditions. In addition, the industrial injury problem was precisely the kind of issue that evoked the bedrock Progressive ideal of faith in expertise. Work-related injuries, as a societal concern, were regarded as highly amenable to rational administrative treatment through preventive measures; specifically, the application of expert systems analysis to the production function. And, at the same time, administrative competence was considered the touchstone to revised strategies of reparation for injury victims, such as rehabilitative measures that were outside the remedial structure of the courts.

Viewed in the broader context, in which the states became laboratories for Progressive experimentation in social welfare, the impulse to engage in fundamental change—outright abolition of the tort remedy, and, concomitantly, rejection of the commitment to liability based on fault—becomes more understandable. If criticism of the treatment of injured workers had been grounded exclusively in concerns about efficiencies in judicial administration and inconsistencies in doctrinal treatment, reform initiatives most likely would have been responsive in kind—unification of doctrine, reassessment of the allocation of power between judge and jury, and similar incremental strategies. But in the context of a far broader re-evaluation of the meaning of freedom of contract in the employment relation, and a corresponding reassessment of the state's police power obligations in the sphere of public health and safety, there was no reason to treat a somewhat arcane judicial commitment to fault liability as an incontrovertible first principle. . . .

Strategies of Reform

It is far from clear that *any* reform strategy based on incremental improvements in the substantive and remedial character of tort law will have a significant impact.

Consider initially the substantive law of tort. The very first recommendation of the Attorney General's Tort Policy Working Group (Working Group), after 60 pages devoted to documenting a tort system out of control, is that the fault principle be retained as the basis for liability. By all accounts, this is an odd opening initiative. Aside from the products liability area, the fault principle in fact continues to reign supreme throughout accident law. Indeed, the most innovative doctrinal developments during the past two decades have been precisely the establishment of a far more robust fault principle, rather than the overthrow of the negligence standard of liability. In one domain after another, the hallmark of recent years has been the enunciation of an unfettered duty of due care.

Perhaps in recognition of this development, the Working Group focused its discussion on the products liability area. But even here, the case for substantive reform is highly suspect. From the outset, there has been unproductive debate over the extent to which the "unreasonably dangerous" requirement in the landmark Second Restatement of Torts, section 402A, created a fault-oriented standard of strict products liability.

In the design defect cases—the *bete noir* of tort critics, including the Working Group—the widely influential holding in *Barker v. Lull Engineering Co., Inc.* has long been recognized as an exceedingly weak foundation for developing a meaningful distinction between a strict liability standard and the fault principle. In that case, the California Supreme Court's "excessive preventable danger" test for whether a product design is defective sounds on close analysis suspiciously like the negligence calculus. The much-maligned ex post test of requisite knowledge of risk spelled out in *Beshada v. Johns-Manville Products Corp.*—namely, a hindsight (time of trial) determination of the knowable risks to users of a product—was, in fact, overturned within two years of the initial decision. In its place, the New Jersey Supreme Court posited an ex ante reasonable forseeability test which, once again, is closely aligned with a due care inquiry.

Product warnings, under generally accepted doctrine, need only be reasonable in character. Indeed, the single area where strict liability for defective products appears to have real meaning—the case of manufacturing defects, in which liability is established solely on the basis that the harm was caused by an atypically dangerous unit of the standard product—is generally acceptable to critics of the tort system such as the Working Group. Thus, although the reaffirmation of the fault principle may not quite be "full of sound and fury," it certainly does not signify very much either.

I cannot hope to survey all of accident law here to demonstrate that doctrinal reform is a barren strategy. But a bit of history may offer a useful perspective, as an exercise in counterpoint. There was a time when substantive law reform did have real significance. When injured consumers were required to establish a contractual bond to a product manufacturer in order to litigate product defects successfully, abolition of the privity doctrine had enormous consequences. When emotional distress was actionable only if linked to a physical injury, the fashioning of an independent duty to protect against intangible harm constituted a breakthrough in victims' rights. As recently as a decade ago, the substitution of comparative for contributory negligence was a notable achievement.

But the time for substantive reform of tort doctrine has largely passed. A robust fault principle occupies the field of accidental harm. There is no turning back to the harsh world of wholesale exemptions from fault responsibility. Correlatively, there is no persuasive indication of a substantial move "beyond" fault, for reasons that I have just indicated. If modest improvements in the system are to be effected, the would-be reformer must look elsewhere than substantive doctrine.

The tendency in the mid-1980s has been to challenge what can be best characterized as the remedial side of tort law. A cluster of issues, loosely tied to the size of damage awards, has been the main focal point of tort critics, as well as state legislative activity: in particular, pain and suffering, punitive damages, joint-and-several liability, and the collateral-source rule. But the full implications of the challenge are rarely, if ever, spelled out.

With the exception of the collateral-source rule, dissatisfaction with the remedial character of tort damages appears to be largely an expression of displeasure about inadequate controls on jury discretion. This is evident in the steady drumbeat of complaints about runaway verdicts and jury nullification of the ground rules of negligence law. But tort reformers have not been willing to meet the issue head-on by, for example, proposing that trial judges instruct the jury on a specified, acceptable range of pain and suffering awards that appear warranted in a given case, or, in the case of punitive damages, by proposing a stringent substantive standard. The source of the reformers' diffidence is readily apparent, even if rarely articulated. Tort reformers of every persuasion are even more suspicious of the good sense of trial judges, generally speaking, than they are of juries. Hence, they express deep-seated antipathy to the entire range of reforms targeted at reallocating authority to decide between judge and jury.

For many of those engaged in the tort reform enterprise, the more attractive option has been a set of cutoff rules: caps on awards, a bar to joint and several liability, or the elimination of collateral-source recovery. Again the implications are rarely articulated in full. Consider the notion of capping awards. The obvious difficulties are the arbitrary nature of any cap and the earlier-mentioned regressive effect of ceilings on recovery: those most seriously injured bear the full burden of the cost minimizing strategy.

But the objection to ceilings cuts deeper. The greater the "bite" achieved through a cap, the more it is tantamount to a form of schedule damages, that is, an approach to reparation that identifies the injury victim as the member of a class, rather than as the bearer of subjectively measured harm. Such an approach, categorical treatment of harm, is the hallmark of a wide array of welfare and social insurance programs. Therein lies the anomaly, however. The proponent of caps professes a continuing attachment to a tort system that makes subjective determinations of liability, at extremely high administrative costs (compared to no-fault and social insurance systems). What is the justification for maintaining this apparatus if an identifiable category of injury victims—arguably the category of victims that provides the strongest reason for case-by-case determinations of harm—is relegated to nonsubjective treatment?

Conversely, of course, a very high cap—a ceiling approaching a million dollars—substantially maintains the tradition of individualized treatment of injury victims. Perhaps such a cap serves symbolic purposes, and even lends a modestly greater degree of predictability to the system—although projecting the impact of a high ceiling on tort and insurance costs involves a leap of faith beyond the existing data. But there is a disquieting aspect to focusing reform efforts on such a limited category of very high awards. Indeed, it is a peculiarity that triggers a general sense of unease over the continuing discussion of runaway verdicts. For juries do not operate in a power vacuum. Trial and appellate court judges retain the long-standing power of remittitur, as well as the ability to order a new trial. Why the excessive timidity and the exaggerated deference if juries are truly out of control? In the final analysis, the question remains whether caps are responsive to a precisely identified malfunction in the tort system.

The fairness objections to joint and several liability and the collateral source rule are of a different kind. Displeasure with the doctrine of joint and several liability has been especially evident in the area of municipal liability. Consider two scenarios. The first involves a pedestrian run down by an intoxicated insolvent driver who fails to notice a stop sign. In an effort to find a solvent defendant, the injury victim joins the municipality, arguing that its failure to trim a hedge adequately at the scene of the accident exacerbated the drunken driver's inability to heed the stop sign. The jury finds the city 5 percent and the driver 95 percent at fault, but under the doctrine of joint and several liability the municipality is responsible for the entire award of damages. In the second scenario, a municipal operator of a hazardous-waste dump is similarly found to be 5 percent responsible for toxic harm, while the generators and transporters of the waste, who are responsible for the remaining 95 percent, are insolvent. Once again, a defendant whose contributory responsibility is slight ends up paying for the entire damage award.

Arguably, full compensation for injury victims ought to be the dominant consideration in these situations, as joint-and-several liability anticipates. Surely, however, a powerful fairness argument can be made for the contrary position: that a defendant's liability ought to correspond roughly to proportionate responsibility for the harm. This is, in essence, the noneconomic fairness rationale for the tort goal of appropriate deterrence. The would-be reformer may well be left in equipose by these competing considerations. Whatever the case, this remedial reform strategy suffers from much more substantial disabilities. There is not a shred of evidence in the data that the joint and several liability rule contributes in any significant way to the growth in claims, award levels, or administrative costs. Indeed, in two of the major areas of perceived crisis, products liability and medical malpractice, it very rarely comes into play. Moreover, in the municipal roadway maintenance scenario, the problem could be eliminated entirely in a far more direct fashion by adopting compulsory motor vehicle liability insurance with adequate exposure limits.

The conscientious reformer is likely to be left with similar lingering doubts about the efficacy of collateral-source rule reform—the last in the package of remedial strategies. From the outset, opponents of the rule have argued that ignoring collateral sources, such as hospitalization insurance and sick pay, in the determination of tort damages results in an unfair windfall, double recovery for an injury victim. The present day response to this claim is that widespread subrogation eliminates most prospects for double recovery and also results in proper cost allocation of liability to risky activities. The tort reformer again is left in a state of uncertainty: the efficiency, as well as the ubiquity, of subrogation is not adequately documented, nor are there data on the extent to which collateral sources have become a return on investment to injury victims. Without this information, the impact of the collateral-source rule—and the case for its abolition—is difficult to assess. What appears, on the surface, to be either a strong argument for fairness to defendants or elimination of excessive litigation may, in fact, be trivial on both counts.

The fundamental problem with both joint-and-several and collateral-source reform initiatives is that they bear no certain relationship to the overriding tort reform concerns of the mid-1980s. The tort system is increasingly burdened by administrative costs for a wide variety of reasons, ranging from the increasing use of experts to the growing complexity of determining due care and cause-in-fact in certain types of cases. In the context of these problems, collateral-source reform or abolition of joint-and-several liability is simply a side show, far removed from the main event.

The irony is that virtually no effort has been directed towards adopting strategies of incremental reform that might directly address the chorus of complaints about steadily increasing administrative costs of the tort system. The substance of this critique is that the most troublesome aspect of the spiralling costs of the system is not excessive litigation per se but too much lawyering—more concretely, the tendency to abuse the torts process through strategic resort to delay and imposition of burdensome costs of trial preparation. The many forms of this abuse include spurious motions practice, excessive deposition taking, unnecessary continuances, frivolous claims and multiple lawyering.

This is not the stuff of breathtaking reform proposals. Few tort critics of any stripe are likely to rally passionately under the banner of a fast-track system, sanctions for abuse of discovery, penalties for spurious claims and defenses, and initiatives to early settlement. Indeed, the centrality of these considerations comes as something of an embarrassment for the would-be tort reformer. Taking reform seriously may well involve the unsettling lesson that the tort specialist has very little to contribute to the effort. . . .

Notes and Questions

1. Would you describe the shift from contributory to comparative negligence as a small or a radical change? What about the historical move toward less sovereign immunity? What explains these changes? Do they suggest anything about the likely direction and magnitude of future reforms?

2. What is the likelihood that, in return for lower prices or taxes, most people would agree in advance to accept caps on their recoveries in the event of a tort suit? Is there a way to let each citizen choose which regime he or she wishes to be governed by?

3. The author bemoans the paucity of reform proposals directed at curtailing the increasing costs (and perhaps abuses) of lawyering. Is it likely that these costs are more closely associated with torts suits than with contracts, antitrust, or other suits?

In the Shadow of the Legislature:
The Common Law in the Age
of the New Public Law

DANIEL A. FARBER AND PHILIP P. FRICKEY

It is a commonplace that we live in a statutory era. A century ago, statutes were considered intrusions into the pristine order of the common law. Today, legislatures are the primary source of law, and the statute books grow exponentially. Nevertheless, the common law has shown great vitality. In the past thirty years, for example, the law of products liability has undergone explosive growth. The common law has also retained its ability to respond to changes in social values. . . .

There is nothing new about change in the common law. But in an era of statutes, the role of the common law in formulating social policy has become problematic. Arguments for innovation in the common law are almost always challenged on the ground that the legislature, not the court, is the proper forum in which to argue for reform. Existing common-law rules may be challenged for reposing too much policymaking discretion in the courts.

In this selection, we explore how modern common-law judges should view their role vis-à-vis the legislature. We suggest that the perspective of the "New Public Law," as we conceptualize it, is surprisingly helpful in considering this problem. . . .

Defining the New Public Law

There are undoubtedly different ways to characterize the public law scholarship for the past decade or so. One of its most distinctive attributes (and the one on which we will focus) is the much more careful and explicit attention granted to political theory. Unlike the pluralist theories of the 1950s, currently influential theories reject the view that the public interest will automatically

Abridged and reprinted without footnotes by permission from 89 *Michigan Law Review* 875 (1991).

emerge from the conflicting efforts of interest groups. In particular, two new movements in the study of political institutions, republicanism and public choice theory, have had a major impact on legal scholarship in public law. . . .

What Is "New" About the New Public Law?

One strand of the New Public Law derives from a communitarian strain in modern political thought that has become known as republicanism. "Republicanism" was hardly a household word even for public law scholars until the 1980s. The term is unfortunately misleading: the political philosophy called "republicanism" has no particular connection with the Republican party. The general obscurity of the term is itself quite meaningful, for the neorepublicans claim to have rediscovered a forgotten yet fundamental strand of American political thought. For the uninitiated, a brief, ruthlessly oversimplified introduction will suffice for our purposes.

The dominant strand of American political philosophy has been liberalism—another misleading term, since philosophical liberalism is at least as much embraced by political conservatives as liberals. Philosophical liberalism begins with the individual rather than the community and posits that individuals have basic human rights that exist independent of any particular political system. Political conservatives may view these rights as involving property; political liberals may stress rights of individual self-expression or equality; but both agree that these rights are constraints on government rather than creations of government.

Liberalism also assumes that individuals have interests that they seek to advance, both in private life and in politics. The government's role is defined in terms of these individual interests while respecting individual rights. One important function of government, therefore, is to provide fair procedures for determining who prevails when individuals or groups conflict.

Philosophical liberalism is the dominant strain in current American thought. In the eighteenth century, however, another political tradition—republicanism—was also highly influential. During the Revolutionary era, many prominent Americans were strongly influenced by the teachings of the seventeenth-century Opposition party in England. The Opposition thinkers had decried the destruction of the old order, the rise of corruption, and the loss of civic virtue. The fate of the nation, they believed, rested on the willingness of individuals to sacrifice private interests to the common good.

Modern reconstructions of republicanism are based on the allure of civic virtue. Political life is not, as liberalism posits, merely an effort to use the machinery of government to further the ends of private life. Rather, politics is a distinct and in some respects superior sphere in which citizens rise above their merely private concerns to join in a public dialogue to define the common good. Indeed, one of the most important tasks of government is to make the citizenry more virtuous by modifying existing individual preferences to further the common good.

According to republicans, courts can play an important role in this process. As forums for public deliberation, courts can identify and promote the acceptance of public values. Courts can also ensure that the decisions of other agencies of the government reflect republican deliberation rather than an equilibrium of private interests.

On the surface, at least, public choice theory stands in stark contrast to republican theory. Public choice usually claims to be positive while republicanism is explicitly normative, and the implications of public choice seem as dismal as those of republicanism seem optimistic. Again, because of its recent prominence, we will present only a brief, simplified overview of public choice for the uninitiated.

As we see it, public choice theory is a hybrid: it applies the economist's methods to the political scientist's subject. Public choice is largely concerned with abstract, axiomatic modeling of political processes. These models often view the legislative process as a microeconomic system in which "actual political choices are determined by the efforts of individuals and groups to further their own interests," efforts that have been labeled "rent seeking." Thus "[t]he basic assumption . . . is that taxes, subsidies, regulations, and other political instruments are used to raise the welfare of more influential pressure groups."

Further, collective action problems make it difficult to organize large groups of individuals to seek broadly dispersed public goods. Public choice suggests that political activity will instead be dominated by small groups of individuals seeking to benefit themselves. The most easily organized groups presumably consist of a few individuals or firms seeking government benefits for themselves, benefits which will be financed by the general public. Accordingly, under this view "rent-seeking" special-interest groups dominate politics.

This vision of politics is disturbing precisely because it suggests that there is an underlying, coherent pattern to political outcomes, but one that most people find normatively unattractive. The implications of another branch of public choice theory are, if anything, even more dismal. Some public choice theorists have suggested that, far from systematic rent-seeking, political processes necessarily lead to entirely arbitrary or incoherent outcomes.

Jerry Mashaw has aptly summarized the potentially dismal implications of the research inspired by Arrow's Theorem:

> The most basic finding of the Arrovian branch of public choice theory might be characterized as indicating that collective action must be either objectionable or uninterpretable. A stable relationship between the preferences of individuals and the outcomes of collective choice processes can be obtained only by restrictions on decision processes that most people would find objectionable. At its most extreme, Arrovian public choice predicts that literally anything can happen when votes are taken. At its most cynical, it reveals that, through agenda manipulation and strategic voting, majoritarian processes can be transformed into the equivalent of a dictatorship. In a more agnostic mode, it merely suggests that the outcomes of collective decisions are probably meaningless because it is impossible to be certain that they are not simply an artifact of the decision process that has been used.

As Mashaw implies, much of the public choice literature makes for fairly grim reading, offering the choice between viewing legislatures as corrupt or mindless. This grimness seems to be diminishing in recent public choice scholarship. Recent studies have shown that legislators are not merely pawns of special interest groups and that a variety of stabilizing devices may allow legislatures to escape incoherence. Nevertheless, it is fair to say that public choice still tends to emphasize the frailties of democratic government.

As we have characterized the New Public Law, its main feature is the self-conscious use of political theory. Yet the two branches of political theory that have been most influential are clearly in tension. Under public choice theory government functions as a mechanism for combining private preferences into a social decision. The preferences themselves remain untouched. In contrast, in republican thought private preferences are secondary; preferences are if anything the products of government action rather than its inputs. Rather than mechanically processing preferences, government involves an intellectual search for the morally correct answer.

It would be hard to imagine a vision of representative government that differs more than republicanism from the rent-seeking models we discussed earlier or from the chaos and cycling of Arrow's Theorem. This contrast forms the basis for much of the appeal of republicanism. Where public choice theorists are prone to see haphazard cycling or rent-seeking by special interest groups, republicans find the possibility of general political dialogue in search of the public good.

This is not the place for a full rendition of our views on republicanism and public choice, but a few brief reflections are relevant to our later discussion. We find some of the lessons of republicanism attractive: ideas as well as pocketbooks matter in politics, civic-mindedness is more than a myth, and government can be a moral teacher as well as a reflection of public opinion. While it is possible to overemphasize these elements of political life, we think it equally wrong to dismiss them. Nevertheless, where public choice theory risks cynicism, republicanism can verge on a dangerous romanticism about politics. Although contemporary republicans concede that the political process is subject to rent-seeking, they may overestimate the extent to which public deliberation elevates political outcomes over prior preferences.

Given that public choice emphasizes the possible pathologies of democratic government while republicanism considers government's possible accomplishments, it is little wonder that the two have appealed to different groups of scholars. Conservatives who are suspicious of government have found in public choice a congenial catalogue of political misdemeanors, while progressive communitarians have been drawn to republicanism. For this reason, the New Public Law has developed along somewhat different left- and right-wing lines. Nevertheless, the two may be used in tandem, with republicanism serving to eliminate the normative possibilities of democratic government and public choice highlighting its potential pitfalls.

One of our purposes here is to suggest that these two very different theories can effectively be combined in legal analysis, although doing so does risk some cognitive dissonance. It would be nice, of course, if the political scientists would present us with some grand synthesis combining public choice and republicanism. But the New Public Law can offer considerable insight even without such a synthesis.

What Is "Public" About the New Public Law?

We suggest that the New Public Law can illuminate some interesting common law problems. For many people, however, "common law" and "public law" are antonyms. How, then, can the New Public Law have anything to say about the common law of torts? To address these concerns, we must first examine the distinction between public and private law. To the extent that this distinction has any remaining vitality, it has little relevance to defining the scope of the New Public Law.

We begin by considering the conventional classification of various fields of law. Torts and contracts are usually considered examples of private law; constitutional and administrative law exemplify public law. But it is hard to know what features of these examples are definitive. Private law is often common law rather than statutory, but the Uniform Commercial Code has not converted Sales into public law, at least in the minds of law professors. Conversely, the use of common-law methodologies does not make constitutional law, section 1983, or Title VII into private law fields. Procedurally, "public law" litigation often involves more parties, more flexible relief, and more judicial initiative than classic private law disputes, but it is not hard to find exceptions on both sides. Often, the government is a party to public law litigation, but many important regulatory schemes rely in whole or in part on private litigation as an enforcement tool. None of these differences seems decisive.

In an earlier age, the distinction between public and private law had deep jurisprudential roots. In nineteenth-century legal thought, the public/private distinction was founded on the idea of an autonomous private order. Private law was designed to protect these pre-political rights. Property law arbitrated disputes about ownership; tort law protected owners from unconsented intrusions; and contract law allowed owners to exchange property. Private law, then, was that part of the legal system protecting the private ordering; public law consisted of government compulsions restricting private freedom.

This vision of a prelegal private ordering was a major target of the legal realists, and today has been attacked by critical legal scholars and neo-republicans. They have argued cogently that the common law itself is based on choices of public policy. Today, in considering issues of contract, tort, or property law, it is almost second-nature to refer to considerations of public policy. Despite the realist critique, it may be possible to salvage something of the distinction between rules that uphold private ordering versus those that override private ordering. Elaborating this distinction would be no easy matter, however, since most fields of law contain some rules of both types. Moreover, the distinction between facilitative and supervening legal rules may itself be problematic. Private law does not endorse all private activity. It attempts to distinguish between "coercive" and "voluntary" transactions, to identify those actions that do further private preferences. These distinctions, however, are themselves often controversial.

If the private law/public law distinction retains any vitality after the realist critique, the line between the two is at best elusive. Fortunately, we need not resolve the issue here. So far as we can see, whether republicanism and public choice are useful analytic tools does not depend on whether a legal rule is ultimately facilitative or supervening.

We suggested earlier that the distinctive attribute of the New Public Law is its explicit reliance on postpluralist political theory. Whether the New Public Law is useful in a particular instance depends, consequently, on whether such political theory helps to solve the problem at hand. Thus, what determines the relevance of the New Public Law is not the substance of the legal issue, but whether an understanding of political institutions seems necessary. Not surprisingly, this most often occurs when the legal issue is formulated from the outset as relating to the proper role of government, or when the interpretation of a governmental act is at stake. But even in other kinds of cases, the perspective of political theory may be important.

In the "market" for legal ordering, the legislature is always a "potential entrant." When a court is asked to make rules to govern a private transaction, it is often relevant that the legislature or its administrative agent are possible alternate decisionmakers. Indeed, today, some legislative act such as the Uniform Commercial Code is often implicated in even the most "private law" case. Once either legislation or the potential of legislation enters the case, the same political theories that have halped shape the New Public Law may become important sources of insight regardless of whether the issue is considered public or private law.

In short, we believe the "New Public Law" is a misnomer. Although it has its origins in fields that were traditionally considered "public law," the applicability of this body of theory is only contingently related to whatever remains of the public law/private law distinction. The New Public Law is potentially useful in analyzing a wide variety of legal problems outside the domain of public law, as classically understood. It is to a couple of brief illustrations of this point that we now turn.

The Viability of the Common Law in the Age of Statutes

In an age in which so much public policy is made by legislatures, the rule of the common law can no longer be taken for granted. When a statute is directly applicable, it of course displaces any conflicting common law rules (constitutional limits aside). Our interest, however, is in situations where the existence of a statute may affect judicial policymaking outside of the statute's domain.

Traditionally, judges viewed the common law as a principled, hermetically sealed system immune from all but clear statutory intrusions. Statutes were "in the law"—they were formal requirements of positive law—but not "of the law" because they were the work of an unprincipled and unpredictable agency, the legislature. With the demise of most aspects of common-law formalism and the rise of the age of statutes, courts began considering statutes as potential sources of worthwhile principles, rather than simply reflections of brute political power. This development conceptualized the legislature as an appropriate potential player in any common-law policymaking dispute, not simply another government agency from which disappointed litigants could seek relief. In essence, common-law courts began viewing themselves as operating in the shadow of the legislature, especially where the legislature had enacted statutes having a penumbral relationship with the common law problem at hand.

Conceptualizing the legislature as a source of public policy has republican connotations, and public choice is implicated when viewing the legislature as a potential entrant into any dispute about public policy. In this section, we illustrate the role that republicanism and public choice might play in analyzing common-law problems in the age of statutes. . . .

The common law and statutes may interact in two distinct ways. In tort law, for example, conduct is often subject to a case-by-case judgment about reasonableness. In these cases, judicial policymaking takes place at the microlevel, when the common-law rule of reasonableness or balancing is applied. If the legislature later adopts a statute regulating conduct subject to this kind of common-law rule, the question becomes whether such microlevel judicial policy decisions have become inappropriate. Here, a legislative scheme may result in a curtailment or even the abandonment of the common law approach. . . .

We consider an instance in which the common law, contrary to its usual approach of case-by-case judgments about reasonableness, has a flat rule barring recovery for injuries arising from certain conduct. Suppose that the legislature later adopts statutes inconsistent with the policies underlying the common-law rules but not directly abrogating it. Should the courts administer the coup de grace to what remains of the common-law rule, in deference to the legislature's policy judgment? Or, on the contrary, should the courts consider what remains of the common law frozen, in deference to the legislature's possible assumption of policymaking responsibility over the subject matter? . . .

Common Law Reform

. . .

The possibilities for common-law creativity in a largely statutory setting are illustrated by *Moragne v. States Marine Lines, Inc.* Understanding *Moragne* requires a brief review of the history of personal injury law. Under British tort law, transplanted to America, there was no cause for action for wrongful death. Accordingly, it was considerably wiser—or at least, more profitable—to kill people than to maim them. In the nineteenth century, American state legislatures abolished this absurd rule by establishing statutory rights to sue for wrongful death. For wrongful death of seaworkers, Congress also enacted remedial legislation in 1920. The Jones Act provides a cause of action for the negligent death of a seaman, and the federal Death on the High Seas Act (DOHSA) establishes a cause of action for wrongful death of workers on the high seas—outside the territorial waters of the United States—"caused by wrongful act, neglect, or default."

To simplify somewhat, in *Moragne* a widow attempted to bring an action for the wrongful death of her husband, who had died from injuries suffered in American territorial waters. Her claim was that the vessel was "unseaworthy," and the common law recognized a well established cause of action for injuries that result from such conditions. Her claim did not fit the federal wrongful-death statutes, however. The Jones Act reaches wrongful death in the territorial waters, but it is limited to negligence claims; DOHSA encompasses unseaworthiness actions, but reaches only wrongful death on the high seas. In short, Ms. Moragne had fallen into a hole in the statutes due to the combination of the nature of the wrongful conduct and the place of the accident. The only other law to apply, federal maritime common law, would preclude her action as well, because of the old rule against recovery for wrongful death.

What was obviously needed in *Moragne* was some way to update federal law. The statutory language was not easily amenable to any construction that would allow her recovery. The more obvious solution was to abandon the old common law rule, which the Supreme Court did in *Moragne* in a well-crafted opinion by Justice Harlan.

Justice Harlan concluded that Congress in 1920 was simply fixing the problems squarely presented to it, not comprehensively addressing—and thus freezing—an entire area of law. Drawing not only from the federal maritime statutes but also from the consistent pattern of state statutes, Justice Harlan found a well-established public value in favor of recovery for wrongful death. Although Justice Harlan had probably never heard of civic republicanism, his decision combined high judicial craftsmanship with republican attention to the legislature's role in articulating public values.

This interpretation is not the only possible reading of *Moragne*. The opinion can also be read, as Judge Posner reads it, in a more purist "legal process" mode. On this reading, the opinion is primarily motivated by a concern for the "neutral principle" of legal coherence; under this view, Harlan's primary concern is eliminating a legal anomaly. While not without support, however, this interpretation is belied by the structure of the opinion. Harlan begins by rejecting *The Harrisburg*'s preclusion of wrongful death recoveries as normatively untenable. He only then turns to 'legal process' issues like coherence in a defensive posture, to determine whether the normatively attractive result is blocked by process arguments.

The *Moragne* setting also illuminates how public choice can provide insights for the judicial role in common-law cases. The interest groups lobbying for statutes like the Jones Act and DOHSA are likely to focus on particular problems, not system-wide inquiries that may complicate passage of legislation. The failure of the various statutes to reach the precise situation of Ms.

Moragne carries no plausible implication of a congressional desire to leave her remediless: it is hard to imagine that the deal underlying this series of statutes was to compensate the families of all victims of tortious conduct except those who were killed within three miles offshore of certain states. Thus, public choice warns against reading negative implications into the precise contours of the complicated statutory scheme.

In *Moragne,* Justice Harlan overtly shaded the inquiry whether a congressional deal precluded adoption of republican values. He noted that legislation sometimes embodies general public policies that courts should apply beyond the statute's domain—the republican insight—and at other times reflects hard-fought compromises between conflicting interests that should not be undone by judicial elaboration—the public choice perspective. Based on the existence of the Jones Act, DOHSA, and a variety of nonmaritime federal wrongful-death statutes, he concluded that "Congress has established a policy favoring recovery in the absence of a legislative direction to except a particular class of cases." He then framed the essential issue in *Moragne* as whether "Congress has given [any] *affirmative* indication of an intent to preclude the judicial allowance of a remedy for wrongful death to persons in the situation of this petitioner." In essence, Justice Harlan presumed the appropriateness of the republican outcome absent a clear statement by Congress. This tactic, usually found in a variety of "public law" cases, places the heavy burden of legislative silence on the "nonrepublican" side of the case.

Like republicanism, public choice also has some implications about the proper placement of the burden of legislative inertia. Ms. Moragne and similarly situated persons have no idea that they are without remedy until they suffer the loss of a loved one; they have no incentive to organize before the fact to lobby for remedial legislation. Such large, diffuse, unorganized groups are, according to public choice, the least likely to lobby successfully for legislative action. Unions may provide a mechanism for overcoming this collective action problem in some contexts. Even within the union membership, however, this group likely is too inchoate to carry any weight. The nature of the defendants in a case like *Moragne* provides a stark contrast. Shipping companies have the problem of compensating work-related injuries every day, in contrast to the one-time tragedy suffered by Ms. Moragne. These companies are small in number, easily identified, and have the resources to lobby Congress—public choice predicts that they can organize and protect themselves in the legislative arena. The industry, in short, is well positioned to obtain congressional relief from any harshness resulting from the application of *Moragne* to future injuries; the Ms. Moragnes of the world are unlikely to obtain legislative relief before their respective losses occur.

Following *Moragne,* The Supreme Court engaged in a tortuous effort to fit the common-law cause of action for wrongful death resulting from unseaworthiness within the broader framework of the maritime statutes. In *Sea-Land Services v. Gaudet,* a sharply divided Court held that, because the *Moragne* cause of action was common law in nature, modern common-law rules defining damages recoverable should apply; recovery could thus include certain nonpecuniary losses such as loss of society. In *Mobil Oil Corp. v. Higgin-botham,* however, the Court confined the *Moragne* cause of action to territo-rial waters—thereby relegating the surviving spouse of a seaworker killed on the high seas to damages only as provided for in DOHSA, which explicitly limits recovery to pecuniary loss. Most recently in *Miles v. Apex Marine Corp.,* the Court held that damages for wrongful loss of society cannot be recovered in a *Moragne* action brought on behalf of a seaman, limiting *Gaudet* to actions for injuries to longshoremen in territorial waters—a limitation that leaves *Gaudet* little or no practical application.

Formalistically, the post-*Moragne* cases make some sense. In *Gaudet* no statute applied, and so (modern) common-law principles controlled; in *Higginbotham* and *Miles,* DOHSA or the Jones Act applied, and thus plaintiff was stuck with the (arguably obsolescent) recovery provided in those statutes. After the two most recent decisions, *Moragne* essentially does little more than eliminate the phrase "high seas" from the title of DOHSA, extending the statutory remedy landward. The policy of uniformity, which was one of the *Moragne* court's concerns, has been fully served, but at the expense of the *Moragne* court's even stronger desire to provide a just remedy for the survi-vors of maritime workers.

A brief evaluation of the post-*Moragne* cases from the perspective of the New Public Law is illuminating. *Moragne* relied upon republican principles— that the common law of torts presumptively embodies modern public values, including recovery for wrongful death. *Gaudet* stressed this same factor in following the "clear majority of States" and "the humanitarian policy of the maritime law to show 'special solicitude' for those who are injured within its jurisdiction." The dissenters in *Gaudet* saw *Moragne* much differently, as simply a gap-filling decision where the precise contours of DOHSA were adopted by analogy to the common law. In essence, the dispute in *Gaudet* was between finding a republican concept in DOHSA to be elaborated upon in defining the modern common law and finding a congressional conception of the appropriate rule of law that, even if outdated, should be applied to pro-mote the legal uniformity. The *Gaudet* dissenters' position is inconsistent with *Moragne*'s public choice insights—that the political dynamics surrounding passage of the Jones Act and DOHSA indicate no congressional intent to prescribe fixed rules for all wrongful deaths on the waters.

Higginbotham is a harder case because DOHSA squarely applied and provided only for recovery of pecuniary loss. Allowing the plaintiff to recover nonpecuniary loss might seem equivalent to Guido Calabresi's proposal that judges simply discard statutes because they have lost majority support or are incompatible with the legal landscape. Although we have serious qualms about Calabresi's approach, we doubt that one need go so far to support the conclusion that *Higginbotham* is at least questionable.

Higginbotham concluded that the court need not engage in republican inquiries about public values because "Congress has struck the balance for us." At most, however, the principle of legislative supremacy is violated only when a court refuses to follow a clear legislative command. *Higginbotham* assumed that such a command was present in DOHSA, but that conclusion is at least overstated, if not wrong.

First, the statutory language at issue granted a "fair and just compensation for the pecuniary loss sustained." Nowhere does it say that nonpecuniary loss cannot be recovered. The majority's conclusion that this section reflected Congress' "considered judgment" to reject nonpecuniary damages is unsupported by reference to any legislative history. Indeed, the legislative history of DOHSA, as interpreted in *Moragne,* suggests that Congress had simply fixed some particular problems brought before it by incorporating the tort values of the era, not necessarily freezing the law to reflect forevermore the 1920s conceptions of how to implement those values. In this respect, the majority's statement in *Higginbotham* that "[t]here is a basic difference between filling a gap left by Congress' silence and rewriting rules that Congress has affirmatively and specifically enacted" seems defensible but also somewhat attenuated. In all events, it conflicts with Justice Harlan's conclusion in *Moragne* that "no intention appears that the Act have the effect of foreclosing any nonstatutory federal remedies that might be found appropriate to effectuate the policies of general maritime law."

In short, *Moragne* and *Gaudet* saw our "general law"—DOHSA, state wrongful-death statutes, and current values—as promoting a case-by-case search for solutions that best effectuate the humanitarian policies of the maritime law. In contrast, *Higginbotham* treated DOHSA like a bill of lading. For us, the appropriate resolution of *Higginbotham* should have turned on a republican evaluation of the policies at issue (for example, is recovery of nonpecuniary loss really necessary to effectuate modern humanitarian policies?) and a public choice inquiry into the "deal"—if any—struck in the 1920 Congress. If the outcome in *Higginbotham* is correct, it is not for the reasons expressed by the majority.

Miles completed the Court's retreat from the creative judicial role contemplated by *Moragne.* The *Miles* Court held that damages for loss of society are unavailable in a *Moragne* action because such damages are unavailable under the Jones Act. It would be "inconsistent with our place in the constitutional scheme were we to sanction more expansive remedies in a judicially created cause of action in which liability is without fault than Congress has allowed in cases of death resulting from negligence." A close reading of the opinion suggests, however, that this judicial modesty is misplaced. If Congress had rejected nonpecuniary damages in the Jones Act, perhaps there would be some basis for judicial deference in fashioning the common-law remedy. In reality, however, there was no legislative decision to which to defer.

The Jones Act itself contains no limitation on the types of damages recoverable. It borrowed its language from the FELA, which provided a federal remedy for injuries to railroad workers. The FELA also contained no express limitation on damages, but the Supreme Court read such a limitation into the FELA in 1913. It was this decision that formed the basis for the ruling in *Miles.* Since Congress said nothing to the contrary, according to the *Miles* Court, it "must have intended" to incorporate this judicial gloss when it passed the Jones Act. In short, what the Court proclaims as judicial deference to the legislature, turns out really to be the 1990 Court's deference to a decision by the 1913 Court under a different statute. Congress is "assumed" to have agreed with the 1913 Court, so the current Court would be exceeding its constitutional role were it to adopt a different rule—not under the FELA, which was the subject of the 1913 decision, but under the federal common law created by *Moragne,* which neither the two earlier statutes nor the 1913 decision had foreseen. This conclusion was especially ironic because *Moragne* was based on a rejection of the grudging attitude toward wrongful-death actions encompassed in the 1913 decision itself.

The *Miles* Court claimed that its decision was controlled by "[t]he logic of *Higginbotham*" and that in *Higginbotham* "Congress made the decision for us." As we have seen, even with repect to *Higginbotham* this is a questionable assertion. It turns out, however, that Congress had never enacted any statutory language bearing in any way on the issue in *Miles.* The only actual decision was made by the Court itself in 1913; everything else is presumed from congressional silence. Deference to a legislative decision is all very well and good, but it is quite another thing to posit a fictional legislative decision first and then "deferentially" extend that fictional decision into a common-law domain. In the end, the *Miles* Court's exercise in deference was little more than an excuse to avoid responsibility for its own decision.

This is not to say that courts should always be in the forefront of law reform. When a court is asked to make a policy judgment . . . there may be good reason to decline. The court may simply be too uncertain of the right answer to intervene, preferring to leave the issue to bodies with greater expertisc or popular accountability. Sometimes the best solution will exceed a court's remedial powers, for example, by requiring a new set of highly integrated rules that do not lend themselves to common law evolution, or by requiring the creation of an administrative agency. In addition, in some circumstances there may be good reason to believe that the legislature approves of the existing rule. Finally, the group adversely affected by a proposed change in rules may be very poorly situated to obtain legislative relief, and the court may be reluctant to foreclose legislative consideration of the issue.

Thus, our endorsement of *Moragne* should not be read as support for wide-open common-law activism. And indeed, it would be highly peculiar to consider Justice Harlan, author of our paradigm opinion, as a judicial activist. We view the legislature as the senior partner in the joint venture of making public policy. But the courts can do a great deal to further the enterprise, and republicanism and public choice are useful tools in defining the border between judicial and legislative prerogatives.

Conclusion

In the first place, we have tried in this selection to debunk the formal categorization of legal problems. In our view, the essential lesson of the New Public

Law—that political theory (and in particular, postpluralist theory) matters—
is applicable as well to the old private law. . . .

More particularly, we have attempted to demonstrate that two important
political theories of the New Public Law, republicanism and public choice,
provide fruitful insights into common law problems. Upon reflection, at least
some of this may seem obvious. The common law has probably always involved
judicial judgments about the ways adjudication can foster the public interest.
Republicanism's focus on civic virtue adds something distinctive to this process, however. For example, by emphasizing the legislature's capacity to define
public values, it strongly reinforces the borrowing of statutory analogies for
common-law purposes, as *Moragne* illustrates. In many instances, the legislature is both more legitimate in theory and more capable in practice of defining
public values than are judges. Yet the institutional insulation of judges and the
deliberative qualities stressed by republicanism—judicial advantages usually
considered only in the public law setting—will sometimes empower the common law judge as well to promote legal change in the pursuit of public values.
Public choice helps these judges consider the institutional issues that arise in an
ever-increasingly interactive world of public policymaking.

It is important, however, not to lose sight of the deeper tension between
republicanism and public choice. Ultimately they embody different perspectives on the nature of politics—as one seeks to remind us of its aspirations, the
other recalls its flaws and thereby calls those aspirations into question. We
doubt if those tensions can ever be fully resolved by incorporating both into
some even more abstract Grand Unified Theory of politics. Rather, this tension may well reflect the ultimate moral ambiguity of politics itself. If the
phrase were not already taken, we might speak here of the "fundamental
contradiction" of the New Public Law. Such contradictions, we suspect, are an
essential source of dynamic tension for law as a living practice, as opposed to
law as abstract theory. In any event, no grand synthesis of these two theories is
yet available, and legal analysis must proceed in the meantime.

Notes and Questions

1. Whose interests seem served by common-law evolution rather than statutory
intervention? If it is the groups that expect to lose out in the legislative process, then
would you expect legislatures to try to supplant the common law as much as possible?

2. The reading refers to "legislative deals," which are often the product of
logrolling—that is, agreements to package a variety of legislative proposals in order to
form a majority coalition. One's attitude toward the new public law is likely to be
related to one's estimation of logrolling as either desirable or troubling. What are some
arguments on both sides of this question? Do the authors tip their hands as to whether
it makes a difference for the analysis whether such deals are normatively good or bad?

3. The tools of the new public law are offered here as a means of understanding
decisions about statutes. Could these tools also be used to predict legal change? What
small-scale changes (or permanent equilibria) in tort law might one predict with these
tools?